高等院校规划教材

基础化学
实验指导
（双语版）

主　编　刘绍乾
副主编　何跃武　钱　频　冯志明
编　委　刘绍乾　王一凡　何跃武　钱　频
　　　　李战辉　肖旭贤　王曼娟　邓凯佳
　　　　李春云　冯志明
主　审　王一凡

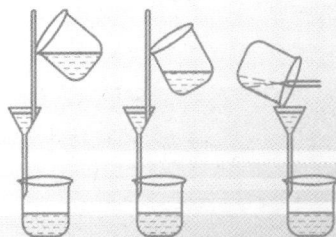

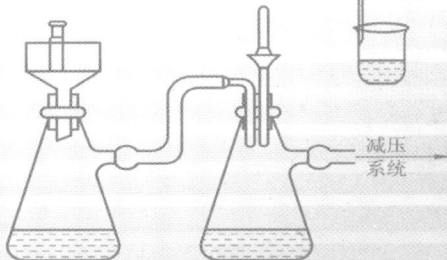

减压
系统

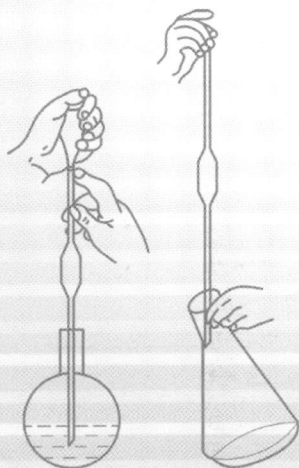

(a)移液管吸液　　(b)移液管放液

中南大学出版社
www.csupress.com.cn

图书在版编目（ＣＩＰ）数据

基础化学实验指导／刘绍乾主编. －－长沙：中南大学出版社，
2013.7

ISBN 978 － 7 － 5487 － 0930 － 5

Ⅰ.①基… Ⅱ.①刘… Ⅲ.①化学实验－高等学校－教学参考资料
Ⅳ.①06 － 3

中国版本图书馆 CIP 数据核字（2013）第 176410 号

基础化学实验指导
（双语版）
JICHU HUAXUE SHIYAN ZHIDAO(SHUANGYU BAN)

主编　刘绍乾

□责任编辑	彭亚非
□责任印制	易建国
□出版发行	中南大学出版社
	社址：长沙市麓山南路　　　　邮编：410083
	发行科电话：0731 － 88876770　　传真：0731 － 88710482
□印　　装	长沙市宏发印刷有限公司

□开　　本	710×1000　1/16　□印张 19.75　□字数 390 千字
□版　　次	2017 年 9 月第 3 版　□2017 年 9 月第 1 次印刷
□书　　号	ISBN 978 － 7 － 5487 － 0930 － 5
□定　　价	39.00 元

前　言

　　基础化学实验是临床医学、预防医学、麻醉医学、口腔医学、护理学等医学相关专业学生的一门重要的必修实验课程，是医学相关专业学生从分子水平上了解生命过程的必备工具。该实验课对培养学生科学的世界观和方法论，提高学生的综合能力和创新意识，以及全面推进素质教育都具有十分重要的意义。

　　该教材自2006年出版以来，受到了教师和学生的好评，也得到了同行们的认可，特别是给医学相关专业学生的基础化学实验课程的学习提供了方便。随着我国高等医学教育事业以及现代化学实验技术的不断发展，医学基础化学实验教学内容、实验方法、实验手段在不断更新，而且，近年来，来我国高校医学相关专业学习的外国留学生呈稳步增长的态势。因此，为了培养适应社会发展需要的高层次医学人才，也为了方便外国留学生基础化学实验课程的学习，根据近年来我们以国家级化学实验示范中心等教育质量工程项目为依托所开展的基础化学实验教学改革实践，结合该教材使用者所反映的意见和要求，参考和借鉴近期国内外出版的同类教材，我们在该教材第二版的基础上，对部分章节内容作了变动或补充，对原书中的一些欠准确之处进行了修正，同时，增加了英文部分。修订后教材的中文部分基本保持该教材第二版的体系，更加突出了基础化学实验内容和生命科学的联系。而增加的英文部分，主要由近年各高校经常选做的一些基础化学实验组成。

　　本书分为中文部分和英文部分：

　　中文部分包括三章：

　　第一章，基础化学实验基本知识。本章较系统和详细地介绍了进行基础化学实验和化学研究所必备的相关基础知识及安全常识。

　　第二章，基础化学实验基本仪器与操作。本章对基础化学实验基本仪器的使用方法以及基础化学实验的基本实验方法、手段作了较系统的总结和详细的介绍。

　　第三章，基础化学实验项目。本章除了选择大量的以训练学生化学实验基本操作能力和基本化学实验方法为目的的典型的基础化学实验外，还引入了一些应用及影响面广，内容上与生命科学结合较为密切的较新的化学实验，比如，"蛋白质含量的分光光度法测定"、"葡萄糖酸钙含量的测定"等实验项目，也对原书中的"磺基水杨酸合铜配合物的组成和稳定常数的测定"等实验内容进行了修订。

英文部分包括两章：

第一章，基础化学实验绪论。本章较系统和详细地介绍了基础化学实验基本仪器与操作方法、基础化学实验的基本实验方法和手段以及化学实验的相关安全知识。

第二章，基础化学实验项目。本章主要是由近年各高校经常选做的一些基础化学实验翻译、编辑而组成，目的是方便教师、学生，特别是留学生对基础化学实验课程的学习。

附录部分，列出了与基础化学实验相关的基本数据、常数和必要的资料，介绍了相关的实验参考书目，以及国内外优秀的化学化工技术类以及生命科学类网站。

本书可作为与生命科学相关的各个专业学生的基础化学实验教材，也可作为从事化学及其相关专业工作者的参考书。

本书由刘绍乾任主编并负责全书修订和校核，何跃武任副主编，王一凡任主审。李战辉、肖旭贤、钱频、王曼娟、邓凯佳、李春云等多位从事基础化学实验教学的老师参加了本书的编写工作。各章和各实验的具体负责编写均列在各章和实验内容之后。

本书的编写和出版得到了中南大学本科生院、化学化工学院和中南大学出版社领导的关心和支持，在此深表谢意。感谢多年来参加基础化学实验教学改革实践和教材编写的中南大学医学基础化学教学团队的同事们。感谢使用该书第一版、第二版的教师和学生，他们的实践和建议使本书得以不断完善。感谢中南大学化学化工学院关鲁雄教授对本书编写工作的大力支持。

本书参考借鉴了兄弟院校教材的一些实验内容，谨表谢意。同时，感谢在网络上提供基础化学实验素材的同行们。

限于编者的学识水平，本次修订中仍难免有疏漏和欠妥之处，恳请同行专家及读者批评指正。

编　者
2017 年 7 月

目 录

上　篇

中文部分

第一章　基础化学实验基本知识

1.1　基础化学实验的目的

化学是建立在实验基础上的科学。基础化学实验是学生化学实验技能与化学素质培养不可缺少的一个重要环节。通过基础化学实验的教学，不仅使学生巩固和加强课堂所学的基础理论知识，更重要的是培养学生的实验操作能力、分析问题和解决问题的能力，养成严肃认真、实事求是的科学态度和严谨的工作作风，培养学生的创新精神和创新能力。它的目的如下：

（1）理论联系实际，使大学化学教学中的重要理论和概念得到巩固和深化，并扩展课堂中所获得的知识。

（2）通过化学实验课程的学习，强化学生的基本操作，注重学生实验技能的训练，使学生掌握一般物质制备的实验方法和技能。掌握研究物质化学性质的实验程序，获得准确的实验数据和结果。熟悉和掌握常用仪器的使用方法，了解、掌握各种实验技能、技术、手段和方法。更重要的是了解各种实验研究方法的应用。

（3）培养学生独立思考和独立工作的能力，学会联系理论知识，独立设计和进行实验，仔细观察和分析实验现象，学会正确处理数据及解释现象，从中得出科学的结论，并且撰写科学报告和论文。

（4）培养严谨科学的工作态度和作风。培养学生的创新能力，为学习其他课程和今后从事化学相关领域的科研、生产打下坚实的基础。诱导学生发散性思维，培养学生的化学素质和创新意识。

1.2　化学实验的要求

（1）实验前要认真预习、领会实验的目的和基本原理，了解实验步骤和注意事项，做到心中有数，有条不紊地做好实验。

（2）预习时，根据实验内容，先写好实验报告的部分内容，画好表格，查出有关数据，以便实验时及时、准确地记录实验现象和有关数据，并进行数据处理。

（3）实验开始前先清点仪器设备，如发现缺损，应立即报告指导教师（或实验室工作人员），并按规定手续向实验员补领。实验中如有仪器破损，应及时报告并按规定手续向实验员换取新仪器，未经教师同意，不得挪用其他人的实验仪器。

（4）实验时应保持肃静，集中精力，要严格按照规范进行认真操作，仔细观察实验现象，及时记录实验结果，积极思考问题，并运用所学理论知识解释实验现象。

（5）实验时应保持实验室和桌面的整洁。实验中的废弃物应倒入废液缸中，严禁投入或倒入水槽内，以防水槽和下水管堵塞或腐蚀。

（6）实验时要爱护国家财产，注意节约水、电、试剂。按照化学实验基本操作规定的方法取用试剂。必须严格按照操作规程使用精密仪器，如发现仪器有故障，应立即停止使用，并及时报告指导教师。

（7）实验室内的一切物品（仪器、试剂和产品）均不得带出实验室。

（8）实验完毕，将玻璃仪器洗涤干净，放回原处。整理桌面，打扫水槽和地面卫生。

（9）认真写好实验报告，对于实验中出现的现象和问题进行认真的讨论。

（10）遵守实验室规则和实验室安全、卫生要求，听从指导教师安排。

1.3　实验室安全知识

1.3.1　安全守则

安全是化学实验工作者要特别注意的大事。安全不仅与个人的安危有关，而且关系到国家财产和其他人的生命安危。在化学实验中，经常使用易燃、易爆、易腐蚀或有毒的化学试剂，大量使用易损的玻璃仪器和某些精密分析仪器，使用水、电、煤气等等，故在化学实验室工作，首先必须在思想上十分重视安全问题，决不能麻痹大意。其次是在实验前应充分了解本实验的原理、步骤和安全注意事项。仔细检查仪器的质量和安装是否良好，并严格遵守操作规程，这样才能避免事故发生。为确保实验的正常进行和人身安全，实验人员应遵守以下实验室的安全守则。

（1）水、电一经使用完毕须立即关闭。离开实验室时应仔细检查水、电、气、门、窗是否均已关好。

（2）实验室内严禁饮食、吸烟。一切化学药品禁止入口。实验完毕后，必须洗净双手。

(3)绝对不允许任意混合各种化学药品,以免发生意外事故。

(4)一切与易挥发的和易燃的物质相关的实验,都应在远离火源的地方进行。产生有刺激性或有毒气体的实验,必须在通风橱内进行。需闻气体气味时,试管口应离面部20 cm左右,用手轻轻扇向鼻孔,不能对着管口去闻。

(5)浓酸、浓碱具有强烈的腐蚀性,切勿溅在皮肤和衣服上,更应注意保护眼睛。稀释它们时(特别是H_2SO_4)应将它们慢慢倒入水中,并不断搅动,而不能反向进行,以避免迸溅。

(6)有毒药品(如重铬酸钾、钡盐、铅盐、砷的化合物,特别是氰化物)不得入口内或接触伤口,剩余的废液须倒入废液缸,不能随手倒入下水道。

(7)加热试管时,不要将试管口指向自己或别人,也不要俯视正在加热的液体,以免溅出的液体把人烫伤。

(8)未经教师许可,不得随意做规定以外的实验。

(9)分析天平、分光光度计、酸度计等均为实验中使用的精密仪器,使用时应严格遵守操作规程。用完后,拔去插头,将仪器各部分旋钮恢复到原来位置。

1.3.2 安全事故的救护措施

(1)玻璃割伤 挑出玻璃碎片,轻伤可抹上龙胆紫或碘酒等药剂包扎。

(2)烫伤 在烫伤处抹上黄色的苦味酸溶液或高锰酸钾溶液,再搽上凡士林、烫伤膏或万花油。受强酸腐伤时,应立即用大量水冲洗,然后搽上碳酸氢钠油膏或凡士林。受浓碱腐伤,请立即用大量水冲洗,然后用柠檬酸或硼酸饱和溶液冲洗,再搽上凡士林。

(3)吸入氯气、氯化氢气体 可吸入少量酒精和乙醚的混合蒸气使之解毒。吸入硫化氢气体而感到不适时,应立即到室外呼吸新鲜空气。

(4)酸(或碱)溅入眼内 立刻用大量水冲洗,再用饱和碳酸氢钠(或硼酸)溶液冲洗,最后再用水冲洗,并立即到医院就医。

(5)毒物进入口内 可把5~10 mL稀硫酸铜溶液加入一杯温开水中,内服后,用手指伸入咽部,促使呕吐,然后立即送医院救治。

(6)触电 先切断电源,然后在必要时进行人工呼吸或送医院。

(7)实验室发生火灾 应根据起火的原因进行针对性的灭火。

①首先防止火势扩展,主要措施如下:

a. 关闭火源,停止加热。

b. 停止通风,减少空气流动。

c. 拉开电闸,以免引燃电线。

d. 把一切可燃物质(特别是有机物质和易爆炸物质)移至远离火源处。

②及时扑灭火焰，实验室一般不能用水灭火，因水能与某些化学药品发生剧烈反应或将可燃物表面扩大而引起重大的火灾。一般的小火可以用湿布、石棉布或沙子覆盖燃烧物，即可灭火。火势大时，可用灭火器。主要方法有：

a. 把沙土或石棉布覆盖在着火物体上（实验室都应备有沙箱和石棉布，放在固定处）。

b. 根据起火原因用不同类型的灭火器进行灭火。不要随便用水灭火，因为水能与某些化学物品（如金属钠）发生剧烈反应，反而会引起更大的火灾。

③实验人员衣服着火时，切勿惊慌乱跑，应赶快脱下衣服或就地打滚或用湿衣服在身上抽打灭火，或用石棉布覆盖在着火处，将火扑灭。

1.4　化学实验的学习方法

要达到实验目的，必须要有正确的学习态度和学习方法。大学化学实验的学习方法，大致可以分为预习、实验和书写实验报告三个环节。

1.4.1　实验前的预习

学生进入实验室前，必须做好预习。实验前的预习，归纳起来是看、查、写三个字。

看：认真阅读实验教材和理论教材的有关内容。预习应解决以下几个问题：明确实验目的和要求，懂得实验的基本原理和操作方法，熟悉实验步骤，预计实验现象，找出实验关键，做相应的思考题。

查：通过查阅书后附录、有关手册以及与本次实验有关的教程内容，了解实验中要用到的或可能出现的基本原理、化学物质的性质和有关理化常数。

写：在看和查的基础上认真写出预习报告，其内容如下：找出本实验的重点、难点和实验成败的关键，要做的步骤，要看的结果，要注意的事项。预习报告附在实验报告上，没有预习者，不允许做实验。

1.4.2　实验中

(1)学生要提前5分钟进入实验室，实验前应清点仪器和药品。如发现有破损或缺少，应申请补足。实验过程中，若有仪器损坏，应报告指导教师，按损坏原因和学校有关规定做出处理后，到供应室领回新仪器补充，不得隐瞒和拿用别的位置上的仪器。

(2)实验中应保持安静和遵守纪律。遵从教师的指导，严格按照操作规程和实验步骤进行实验。集中思想，周密思考，正确操作，细致观察，认真记录。

不得擅自离开实验室。

(3)保持实验室的整洁。实验时，应注意合理安排实验仪器，做到实验台上有条不紊，高低有序。

(4)应备有实验原始记录纸，附在实验报告上，随时将实验中的现象、数据、计算和结果等记录下来。此外，还应把预习中不懂或需要重点观察的现象记录下来，以便实验时能抓住关键。原始记录要求真实、及时、清楚(条理、字迹)、准确、持久，不准用铅笔记录，不准随意涂改，应实事求是地记录，否则无效。

实验原始记录中应包括以下内容：

①每一步操作所观察到的现象，如是否放热、颜色变化、有无气体产生、有无沉淀产生等。尤其是与预期结果相反或与教材、文献资料所述不一致的现象更应如实记载。

②实验中测得的各种数据，如重量、体积、吸光度、pH 等。

③产品的色泽、晶形等。

④实验操作中的失误，如抽滤中的失误、粗产品或产品的意外损失等。

(5)爱护公物，节约用水、用电，按规定取用药品。公用仪器及药品用后应立即归还原处。火柴棍、废纸屑等应投入废纸篓。废液、废金属、残渣应倾入废液缸中。以上物质都不得倒入水槽，以免下水道堵塞和腐蚀金属管道。

1.4.3　实验后

(1)洗涤用过的仪器全部复原，精密仪器用完后应签名，实验台面应擦拭干净，同学应轮流做好实验室卫生，包括以下职责：整理公用的试剂和仪器，打扫卫生，清倒废物，关好水、电、门、窗。

(2)原始记录须经指导老师签字后才可离开。

(3)做完实验后应及时完成实验报告，包括解释实验现象，并做出结论，或根据实验数据进行计算，在规定时间交指导教师评阅。

实验报告是实验的重要组成部分，是分析问题解决问题的过程，也是综合运用知识的过程。实验报告应写得简明扼要、整齐洁净，实验报告一般应包括下列几个部分。

实验名称：该部分应该包括实验名称、实验日期、气温、操作者。

实验目的：实验目的通常包括以下三个方面。

①了解本次实验的基本原理；

②掌握哪些基本操作；

③进一步熟悉和巩固已学过的某些操作。

实验原理：本项内容在写法上应包括以下两部分内容。

①文字叙述：要求简单明了、准确无误、切中要害。

②主、副反应的反应方程式。（配平）

实验试剂与仪器：实验中使用的主要试剂和仪器。

实验内容：尽量用简图、表格、化学式、符号等表示。例如性质实验用表格表示内容、现象、解释（文字或反应方程式）。

实验数据与处理：把测得的各种原始数据记录下来，按照误差要求及有效数字修约规则进行数据处理。

实验结果及讨论：对于定性实验，根据实验现象对结果进行分析、解释，并写出相应的化学反应方程式；对定量实验，将计算结果与理论值或文献值进行比较，分析产生误差的原因。

实验结论：根据实验现象进行分析、解释，得出正确的结论。

1.5　我国的法定计量单位、化学测量的误差与数据处理

1.5.1　法定计量单位

1971 年十四届国际计量大会（CGPM）决定：国际单位制共有 7 个基本单位，并用国际符号"SI"表示。这 7 个基本单位如表 1－1：

表 1－1　SI 基本单位及符号

量的名称	长度	质量	时间	电流	热力学温度	物质的量	发光强度
单位名称	米	千克	秒	安培	开尔文	摩尔	坎德拉
单位符号	m	kg	s	A	K	mol	cd

国际单位制由 SI 单位和 SI 单位的倍数单位组成。其中 SI 单位分为 SI 基本单位和 SI 导出单位两大部分。SI 单位的倍数单位由 SI 词头加 SI 单位构成。在实际应用中，基本单位、导出单位以及它们的倍数单位是单独或交叉或混合或组合使用的，构成了可以覆盖整个科学技术领域的计量单位体系。一切属于国际单位制的单位都是我国法定计量单位。

1.5.2　测量的误差

在测量过程中不可避免地会产生误差，有些误差是不可避免的，有的误差可加以校正，有的误差则是由于实验过程中的错误造成的。即使采用最可靠的

分析方法，使用最精密的仪器，由很熟练的分析人员操作，也不可能得到绝对准确的结果。

为了使测定结果尽可能接近客观真值，必须了解误差产生的原因及误差出现的规律，并采取相应措施以减小误差。在进行数据处理时，需要对误差的大小做出正确表述，并对分析结果的可靠性和精确程度作出合理判断。

误差产生的原因和分类

根据误差产生的原因，可以大致将误差分为两大类。

1. 系统误差

系统误差是由某种固定原因造成的，它具有单向性、重复性，即正负、大小都有一定的规律性，当重复进行测定时会重复出现，若能找出原因，并设法加以测定，就可以消除，因此也称为可测误差。

产生系统误差的原因主要有以下几个方面：

（1）方法误差　分析方法本身所造成的误差；

（2）仪器误差　仪器本身不够精确引起的误差；

（3）试剂误差　试剂不纯，含有被测组分或干扰物质等所引起的误差；

（4）操作误差　操作人员的主观原因造成的误差。

系统误差的存在影响测定结果的准确度。为了检查分析过程中有无系统误差存在，需做对照试验（与标准样品对照，与标准方法对照），对实验结果用统计检验方法确定有无系统误差。

2. 偶然误差

偶然误差是由某些难以控制、无法避免的偶然因素造成的，其大小、正负都不固定。偶然误差不能通过校正而减小或消除，但统计规律表明，增加测定次数，可以使分析结果的平均值更趋近于真实值。一般的分析结果总是平行测定 4~6 次，用平均值报告结果。

偶然误差的大小决定分析结果的精密度。

由于分析工作人员粗心大意或违反操作规程所引起的误差称为过失，结果应弃去重做。

误差的表示方法

1. 误差与准确度

误差：测定值（x）与真实值（T）之间的差值。

准确度：测定值与真实值符合的程度。

误差的大小是表示准确度高低的尺度。

绝对误差 = 测量值 − 真实值，即 $E = x - T$，有正值或负值，表示测定结果偏高或偏低。

相对误差＝绝对误差/真实值，即 $E_r = E/T$，相对误差能更好地反映误差在真实值中所占的比例，对于比较测定结果的准确度更为合理。

例如称量，绝对误差相同时，称取质量较大的样品的相对误差较小，即该样品称得的准确度较高。

2. 偏差与精密度

在成分分析实验中，一般待测物含量的真实值是未知的，为了得到可信赖的结果，往往在相同的条件下，对试样进行重复多次测定，取其平均值，将任一测得值与平均值进行比较，其差值称为偏差。多次测定值之间相互接近的程度，则称为精密度，表示测定结果重现性的优劣。

偏差的大小表示精密度的高低。

偏差是单独测定值与平均测定值的差异。

绝对偏差：单个测量值与测量平均值之差。绝对偏差的值可正可负。若令 $\bar{x}$ 代表一组平行测量的平均值，则单个测量值 x_i 的偏差 d 为：

$$d = x_i - \bar{x}$$

平均偏差：各单个偏差绝对值的平均值，称为平均偏差，以 $\bar{d}$ 表示，n 次测量结果的平均偏差定义式如下：

$$\bar{d} = \frac{\sum_{i=1}^{n} |x_i - \bar{x}|}{n}$$

相对平均偏差：平均偏差 $\bar{d}$ 与测量平均值 $\bar{x}$ 的比值称为相对平均偏差，定义如下式：

$$相对平衡偏差(\%) = \frac{\bar{d}}{\bar{x}} \times 100\%$$

$$= \frac{\sum_{i=1}^{n} |x_i - \bar{x}|/n}{\bar{x}} \times 100\%$$

相对平均偏差越小，表示实验结果的精密度越高。

3. 准确度与精密度的比较

准确度：表示测量结果与被测组分的真实值接近的程度，用误差大小来衡量。

精密度：表示几次平行测定结果相互接近的程度，用偏差大小来衡量。

1.5.3　有效数字及其运算规则

1. 有效数字

实验数据不仅表示测量值的大小，也反映出测量的精确程度，过多或过少地使用有效数字都会对分析结果的精密度产生误解,因此必须正确记录数字的位数。

有效数字就是实际上能够测量得到的数字,它包括所有确定的数字和第一位可疑数字。有效数字的位数是由实验方法和仪器准确度决定的。

如常用容量仪器,容量分析一般应保证四位有效数字。而量筒的精度为1 mL,故读数可取10.3 mL。台秤的精度为0.1 g,故读数可取0.56 g。分析天平读数可达小数点后5位。

2.有效数字的使用

(1)0的定位认定

数字末位和中间的0有效,第一个非0数字前的0无效。

(2)计算规则

数据处理时,经常遇到一些有效数字位数不相同的数据,因此必须按一定规则进行计算,以节省时间,减少计算错误。

①在记录测定数据时,只保留一位可疑数字。

②加减法。会造成各个数据绝对误差的传递,以小数点后位数最少的数为依据,保留和(或差)的位数。

③乘除法。各个数值相对误差会传递,以有效数字位数最少的数为依据,保留积(或商)的位数。

④对数运算。整数部分代表该数的方次,其有效数字的位数仅取决于尾数部分的位数,而且尾数部分的所有"0"都为有效数字。

⑤非测量数。不是测量所得数字,其有效数字位数可视为无限。

⑥准确度和精密度。多数情况下,只取1~2位有效数字。

⑦用计算器计算时,只对最后的结果进行修约。

(3)修约规则

数据处理时,常遇到一些准确度不同即有效数字位数不同的数字,每一个测量值的误差都要传递到结果上面去,对于这些数据,必须按一定规则修约。当有效数字位数确定后,多余的尾数应弃去。规则是"四舍六入五成双"

当尾数≤4时,舍去;当尾数≥6时,进位。

当尾数=5时,5后面为非0数时进位。

5后面为0时,前位数为奇数时,进位。

前位数为偶数时,舍去。

1.5.4　数据处理

数据是表达实验结果的重要方式之一,因此要求实验者将测量的数据正确地记录下来,加以整理、归纳、处理,并正确地表达由实验结果所获得的规律。在基础化学实验课程中,主要用到列表法和图解法。

1. 列表法

化学实验中,多数测量至少包括有两个变量。在实验数据中,选出自变量和应变量,将两者的对应值尽可能整齐地、有规律地列成表格表达出来,使得全部数据一目了然,便于处理、运算,容易检查而减少差错。

列表时应该注意以下几点。

(1)每一个表的开头都应写出表的序号及简明而又完备的表的名称。

(2)在表的每一行或每一列应正确写出表的栏头,即名称和单位。

(3)表中的数值应用最简单的形式表示,公共的乘方因子应放在栏头注明。

(4)每一行的数字要排列整齐,小数点应对齐。

(5)直接测量的数值可与处理的结果并列在一张表上。必要时应在表的下面注明数据的处理方法或数据的来源。

(6)表中所有数值的填写都必须遵守有效数字规则。

2. 图解法

1)作图工具

处理化学实验数据时,作图工具主要有铅笔、直尺、曲线板、曲线尺、圆规等。

2)坐标纸

用得最多的是直角坐标纸,半对数坐标纸和全对数坐标纸(对数—对数坐标纸)也常用到。

3)坐标轴

用直角坐标纸作图时,以自变量为横轴,应变量(函数)为纵轴,横轴和纵轴的读数不一定从0开始,视具体情况而定。在坐标轴旁应注明该轴变量的名称和单位。在纵轴的左边和横轴的下边每隔一定距离写上该处变量应有的值,以便作图及读数。

4)代表点

代表点是指坐标系中与测得的各数据相对应的点,代表点反映了测得的数据的准确度和精密度。将测得数量的各点绘于图上,在点的周围画上圆圈、方块或其他符号,其面积大小应代表测量的精确度。若同一坐标纸上有几组不同的测量值,则各组测量值的代表点应用不同的符号表示,以示区别,并须在图上注明。

5)曲线

在图纸上画好代表点后,按代表点的分布情况,用曲线板或曲线尺,连出尽可能接近于诸实验点的曲线。曲线应光滑均匀,细而清晰。曲线不必也不可能通过所有的点,但各点在曲线两旁的分布,在数量和远近程度上应近似于相等。

6)图名及图坐标的标注

每个图应有序号和清楚完备而简明的标题(即图名),有时还应对测试条件等作简要说明,这些一般都放置在图的下方(如写实验报告也可在图纸的空白地方写上实验名称、图名、姓名、日期)。

1.6　水的制备与检验

在化学实验中,根据具体任务及要求的不同,对水的质量也有不同的要求,对于一般的化学分析工作,采用蒸馏水或去离子水即可;而对于超纯的物质的分析或有特别要求的分析,则要求使用纯度更高的"高纯水"。

空气中的二氧化碳可溶于水,因而纯水的 pH 一般小于 7.0,在 6.0 左右;特殊情况下,应煮沸赶走二氧化碳之后再使用。

1.6.1　纯水的制备

1. 蒸馏法

目前使用的仪器有玻璃、铜、石英等蒸馏器,只能除去水中不能挥发的杂质,而溶解在水中的气体并不能除去。若需要更纯净的水,则可将一次蒸馏水用石英或硬质玻璃蒸馏器重新蒸馏,获得二次蒸馏水。

2. 离子交换法

利用阴、阳离子交换树脂上的 OH^- 和 H^+ 基分别与天然水中的阴、阳离子进行交换,将水中的杂质离子滞留在交换树脂上,达到纯化水的目的。

3. 电渗析法

这是在离子交换技术的基础上发展起来的一种方法,它是在外电场的作用下,利用阴、阳离子交换膜使天然水中的离子选择性地透过,从而达到溶质与水分离,获得净化水的目的。

1.6.2　纯水水质的检验

纯水的检验有物理方法(测定水的电阻率)和化学方法两类,根据一般分析实验室工作的要求,纯水检验通常有以下几个项目。

电阻率(电阻率越高表示水中的离子越少,水的纯度越高),pH(一般为 6 左右),硅酸盐,氯化物,Cu^{2+}、Pb^{2+}、Zn^{2+}、Fe^{3+}、Ca^{2+}、Mg^{2+} 等金属离子。

1.7　实验仪器安装

实验室常用的玻璃仪器装置，一般是用铁夹将各种仪器依次固定于铁架上而组成。铁夹的双钳应贴上橡皮、绒布等软性物质，或缠上石棉绳、布条等，以避免铁钳直接夹持玻璃仪器不能拧紧或将仪器夹坏。

安装时，先用左手将双钳夹紧，再拧紧铁夹螺丝，做到夹物不松不紧。

总之，仪器安装从下到上，从左到右，做到正确、整齐、稳妥、端正，其轴线与实验台边沿平行。

1.8　溶液的组成与溶液浓度的测定

1.8.1　溶液的组成

两种或两种以上的物质均匀混合，彼此呈分子或离子状态分布的体系称为溶液。溶液的组成是指溶液中溶质(B)和溶剂(A)的具体物种及其相对含量。溶液的组成又称为溶液的浓度，其表示方法很多，现分别介绍：

(1)物质的量浓度　单位体积的溶液中所含溶质的物质的量。

$$c_B = n_B / V \ (\mathrm{mol \cdot L^{-1}})$$

(2)质量浓度　单位体积的溶液中所含溶质 B 的质量。

$$\rho_B = m_B / V \ (\mathrm{g \cdot L^{-1}})$$

(3)体积分数　某气体的分体积与气体混合物总体积之比。

$$\varphi_B = V_B / V$$

(4)质量分数　溶质的质量与溶液总质量之比。

$$w_B = m_B / m$$

(5)物质的量分数　溶质的物质的量与溶液的总的物质的量之比。

$$x_B = n_B / n$$

(6)质量摩尔浓度　单位质量的溶剂中所含溶质的物质的量。

$$b_B = n_B / m_A (\mathrm{mol \cdot kg^{-1}})$$

(7)摩尔比　溶质的物质的量与溶剂的物质的量之比。

$$r_B = n_B / n_A$$

1.8.2　溶液浓度的测定方法

根据测定的手段不同，可分为化学分析法和仪器分析法两大类。化学分析法可通过测量反应终点时标准溶液的体积进行计算。

1.9　物质的酸碱性与溶液 pH 的测定

1.9.1　溶液的酸碱性

如果已知溶液中的 H^+ 或 OH^- 浓度，就可定量地表达出溶液的酸碱度，一般用 $[H^+]$ 表示。室温时：

中性溶液中　　$[H^+] = [OH^-] = 1.0 \times 10^{-7} \, mol \cdot L^{-1}$

酸性溶液中　　$[H^+] > 1.0 \times 10^{-7} mol \cdot L^{-1} > [OH^-]$

碱性溶液中　　$[H^+] < 1.0 \times 10^{-7} mol \cdot L^{-1} < [OH^-]$

考虑到许多溶液的 $[H^+]$ 很小，为了使用方便，常用 pH 表示溶液的酸碱性，pH 的定义为氢离子活度的负对数：$pH = -\lg a(H^+)$。

在稀溶液中，浓度和活度的数值很接近，在实际工作中通常用浓度代替活度，故

$$pH = -\lg[H^+]$$

这样，室温时：

中性溶液中　pH = 7.00

酸性溶液中　pH < 7.00

碱性溶液中　pH > 7.00

1.9.2　溶液 pH 的测定

1. 酸碱指示剂

将指示剂涂在试纸上制成 pH 试纸，使得溶液酸碱性的测定变得更为方便。

2. 酸度计

将玻璃电极、甘汞电极与待测溶液构成原电池，通过测定电池的电动势来确定溶液的 pH，从酸度计上可直接读出溶液的 pH，其测量精度可达 0.001pH 单位，是一种定量测定溶液酸碱性的方法。

3. 酸碱滴定

用已知准确浓度的标准强酸或强碱溶液滴定碱性或酸性未知溶液，以酸碱指示剂颜色变化来确定滴定终点，再根据消耗的标准溶液的体积来计算待测液中的 OH^- 或 H^+ 浓度，这是一种经典和常用的定量测定溶液酸度的方法。

1.10　物质溶解度的表示方法

溶度积和溶解度都可以表示难溶电解质的溶解能力,但两者既有联系又有区别。溶度积是指在一定温度下,难溶电解质饱和溶液中各离子浓度以其化学计量数为指数的乘积;而溶解度是指在一定温度下物质饱和溶液的浓度。在同一温度下,溶度积与溶解度之间一般可进行互相换算。在换算时,应注意浓度应以 $mol \cdot L^{-1}$ 为单位。由于难溶电解质的溶解度很小,即溶液很稀,可以近似地认为它们饱和溶液的密度和纯水一样,为 $1\ g/cm^3$。

设难溶电解质 A_mB_n 的溶解度为 $S(mol \cdot L^{-1})$,在其饱和溶液中

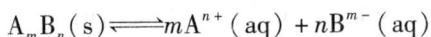

$$A_mB_n(s) \rightleftharpoons mA^{n+}(aq) + nB^{m-}(aq)$$

因为 $[A^{n+}] = mS$,$[B^{m-}] = nS$,

在一定温度下 $[A^{n+}]^m[B^{m-}]^n = K_{sp}$,$K_{sp}$ 为一常数,称为溶度积常数

$$K_{sp} = [A^{n+}]^m[B^{m-}]^n = (mS)^m \cdot (nS)^n$$
$$= m^m \cdot n^n \cdot S^{(m+n)}$$

或

$$S = \sqrt[m+n]{\frac{K_{sp}}{m^m \cdot n^n}}$$

上式就是难溶电解质的溶解度与其溶度积的定量关系式。

1.11　实验室三废治理方案

化学实验不可避免地会产生"三废"(废水、废气、废渣),因此化学实验室也是不可忽视的环境污染源,要解决实验室的污染问题,首先应对传统的实验方法进行改革,以减少"三废"的产生,在此基础上再进行"三废"的治理。治理的原则是:化害为利、变废为宝、综合利用。

1.11.1　废水的处理

实验室排放的废水,会对水源造成严重的污染,危害人体健康,并使自然环境受到破坏,同时对渔业、畜牧、农业和林业也会带来极大的危害,此外还会使金属制品遭受腐蚀、建筑材料受到损害。对废水的处理,常用以下方法:

1. 物理方法

(1)沉淀法　废水中的悬浮固体一般采用沉淀法除去。此法是利用固体与水两者密度上的差异,使固体和液体分离。该法被广泛用作废水的预处理。

沉淀法又分为自然沉淀法和混凝沉淀法两种。自然沉淀法是依靠废水中固体颗粒自身重量进行沉淀，这种方法只能除去废水中的较大颗粒，属于物理方法。混凝沉淀法是在废水中加入特定的混凝剂，使废水中的微小颗粒与混凝剂结成较大的胶团，在水中加速沉降。该方法实质是一种化学方法。

（2）过滤法　废水中含有的微粒物质和胶状物质，可以采用机械过滤的方法加以清除。过滤操作的方式很多，常压过滤主要是依靠自身的重力作用，可以在敞开式的过滤池中进行，而加压过滤和真空过滤则需要在密闭的容器中进行。

（3）离心分离法　利用高速旋转所产生的离心力，使废水中的悬浮颗粒分离。由于悬浮颗粒的密度与液体的密度不同，密度大的固体颗粒在高速旋转的过程中所受到的离心力较大，被甩到外圈，沿器壁下沉，从而达到分离。

2．物理化学法

废水经过物理方法处理后，仍含有某些细小的悬浮物以及溶解的有机物、无机物。为了进一步除去残存在水中的污染物，可以采用物理化学方法处理，常用的方法有吸附法、浮选法、电渗析法、反渗透法等。

（1）吸附法　利用多孔固体物质作为吸附剂，以吸附废水中的某些污染物的方法。常用的吸附剂有活性炭、硅藻土、铝矾土、磺化煤、矿渣以及树脂等。其中以活性炭最为常用，因为活性炭具有很好的吸附性能、机械性能和化学稳定性，而且可再生利用，价格便宜，来源方便。

（2）浮选法　向废水中加入浮选剂或凝聚剂、表面活性剂、调整剂等，然后通入空气，当空气通入到含有细小颗粒物质的废水中时，废水中会产生泡沫，颗粒黏附于气泡上，然后将泡沫排除达到分离的目的。

（3）反渗透法　这是一种利用半渗透膜进行分子过滤来处理废水的新方法，半渗透膜可以使水通过，但不能使水中的悬浮物及溶质通过。由于膜两边水分子浓度差的关系，水分子可以从稀溶液一边向浓溶液一边渗透，如果在浓溶液一边施加压力，迫使水分子逆向渗透，则称为反渗透。结果使废水中的悬浮物及溶质被分离，达到废水净化的目的。

电渗析法能耗较高，仅用于废水中有效成分的回收。

3．化学方法

化学法是利用物质之间的化学反应，进行废水处理的方法，主要有中和法、氧化还原法、化学絮凝法等。

（1）中和法　主要用于处理含酸或含碱废水。对含酸或含碱浓度高的废水，首先必须考虑回收，综合利用。浓度在4%以下时，没有回收利用价值，则应中和至中性才可排放。通常，应尽量选用碱性废水来中和酸性废水，达到以废治废的目的。

（2）混凝沉淀法　在废水中加入混凝剂，混凝剂为电解质，在废水中形成胶团，与废水中的胶体物质产生电中和，形成绒粒沉降。

（3）氧化还原法　废水经过化学氧化处理，可使废水中所含有的有机物质和无机还原性物质进行氧化分解，达到净化的目的，此外还可以达到去臭、去味、去色的效果。废水处理中使用最多的氧化剂是臭氧、次氯酸、氯和空气。

4. 生物化学方法

生物化学处理方法是利用自然界大量存在的各种微生物来分解废水中的有机物质的一种方法。其原理是：在微生物酶的催化作用下，依靠微生物的新陈代谢作用使废水中的有机物质氧化、分解，最终转化为稳定的、无毒的无机物而被除去。根据生化处理过程中起主要作用的微生物种类的不同，又分为以下两大类。

（1）好气生物处理　在有氧条件下，利用好气微生物的作用来处理有机物废水的方法。

（2）厌气生物处理　在无氧条件下，利用厌气微生物的作用来处理有机物废水的方法。

实践证明，生物化学处理法处理废水效率高、费用低、设备简单，是一种比较经济、实用的方法。

1.11.2　废气的处理

大气污染，是指由于有害气态物质进入大气，使大气的成分、气味、颜色和性质等发生变化，危害生物的生活环境，影响人体健康和动植物的生存。

化学实验室排放的废气中主要污染物质有二氧化硫、氮氧化物、氟化物、碳化物和各种有机气体。目前处理气态污染物的方法主要有吸收、吸附、冷凝和燃烧等方法。

吸收法是普遍采用的方法，一般可归纳为以下几种类型：

水吸收法；酸吸收法（如硫酸法、稀硝酸法等）；碱性溶液吸收法（如烧碱法、纯碱法、氨水法等）；还原吸收法（如氯氨法、亚硫酸盐法等）；氧化吸收法（如次氯酸钠法、高锰酸钾法、臭氧氧化法等）；生成配合物吸收法（如硫酸亚铁法等）；分解吸收法（如酸性尿素水溶液吸收法等）。

上述各方法对不同的气态污染物有不同的吸收效果，因此这些方法既可以单独使用也可以组合作用。

吸收气态污染物后的溶液可按照废水处理办法进一步进行处理。

1.11.3 废渣的处理

化学实验过程中产生的废渣如果作为废弃垃圾，也同样会对环境造成污染。由于实验过程产生的废渣量小，一般情况下，都是收集储存，然后集中处理。

处理废渣的方法大致有以下几种：再生处理法、热分解法、焚烧法、湿式氧化法和化学处理法等。

（编者：何跃武　刘绍乾　王曼娟）

第二章　基础化学实验基本仪器与操作

2.1　化学实验室常用的基本仪器

化学实验室使用的玻璃器皿，按它们的用途大体可分为容器类、量器类和其他常用器皿三大类。下面列出的是常用的玻璃仪器（表2-1）。

表2-1　常用的玻璃仪器

仪　器	规　格	主要用途	注意事项
试管　离心试管	分为硬质试管、软质试管、普通试管、离心试管。普通试管以管口外径×长度（mm）表示，如：15 mm×75 mm。离心试管以容积（mL）表示	用作少量试剂的反应容器，便于操作和观察。离心试管还可用于定性分析中的沉淀分离	可直接用火加热。硬质试管可以加热至高温。加热后不能骤冷，特别是软质试管更容易破裂。离心试管只能用水浴加热
量筒	以所能量度的最大容积（mL）表示。上口大，下口小的叫量杯	用于液体体积的量度	不能加热，不能用作反应容器
烧杯	以容积（mL）大小表示。外形有高、低之分，并分为带有刻度的和无刻度的	用作反应物较多时的反应容器，使反应物易混合均匀	加热时应放置在石棉网上，使受热均匀，反应液体不得超过容积的2/3

续上表

仪　器	规　格	主要用途	注意事项
滴瓶　细口瓶　广口瓶	以容积（mL）大小表示。分为棕色、无色两种	广口瓶用于盛放固体药品，滴瓶、细口瓶用于盛放液体药品，不带磨口塞子的广口瓶可作集气瓶	不能直接用火加热。瓶塞不要互换，如盛放碱液时，要用橡皮塞，不能用磨口瓶塞，以免玻璃磨口瓶塞被腐蚀黏结
表面皿	以口径（mm）大小表示	用作烧杯盖，防止流体迸溅或其他用途	不能用火直接加热
容量瓶　E20°　100mL	以刻度线以下的容积（mL）大小表示	用于配制准确浓度的溶液，配制时液面应恰好与刻度线相切	不能加热，瓶塞不能互换并防止打碎
锥形瓶	以容积（mL）表示。分为有塞的和无塞的	反应容器，适用于滴定操作，便于摇荡	加热时应放置在石棉网上，使受热均匀，盛液不能太满
称量瓶	以外径（mm）×高（mm）表示。分扁形和高形两种	用于准确称取一定量的固体物质	不能直接用火加热，盖子不能互换
漏斗　长颈漏斗	以口径（mm）大小表示	用于过滤等操作，其中长颈漏斗适用于定量分析中的过滤操作	不能直接用火加热

续上表

仪　器	规　格	主要用途	注意事项
吸滤瓶　布氏漏斗	布氏漏斗为瓷质的，以容量(mL)或口径(cm)大小表示，吸滤瓶以容积大小表示	两者配套使用，用于无机制备中晶体或沉淀的减压过滤，利用水泵或真空泵降低吸滤瓶中压力以加速过滤	不能直接用于加热，滤纸要略小于布氏漏斗的内径
圆形分液漏斗　梨形分液漏斗	以容积(mL)大小和形状(球形、梨形)表示	用于互不相溶的液—液分离，也可用于气体发生器装置中滴加液体	不能用火直接加热，漏斗的磨口塞子不能互换，活栓处不能漏液
蒸发皿	以口径(cm)或容积(mL)大小表示，有瓷、石英、铂等不同材质的，并有有柄和无柄的	蒸发液体用，随液体性质不同可选用不同材质的蒸发皿	能耐高温，但不宜骤冷，蒸发溶液时，一般应放在石棉网上加热，也可直接用火加热
坩埚	以容积(mm)大小表示。有瓷、石英、铁、镍或铂等不同材质的	灼烧固体用，随固体性质不同可选用不同材质的坩埚	可用火直接灼烧至高温，灼热的坩埚不要直接放在桌上(可放在石棉网上)
研钵	以口径大小表示。有瓷、玻璃、玛瑙或铁等材质的	用于研磨固体物质。按固体的性质和硬度选用不同材质的研钵	不能用火直接加热，不能敲击，只能挤压研磨
水浴锅	铜或铝质制品	用于间接加热，也用于控温实验	使用时要及时补充水，防止将水烧干

续上表

仪　器	规　格	主要用途	注意事项
干燥器	以外径（mL）大小表示。 分普通干燥器和真空干燥器	内放干燥剂，可保持样品或产物干燥	防止盖子滑落而打破，红热的物品待冷到室温后才能放入
短颈平底烧瓶　长颈圆底烧瓶	以容积（mL）表示，烧瓶有平底的和圆底的	常用作反应容器	加热时放置在石棉网上使加热均匀
蒸馏烧瓶	以容积（mL）表示	用于液体蒸馏，也可用于少量气体的发生	加热时放置在石棉网上使加热均匀

2.2　常用玻璃仪器的洗涤和干燥

2.2.1　仪器的洗涤

化学实验室经常使用各种玻璃仪器，而这些仪器是否干净，常常影响到结果的准确性，所以应保证所使用的仪器是很干净的。"干净"两字的含义决不是我们日常所说的干净，而是具有纯净的意思。

洗涤玻璃仪器的方法很多，应根据实验的要求、污物的性质和玷污的程度来选用。一般说来，附着在仪器上的污物既有可溶性物质，也有尘土和其他不溶性物质，还有油污和有机物质。针对不同情况，可以分别采用下列洗涤方法：

1. 用水刷洗

用毛刷就水刷洗，既可以使可溶物溶去，也可以使附着在仪器上的尘土和不溶物质脱落下来。但往往洗不去油污和有机物质。

2．用去污粉、肥皂或合成洗涤剂洗

肥皂和合成洗涤剂的去垢原理已众所周知，不必再重述。去污粉是由碳酸钠、白土、细沙等混合而成的。使用时，首先把要洗的仪器用水湿润（水不能多），洒入少许去污粉，然后用毛刷擦洗。碳酸钠是一种碱性物质，具有强的去油污能力，而细沙的摩擦作用以及白土的吸附作用则增强了仪器清洗的效果。待仪器的内外器壁都经过仔细的擦洗后，用自来水冲去仪器内外的去污粉，要冲洗到没有微细的白色颗粒状粉末留下为止。最后，用蒸馏水冲洗仪器三次，把从自来水中带来的钙、镁、铁、氯等离子洗去，每次的蒸馏水用量要少一些，注意节约（采取"少量多次"的原则）。这样洗出来的仪器的器壁就完全干净了，把仪器倒置时就会观察到仪器中的水可以完全流尽而没有水珠附在器壁上。

3．用铬酸洗液洗

这种洗液是由等体积的浓硫酸和饱和的重铬酸钾溶液配制成的，具有很强的氧化性，对有机物和油污的去污能力特别强。在进行精确的定量实验时，往往遇到一些口小、管细的仪器很难用上述的方法洗涤，这可用铬酸洗液来洗。

往仪器内加入少量洗液。将仪器倾斜并慢慢转动，让仪器内壁全部被洗液湿润；转几圈后，把洗液倒回原瓶内，然后用自来水把仪器壁上残留的洗液洗去；最后用蒸馏水洗三次。

如果用洗液将仪器浸泡一段时间，或者用热的洗液洗，则效果更好。但要注意安全，不要让热洗液灼伤皮肤。

能用别的方法洗干净的仪器，就不要用铬酸洗液洗，因为后者成本较高。但实验要求高的仪器除外。

洗液的吸水性很强，应随时把装洗液的瓶子盖严，以防吸水，降低去污能力。当洗液用到出现绿色（重铬酸钾还原成硫酸铬的颜色），就失去了去污能力，不能继续使用。

4．特殊物质的去除

应根据器壁上的物质的性质，对症下药，采用适当的药品来处理。例如沾在器壁上的二氧化锰用浓盐酸来处理时，就很容易除去。

凡是已洗净的仪器，决不能再用布或纸去擦拭。否则，至少布或纸上的纤维将会留在器壁上而玷污仪器。

2.2.2　仪器的干燥

仪器干燥可以采用下列方法。

1．晾干

不急用的仪器洗净后可倒置于通风干燥处，或仪器架上，任其自然晾干。

2．吹干

用压缩空气或吹风机把仪器吹干。

3．用烘箱加热烘干

洗净的仪器可放在电烘箱(图2－1)内烘干。烘箱内的温度根据需要调节(例如控制在105℃左右)，注意仪器口朝下(倒置后不稳的仪器应斜放)，在烘箱的最下层放一搪瓷盘，以接受从仪器上滴下的水珠，防止水滴到电炉丝上，以免损坏电炉。

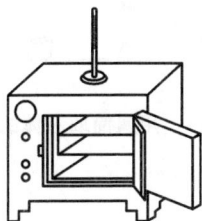

图2－1　电烘箱

图2－2　烤干试管

4．用火加热烤干

烧杯或蒸发皿可以放在石棉网上，用小火烤干(在用火烤之前，容器外壁的水要先擦干)。

试管可以用火直接烤干。操作时，先将试管外壁揩干，然后将试管略为倾斜，管口朝下，以免水珠倒流炸裂试管(图2－2)，火焰从底部开始缓慢向下移至管口，如此反复烘烤直到不见水珠后再将管口朝上，以赶尽水汽。

5．用有机溶剂干燥

带有刻度的计量仪器，不能用加热的办法干燥，因为这会影响仪器的精密度。可以用易挥发、易与水混溶的有机溶剂(最常用的是酒精或酒精与丙酮的混合液)加到洗净的仪器里，倾斜倒出，少量残留在仪器中的混合液很快挥发而干燥。若用吹风机或压缩空气吹仪器则干燥得更快。

2.3　基本衡量仪器及其使用方法

2.3.1　容量仪器及其使用方法

量筒、量杯、移液管、吸量管、滴定管、容量瓶是实验室中常用的度量液体体积的容量仪器。在读取液体体积时，常以液体弯月面的最低点为准，即仪器垂直时，使视线与液体的弯月面的低点处保持同一水平面，弯月面最低点与刻度线水平相切的刻度就是液体体积的读数(图2-3)。

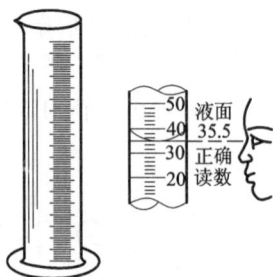

图2-3　量筒及其读数法

图2-4　用量筒取液体的操作

1. 量筒和量杯

量筒和量杯常用于对液体体积要求不十分精确的液体的量度，量筒和量杯有 10 mL、20 mL、100 mL、1000 mL 等多种规格，使用时可按具体情况选用较合适规格或几种规格的量筒。

用量筒量取液体时，用左手持量筒，大拇指指尖指示所需体积的刻度处，右手持药瓶(注意标签应朝手心处)，瓶口紧靠量筒口边缘，慢慢注入液体(图2-4)到所需刻度。

2. 移液管和吸量管

(a)移液管　　(b)吸量管

图2-5　移液管与吸量管

移液管、吸量管是用于准确移取一定体积的液体的仪器。移液管的中间有一膨大部分，管颈上部刻有一条刻度线，如图2-5(a)，而吸量管是一根内径均匀的、管上刻有分刻度的玻璃管，如图2-5(b)。移液管有 5 mL、10 mL、20 mL、25 mL、50 mL 等规格。吸量管有 0.2 mL、0.5 mL、1 mL、2 mL、5 mL、10 mL、20 mL、25 mL 等规格，最小分度为 0.1 mL、0.02 mL、0.01 mL 等。

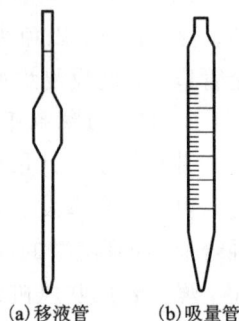

量液时先将移液管洗干净，然后用少量被量取的液体淌洗三次，以免量取的液体被残留在移液管内壁的蒸馏水所稀释。

移液管的取液操作是用右手大拇指和中指持住移液管的上部位，将移液管下尖端插入到液面下适当的深度(不能太浅，以免液面下降时吸入空气)。左手拿洗耳球，先挤出球内空气，洗耳球的尖嘴对准移液管的上口[图2-6(a)]，慢慢松开左手手指，使液体吸入管内。当液面上升到刻度以上时迅速拿开洗耳

球,同时用右手食指迅速按住管口,然后用
拇指和中指轻轻转动移液管使食指稍微松
动,此时管内液面慢慢下降,直至弯月面最
低点与刻度线相切,食指即用力按紧管口,
使液体不再流出。提取移液管,使管的下
尖端与盛液容器的内壁靠一靠去掉悬挂在
移液管口的液滴,取出移液管,移至已准备
好的接受液体的容器中,让移液管的下尖
端紧靠容器内壁[图2-6(b)],松开食指,
让液体自然流出,待液体全部流出后,停留
片刻(约15s),取出移液管。最后留在尖端
的液滴不能吹出,因为移液管在容积校正
时,已不计残留的液体。也有少数移液管
上面标有"吹"字,使用这种移液管时必须
用洗耳球将管内的残液吹入接收器中。

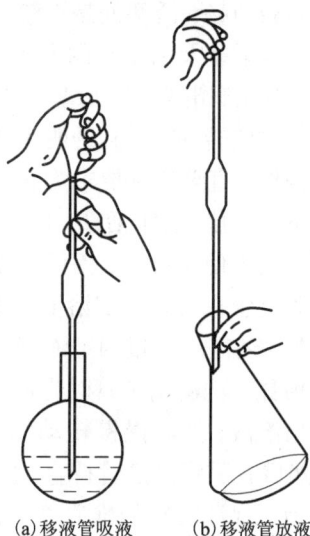

(a)移液管吸液　　(b)移液管放液

图2-6　移液管的使用

3. 滴定管

滴定管是具有刻度的细长玻璃管,内径均匀,精确度高,可准确地连续量
取不同体积的液体,用于滴定分析,使用方便。

滴定管分为酸式滴定管与碱式滴
定管两种(图2-7),有5 mL,10 mL,
25 mL,50 mL等规格。

碱式滴定管用于盛放碱性溶液,
它的下端用橡皮管连接一个带有尖嘴
的小玻璃管,橡皮管内装有一玻璃珠,
玻璃珠起活塞阀门的作用。使用时只
要用大拇指和食指紧捏玻璃珠处的橡
皮管,使玻璃珠往一边挤压,橡皮管内
会形成一条狭缝,溶液便会流出,通过
手指用力的大小可以控制液体流量。
若碱式滴定管漏液则应更换橡皮管或
玻璃珠。

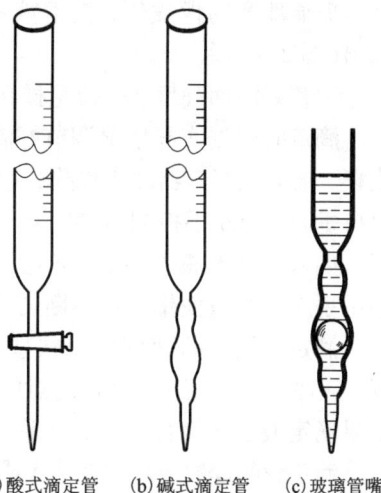

(a)酸式滴定管　(b)碱式滴定管　(c)玻璃管嘴

图2-7　移液管与吸量管

酸式滴定管用于盛放酸性溶液,
它的下端有一玻璃活塞,其作用是控制滴定过程中溶液的流出速度。滴定管使
用前应检查是否漏液,检查的方法是在滴定管内注入半管水,转动活塞,观察

水流是否通畅，活塞处是否渗漏。若转动不灵活或渗水，可将活塞取出，用滤纸擦干净活塞及活塞孔的内壁，然后在活塞大小两端涂上一层凡士林(不能过多，恰好能润滑为宜，活塞中段有孔的一圈不要涂抹)。小心插回活塞，将活塞沿一个方向转动，直到透明为止。再一次检查是否漏液，若达不到要求，重复上述操作，直至达到要求为止。最后用一小橡皮圈套在活塞的小端(软橡皮管上剪下一小段)套牢活塞，以免活塞脱落打碎(活塞打碎后，这支酸式滴定管就完全失去了使用价值，因而报废，造成浪费)。

　　滴定管在装入滴定溶液前要洗涤，除了用铬酸洗液、自来水及蒸馏水依次洗涤外，还需用少量滴定液(约10 mL)淌洗三次，以免滴定液被管内壁残留的水所稀释。洗涤滴定管时先注入洗涤液或蒸馏水，再将滴定管接近平持，但不要倒出溶液，然后慢慢转动滴定管，使洗涤液或蒸馏水与内壁各个部分都接触，然后右手将滴定管持直，左手打开活塞，放出洗涤液或水。

　　为了保证实验的准确度，滴定前须驱赶出滴定管内的气泡，滴定管中的气泡的清除方法如下：酸式滴定管中气泡的清除，先在滴定管中装满溶液，然后将滴定管倾斜成约30°角，左手迅速打开活塞，使溶液冲出，利用溶液的急流把气泡冲出。碱式滴定管中气泡的清除，则把橡皮管向上弯曲，玻璃小管嘴尖也斜向上，两指用力挤压玻璃珠旁的橡皮管，使溶液从嘴尖冲出，气泡随之逸出(图2-8)。

图2-8　排除气泡

　　调整好液面高度，去掉悬挂在尖嘴端的液滴，记下开始时的读数，开始滴定。滴定时根据实际情况调节好滴定管在铁架台上的最适宜的高度，原则上是烧杯、锥形瓶等容器便于操作，运用自如。酸式滴定管的活塞柄向右，滴定管保持垂直，用左手控制活塞，卡于左手的虎口处，用拇指、食指和中指转动活塞。如果在锥形瓶内滴定，滴定管下嘴尖伸入瓶口内，右手持瓶颈，手心要空，利用手腕的力量使锥形瓶不断转动(图2-9，图2-10)。

　　滴定开始时的速度可以快些，但要成滴不能成柱，接近终点时应减慢速度，应逐滴滴入瓶中，接近终点的最后几滴溶液更应该小心慢放，以防止过量。如果滴定液滴入烧杯中，则右手用玻璃棒不断轻轻搅动溶液。

　　滴定管的读数应注意在注入或放出溶液后等待1~2 min，待附着于内壁上的溶液完全流下后再读数。25 mL和50 mL的滴定管的刻度一般细分至0.1 mL。读数应读到小数点后第二位数值，如0.02 mL、15.34 mL、27.58 mL等，末尾一位数字是估计值。为了方便观察，读数时可在滴定管后衬上一黑色纸卡片，将卡片上移至弯月面下1 mm左右处，使弯月面的反射层染上黑色，便于读数(图2-11)。

图 2 - 9　酸式滴定管操作

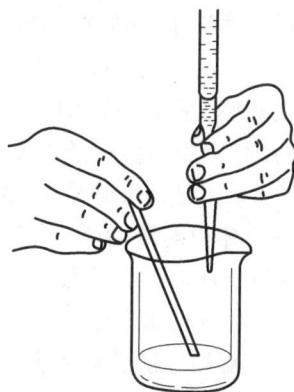

图 2 - 10　碱式滴定管操作

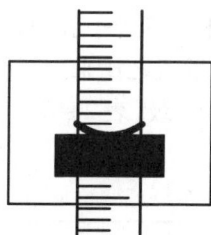

图 2 - 11　滴定管读数

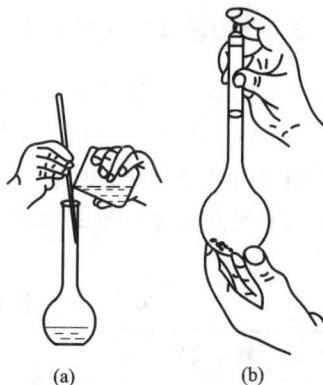

(a)　　　　　(b)

图 2 - 12　容量瓶的使用

4. 容量瓶

如图 2 - 12 所示，容量瓶是细颈的平底玻璃瓶，瓶口和瓶塞带磨口，颈上有刻度线的容量仪器。瓶身上通常注明指定温度下的容积。容量瓶通常用来配制准确浓度的溶液。容量瓶的使用方法如下：

(1)使用前应检查容量瓶是否漏水，瓶中加入自来水盖好塞子，右手(或左手)按住塞子，左手(或右手)托住瓶底部分，将瓶子倒立数分钟，观察瓶塞与瓶口处是否漏水，不漏水才可供使用。用细绳把配套的容量瓶和塞子相互连接起来，以免容量瓶使用中瓶塞错配，引起漏水，或不小心打破。

(2)按玻璃仪器的洗涤方法洗净容量瓶，编号备用。

(3)用固体物质配制溶液时，将已准确称量的物质在烧杯中用少量水溶解，然后将溶液转移到容量瓶中，如图 2 - 12(a)所示，用蒸馏水少量多次冲洗烧杯

（约 3 次），洗涤水也转移到容量瓶中，这时溶质已全部转移到容量瓶中。加入蒸馏水，直到刻度线。最后盖好塞子，一手握瓶颈，食指按住瓶塞，另一只手五指托住瓶底，将容量瓶倒置，待气泡上升后，加以摇动。如此反复倒转数次，就能使瓶内的溶液混合均匀，如图 2 - 12（b）所示。

（4）如果用已知准确浓度的浓溶液稀释成准确浓度的稀溶液，则可用移液管移取一定体积的浓溶液于容量瓶中，然后按上述方法加蒸馏水至刻度，摇匀，即配制成所需的稀溶液。

（5）热溶液不能直接倒入容量瓶中，固体经加热溶解成的溶液，必须待溶液冷却到室温后才能转移到容量瓶中，否则溶液体积会造成较大的误差。

（6）实验精度要求很高的实验，所用的容量瓶的容积都必须用天平称量的方法进行校正，以便得到所用的容量瓶的精确容积。

2.3.2　衡器及其使用方法

天平是实验室衡量质量用的衡器。天平的种类很多，一般可按其结构、精度、用途或称量范围等分类。通常所说的天平是指杠杆天平，我国目前采用的方法是相对精度分类法，即以天平分度值与最大称量值之比来划分精度级别，规定按天平名义分度值与最大载荷之比将天平分成 10 级（表 2 - 2）。

表 2 - 2　天平的等级

精度级别	1	2	3	4	5
名义分度值与最大载荷的比值	1×10^{-7}	2×10^{-7}	5×10^{-7}	1×10^{-6}	2×10^{-6}
精度级别	6	7	8	9	10
名义分度值与最大载荷的比值	5×10^{-6}	1×10^{-5}	2×10^{-5}	5×10^{-5}	1×10^{-4}

1. 托盘天平

托盘天平又称台秤，是实验室中常用的称量仪器。托盘天平多为双盘，一般能准确到 0.1 g。托盘天平的构造如图 2 - 13 所示。横梁架在托盘天平座上，横梁左右各有一个盘子用来承重物和砝码。横梁中部的上面或下面有指针，用以指示天平的平衡状态。

使用托盘天平前，先将游码拨至刻度左端刻度"0"处，检查指针是否停在刻度盘上中间位置，如果指针停在刻度盘中间位置，则无须调节便可使用。否则，须调节托盘下面的螺旋（可进可出），使横梁重心改变位置，使指针停在中间位置（称为零点）。称量时，左边放称量物，右边放砝码，5 g 或 5 g 以上的砝码放在砝码盒内，5 g以下的砝码通过移动游标尺上的游标来调节。当添加砝码使托盘天平两边平衡时，指针停在近中间位置，称为停点，停点和零点之间允许偏差在1小格之内，此时砝码所示的质量就是称量物的质量。称量应注意以下几点：

（1）称量药品时不能直接放在托盘上，应在左盘先放上纸片或表面皿、小烧杯、称量瓶等盛器，药品再放在盛器上称量。易潮解或具有腐蚀性的药品必须放在玻璃容器内，不可放在纸片上称量。

（2）不能称量热的物体，热的物体需冷却到室温后再称量。

（3）称量完毕后，砝码应放回砝码盒中，并将游砝退回到刻度"0"处。

（4）台秤应经常保持清洁，托盘上撒有药品时，必须立即清除干净。

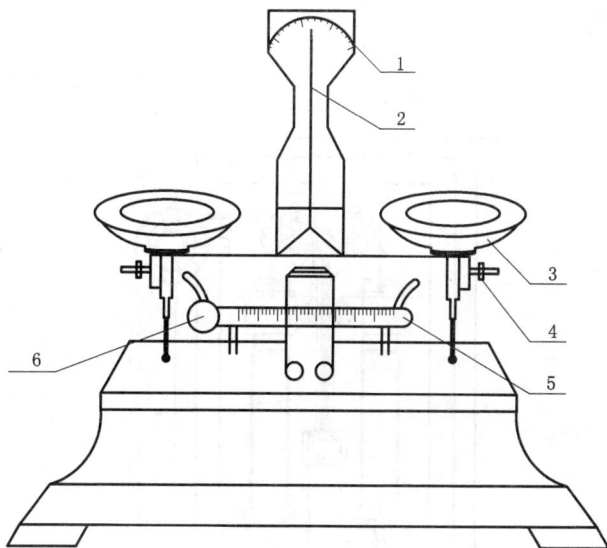

图2-13　托盘天平

1—标尺；2—指针；3—托盘；4—螺旋；5—刻度尺；6—游码

2. 分析天平

（1）分析天平的工作原理

分析天平是实验室最常用的一种精密称量仪器。其称量的精确度一般为万分之一克，与托盘天平一样是根据杠杆原理设计的，如图2-14所示。设有一杠杆 ABC，B 为支点，A、C 两端所受的力为 W_1、W_2，当达到平衡时，支点两边的力矩相等，即

$$W_1 \times AB = W_2 \times BC$$

因为天平是等臂的，$AB = BC$，故 $W_1 = W_2$

而 $W = mg$（g 为重力加速度），有 $m_1 g = m_2 g$，$m_1 = m_2$ 即在等臂天平中被称物的质量 m_1 等于砝码的质量 m_2。

当分析天平使用不当时，会造成支点刀 B 的磨损，致使支点两边的力臂不相等，精度下降，因此应注意保护好支点刀。

图 2 – 14　杠杆原理示意图

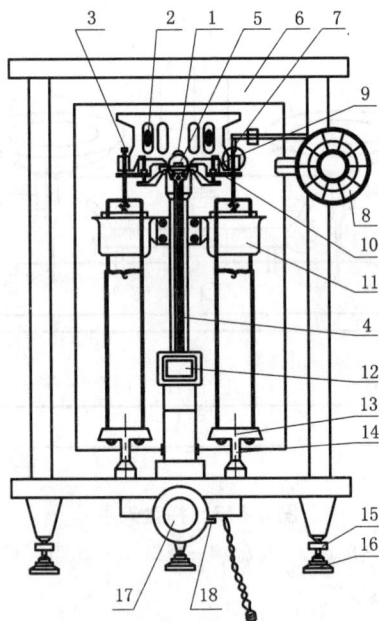

（2）分析天平的结构

半自动机械加码电光天平是一般化学实验室中使用较多的一种分析天平，其结构如图 2 – 15 所示。它主要由天平梁、升降枢、光学读数装置、机械加码装置等部件组成。

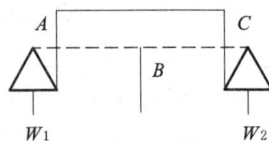

图 2 – 15　半自动电光分析天平

1—横梁；2—平衡螺丝；3—吊耳；4—指针；5—支点刀；6—框罩；
7—环码；8—指数盘；9—支力销；10—托叶；11—阻尼筒；12—投影屏；
13—称盘；14—托盘；15—螺旋脚；16—垫脚；17—升降枢；18—微动拨杆

①天平梁　天平梁是天平的主要部件，由铝合金制成。梁的中间和两端装

有三个三棱形的玛瑙刀，中间的玛瑙刀的刀口向上，放在天平柱的玛瑙刀面上，天平两端的两个玛瑙刀的刀口向下，通过蹬(或称吊耳)上的玛瑙平板与刀口接触，蹬上有挂托盘与空气阻尼器内筒的悬钩。中间的玛瑙刀称为支点刀，两端的玛瑙刀称为承重刀。支点刀与两端承重刀口的尖锐程度决定着天平的灵敏度，应注意保护刀口防止损伤。另外在梁的正中装有细长的指针，指针的偏转可以指示梁的平衡情况。梁的两边各装有一个平衡螺丝，用来调节天平的平衡位置。

②升降枢　为了保护刀口，当天平不工作时，由两个托叶将天平梁支住，使刀口离开天平柱上的承刀面；当天平工作时，则由升降枢将托叶放下，刀口缓缓落在天平柱的刀承面上，使天平处于工作状态。另外，升降枢还控制着电光读数装置的开关，升降枢放下，电源开关打开，可在投影屏上看到缩微标尺的投影；升降枢上升，则电源切断。

③托盘和阻尼装置　天平有两个托盘挂在吊耳下，可分别用于放置砝码和被称物品。盘下装有盘托，在支起托叶的同时，盘托支持托盘，防止其摆动。为了尽快地使天平静止下来，提高称量速度，吊耳下面安装了阻尼器，它是由两只内外相互套合而彼此不接触的铝盒构成。外盒固定在天平柱上，内盒挂在吊耳上面，利用空气阻尼作用使天平很快停止摆动而达到平衡。

④砝码和机械加码装置　砝码放在砝码盒(图2-16)内。最大质量为100 g，最小质置为1 g。1 g以下10 mg以上是用金属丝做成的环码，分别挂在环码钩上，利用机械加码装置(图2-17)来加减环码，可在10~990 mg的范围内调节质量，其中外轮调节范围为100~900 mg，内轮调节范围为10~90 mg的环码。10 mg以下则由电光读数装置直接读出。

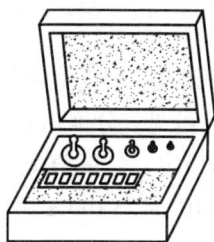

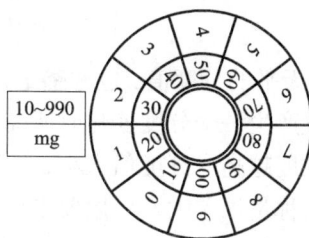

图 2-16　砝码盒　　　　　　　　　图 2-17　机械加码指数盘

⑤光学读数装置　光学读数装置如图2-18所示，由升降枢控制电路的开关，小灯泡的光线将固定在指针下端的微分标尺经放大、反射，在投影屏上显示出来，并能准确地读出10 mg以下的质量。在升降枢下面有一微动调节杆，可使投影屏左右移动，用于天平零点调节。

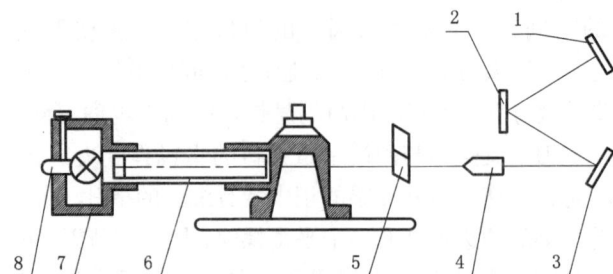

图2-18　光学读数装置示意图

1—投影屏；2、3—反射镜；4—物镜筒；
5—微分标牌；6—聚光管；7—照明筒；8—灯头

⑥天平橱和水平仪　为了防止灰尘、酸气、水气的侵蚀以及空气对流对天平称量的影响而设置了天平橱。橱上有三个门，前面的门是清理、维修时才打开，左右两个门供取放被称物和砝码用。称量读数时，三个门都要关闭。天平要保持水平，才能进行准确称量，因此在天平柱上装有气泡水平仪，指示天平是否水平，当气泡在水平仪的正中央时天平才水平，否则需用天平橱下的两只螺旋脚来调节。一般是由安装天平者进行调节。

（3）分析天平的灵敏度

天平的灵敏度通常是指在天平的一个盘上增加1 mg质量所引起的指针偏转的程度，指针偏转程度愈大，天平的灵敏度愈高。灵敏度（E）的单位为分度/毫克。在实际工作中，常用灵敏度的倒数来表示天平的灵敏度，即

$$S = \frac{1}{E}(\text{mg/分度})$$

S称为天平的分度值，也称感量，单位为mg/分度。因此，分度值是使天平的平衡位置产生一个分度变化时所需的质量值（mg）。可见，分度值越小的天平，其灵敏度越高。分度值太小，灵敏度太高则天平不稳定；分度值太大，灵敏度太低则称量误差大。一般要求分析天平的分度值为10 mg/（100±2）分度。

（4）称量方法

①直接称量法

称量物体前，先测定天平零点，然后把物体放在天平左盘上，在右盘上加砝码，使其平衡点与零点重合，此时砝码所表示的质量就等于物体的质量。

②固定质量称量法

这种方法是为了称取指定质量的物品，一般为粉末或液体。要求被称量物品本身不吸潮并在空气中性质稳定，其方法如下：

a. 先称容器(如表面皿、称量瓶、铝铲)的质量,并记录平衡点。

b. 如指定称取 0.4000 g 时,在右边称盘增加 0.400 g 砝码,在左边称盘的容器中加入略少于 0.4 g 的物品,然后再往容器中慢慢补加物品,直至平衡点与称量容器时的平衡点刚好一致。

这种方法的优点是称量过程简单,结果计算方便。因此,在工业生产分析中,广泛采用这种称量方法。

③递减称量法

这种方法对称出物品的质量不要求固定在某一数值,只需在要求的称量范围内即可,适于称取多份易吸水、易氧化或易与 CO_2 反应的物质。将此类物质盛在带盖的称量瓶中进行称量,既可防止吸潮和防尘,又便于操作。其方法于下:

a 在称量瓶中装适量待称物质(如果是曾经烘干的固体物质,应放在干燥器中冷却到室温),用洁净的小纸条或塑料薄膜条,套在称量瓶上,拿取放在天平盘上,设称得其质量为 $m_1(g)$。

b. 将称量瓶取出,并从右盘取出与某一数量物质质量相当(最好是略小)的砝码。在准备承装的容器上方打开瓶盖,用称量瓶盖轻轻地敲瓶的上部,使瓶内物质慢慢地倾出一部分于承装的容器中,然后慢慢地将瓶竖起,用瓶盖轻敲瓶口的上部,使沾在瓶口的物质落入瓶中,盖好瓶盖。再将称量瓶放回天平盘上称量,如此重复操作,直到倾出的物质质量达到要求为止。设其质量为 m_2(g)。则第一份物质重 = $m_1(g) - m_2(g)$。

c. 同上操作,逐次称量,即可称出多份物质,例如:

瓶 + 物品(1)质量为 20.3720 g;

瓶 + 物品(2)质量为 20.1237 g,差为 0.2483 g……第一份物品的质量;

瓶 + 物品(3)质量为 19.8937 g,差为 0.2300 g……第二份物品的质量;

瓶 + 物品(4)质量为 19.6327 g,差为 0.2410 g……第三份物品的质量。

3. 电子天平

(1)电子天平的特点

电子天平又称电磁力式天平,它与传统的杠杆式机械天平比较主要有如下特点:

①传感器的反应速度快,从而可以提高称量速度;

②结构简单,体积小,重量轻,受安装地点的限制小;

③称量信号可以用计算机进行数据处理,自动显示、记录称量结果;

④称重传感器密封性好,因而有优良的防潮、防腐蚀性能;

⑤它没有作为支点的刀承和刀口,稳定性好,机械磨损小,减轻了维修保养工作,使用方便,寿命长;

⑥精度高。

所以电子天平在目前已成为衡器发展的主流。

(2)电子天平的构造及工作原理

电子天平的结构及工作原理如图 2 – 19 所示。

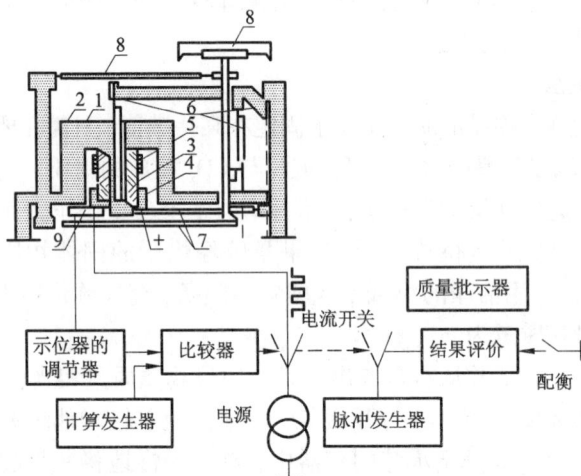

图 2 – 19　电子天平结构示意图

1—磁轭；2—磁钢；3—极靴；4—补偿线圈；5—温度补偿；
6—挠性轴承；7—秤盘；8—罗伯威尔机构；9—示位器

根据称重传感器的工作原理，电子天平又可分为电容式、电感式、磁电式、电磁式和电阻应变式几种类型。无论哪种类型的电子天平都是利用称重传感器作为变换元件，把被称物体的重量按一定的比例关系转换成与其相应的电信号，然后用电子仪表进行测量和显示。

电子天平的外观如图 2 – 20 所示。

①底座部

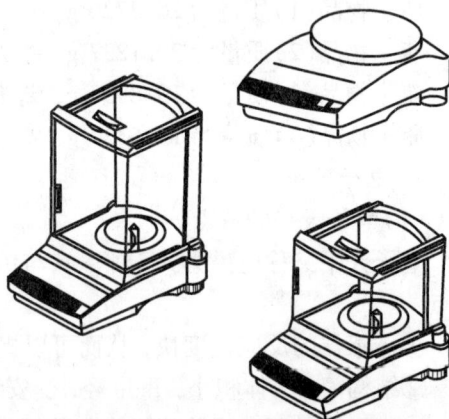

图 2 – 20　电子天平外观图

底座部是安装电子天平全部部件的基础。天平的磁轭、磁钢、极靴、罗伯威尔机构支架、水准器以及天平绝大部分的测量、控制、显示部分的电器件及其线路都安装在底座上或固定于与底座相连的支架上。在底座的下表面，装有三只脚，其中两只是用来调整水平的底脚螺丝。绝大多数的电子天平，显示部分通常和天平主机装在一起，此时，常安装在底座上表面的前端部位。

②载荷接受、传递部

载荷接受、传递部是电子天平按受外界载荷，并将该载荷以力或力矩的形式传递给载荷测量、控制部分的机械装置。通常它由称量盘（衡量盘）、护环、护板、罗伯威尔机构等零部件所组成。

③载荷的测量、控制、显示部

载荷的测量、控制、显示部是电子天平测量载荷的质量值，对测量过程实行开环或闭环控制，并将测量结果显示出来的装置。通常也将天平的开关、校准、配衡（除皮）、调零、打印等若干功能键也归此部。

④框罩部

框罩部是安装在底座上的部件。它是起着固定天平外形，防止外界灰尘、气流对传感器和天平主机内其他电器件或系统产生干扰的装置。框罩部也可作为天平主机内的个别零部件空间定位用（只限无精确度定位要求）的框架。精确度高的天平，通常框罩部备有门、窗，称量盘在天平罩内。对于精确度要求不太高的天平，往往不带天平罩。

4. 半自动电光分析天平的使用方法

（1）天平检查　称量前首先应检查天平是否处于正常工作状态、天平是否水平；天平梁是否套在托叶上；砝码是否齐全；环码有无脱落；机械加码装置是否在零位置；秤盘是否干净；灯泡是否亮；投影屏上显示刻度是否清晰。

（2）零点调整　当天平空载（即不加砝码和重物）时，用左手顺时针旋转升降旋钮接通电源，此时可看到标尺的投影在摆动，当投影稳定后，观察投影屏上的刻线与标尺的零点是否重合，通过微调拨杆的移动使其重合，否则则应关闭升降枢，调节天平梁上的平衡螺丝（一般由实验室老师处理）。零点调整好后，关闭升降枢，然后，再打开升降枢看刻度线是否指零，否则需再次调整，直到两次打开升降枢，投影屏上刻度部指零为止。

（3）预称物体质量　如果被称物体的质量较大，则应先在托盘天平上预称，估计到 0.1 g，若发现被称物体超过 200 g，则不能在这种型号的分析天平上称量。

（4）称量物体　将被称物体从左侧门放在左盘中心，从右侧门加砝码于右盘。然后轻轻转动升降枢，观察指针移动方向，判断砝码或环码是太重还是太轻。若指针迅速向左移动，表示砝码或环码太重，反之则太轻，应随即轻轻地关闭天平，加减砝码（注意每次加减砝码或环码时，都必须预先关闭升降枢以保护天平梁上的刀口），加减砝码或环码时要由大到小顺序加减。只有当天平启动后，指针偏转不明显时，才能从投影屏上观察。屏上刻线往"＋"方向迅速移动，表示环码太轻，反之则太重，应加减环码。当屏上刻线往"＋"方向移动缓慢时，则应将升降枢全部打开，观察屏上刻线停止的位置，此点称为载重时的平衡点也称停点，记下停点值，这时被称物体的质量＝盘中砝码质量＋指数盘读数＋投影屏读数。

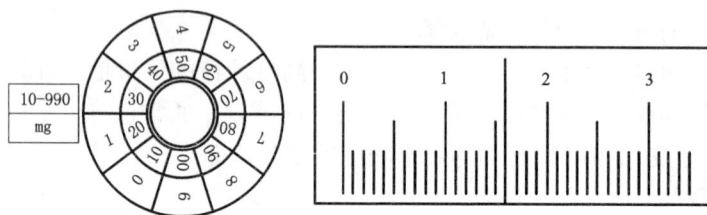

图 2 – 21   光学读数装置示意图

例如：先在天平右盘上放置 16 g 砝码，然后旋动机械加码指数盘旋钮，如图 2 – 21。指针停止后投影屏上刻线指在 图 2 – 21 所示位置，这时物体的质量为

右盘砝码质量：10 g ＋5 g ＋1 g ＝16 g

指数盘读数：200 mg ＋30 mg ＝230 mg

投影屏读数：1.6 mg

被称物体质量 ＝16 g ＋0.230 g ＋0.0016 g ＝16.2316 g

（5）结束工作   称量完毕，取出被称物体和砝码，指数盘旋至 0，同时校对一下读数记录。在空载条件下检查天平零点，看是否与称量前的零点重合，如不重合则说明在称量过程中操作出现错误，应重称。检查天平是否复原，复原后罩好天平罩，并登记使用情况。

5. 半自动电光天平的使用规则

（1）电光天平是一种精密的称量仪器，应注意爱护，不准任意移动天平，应保持天平内的清洁。

（2）在加减砝码和取放被称物体时，一定要预先将升降枢关上，以免损坏刀口、降低天平的灵敏度。开启升降枢时动作要轻缓，发现光标移动很快应立即轻轻关闭升降枢。

（3）天平不能超载（＞200 g），也不能称量过冷或过热的物体。化学药品不能直接放在天平盘上，一般应装在称量瓶中。

（4）砝码一定要用镊子取放，较大的砝码和物体应尽量放在托盘中央。使用机械加砝装置一定要轻轻地逐格转动。

（5）称量完毕，将天平复原，并检查天平是否休止；砝码是否齐全；机械加码装置指数盘是否置零；被称物是否取出，关好天平门，登记使用情况，请教师检查签字后才能离开天平室。

2.4　常用化学试剂及其取用方法

2.4.1　常用化学试剂的等级规格

我国常用试剂等级的划分如表2－3。

表2－3　常用试剂等级的划分

国家标准	优(质)级纯 保证试剂 G.R	分析纯 A.R	化学纯 C.P	实验试剂 L.R
等级	一级品（Ⅰ）	二级品（Ⅱ）	三级品（Ⅲ）	四级品（Ⅳ）
标志	绿色标签	红色标签	蓝色标签	棕黄色标签
用途	精密的分析工作和科研工作	一般的分析工作和科研工作	厂矿日常控制分析和教学实验	实验中的辅助试剂及制备原料

除上述四个等级外，还根据特殊需要而定出了相应的纯度规格。如供光谱分析用的光谱纯，供核实验及其分析用的核纯等。

对于各种规格不同的试剂，其要求标准不同。但总的来说，优级纯试剂杂质含量最低，实验试剂杂质含量较高。所以应根据实验工作的需要，选用适当等级的试剂，既满足工作要求，又要符合节约原则。

2.4.2　常用化学试剂的纯化与干燥

2.4.2.1　试剂的纯化

根据需要，试剂在使用前，往往需要纯化处理后，再用于化学实验。

1. 常见无机试剂的纯化处理方法

常见无机试剂的纯化处理方法见表2－4。

表 2 - 4　常见无机试剂的纯化处理方法

试剂名称	纯化处理方法	备注
盐酸	(1)除去一般杂质 　将 G. R 级盐酸与高纯水以 7 + 3 的体积比稀释。取此溶液 1.5 mL 装入一只 2 L 的石英或硬质玻璃双重蒸馏器中，调节加热器，将馏速控制在 200 mL/h。弃去开始馏出的 150 mL 后，收集中间馏出液 1 L，即得纯的 6.5 ~ 7.5 mol/L 的盐酸 (2)除砷 　将商用纯水与 G. R 盐酸按上述比例稀释，加入适量氧化剂(按体积加入 2.5% 硝酸或 2.5% 过氧化氢或0.2 g/L的高锰酸钾)。取此溶液 1.5 L 装入一只 2 L 的石英或硬质玻璃双重蒸馏器中，放置 15 min 后，以 100 mL/h 的馏速进行蒸馏。弃去开始馏出的 150 mL 后，收集中间馏出液 1 L (3)等温扩散提纯盐酸的方法 　在一只内壁涂有石蜡层(防止污染)的口径为 300 cm 的干燥器中，注入 3 kg G. R 盐酸，在干燥器的瓷托板上放一只聚乙烯或石英容器，内装有 300 mL 高纯水，盖好盖子，在室温下放置 7 ~ 10 天(在 20 ~ 30℃的气温下放置 7 天，15 ~ 29℃的气温下放置 10 天)，可得浓度为 9 ~ 10 mol/L 的盐酸 　用"等温扩散"法，也可以制备某些高纯度的易挥发的试剂	(1)提纯的盐酸，铁、铝、钙、镁、铜、铅、锌、钴、镍、锰、铬、锡的含量可控制在 $5 \times 10^{-6}\%$ ~ $10^{-7}\%$ 之间。 (2)提纯的盐酸，砷的含量控制在 $1 \times 10^{-6}\%$ 以下。 (3)制得的盐酸，铁、铝、钙、镁、铜、铅、锌、钴、镍、锰、铬的含量可控制在 $2 \times 10^{-7}\%$ 以下
硝酸	(1)除去杂质 　将 1.5 L C. R 级的硝酸放入一只硬质玻璃蒸馏器中，控制蒸馏温度，馏速保持在 200 ~ 400 mL/h，弃去开始馏出的 150 mL 溶液，收集中间馏分 1 L。 　将上述中间馏分 2 L，装入一只 3 L 的石英玻璃蒸馏器内，在石蜡浴中进行蒸馏，馏速保持在 100 mL/h，弃去开始馏出的 150 mL 溶液，收集中间馏分 1600 mL (2)制取无水硝酸 　将高浓度的 G. R 硝酸蒸馏后，通入预热的无尘惰性气体气流，以驱除氮的氧化物。随后加入少量五氧化二磷，再进行蒸馏，并测定其中是否有游离的五氧化二氮存在，若有游离的五氧化二氮时，可加入与过剩的五氧化二氮量相当的含水的硝酸 (3)制取无卤素的硝酸 　在 G. R 硝酸中加入少量硝酸银，于石英或硬质玻璃容器中在通风柜内进行蒸馏，收集中间馏分	制得的硝酸，铁、铝、钙、镁、铜、铅、锌、钴、镍、锰、铬含量可控制在 $3 \times 10^{-7}\%$ 以下

续上表

试剂名称	纯化处理方法	备注
氨水	（1）蒸馏吸收法 　　将 3 L A.R 级氨水注入一只 5 L 硬质玻璃烧瓶中，加入少量 1% 高锰酸钾溶液至呈微红紫色，烧瓶上端装一支回流冷凝管，冷凝管的上端串接三只洗气瓶（第一只瓶盛有 1% 的 EDTA 二钠盐溶液，另两只洗气瓶装高纯水），最后一只洗气瓶与装有 1.5 L 高纯水的有机玻璃接收瓶相连，此接收瓶置于冰盐浴中。将加热温度控制在 40℃，氨气即通过洗气瓶而被接收瓶中的高纯水所吸收，当大部分氨挥发后，最后可升温至 80℃，使氨全部进入吸收瓶。 （2）等温扩散法 　　将 2 L A.R 级氨水注入一只干燥器中，在干燥器的瓷板上放置 3 ~ 4 只盛有 200 mL 高纯水的聚乙烯或石英广口容器，再通过瓷板向氨水中加入 2 ~ 3 g 氢氧化钠，迅速盖上干燥器，放置 5 ~ 6 天，并要每天摇动一次，可得浓度为 10% ~ 12% 的高纯氨水。 （3）制备无碳酸盐的氨水 　　将钢瓶中的氨气，通过一支装有活性炭的管子和一支装有碱石灰的管子后，通入煮沸过的商品纯水中（盛高纯水的容器要冷却），直到高纯水中的氨气饱和为止	制得的氨水浓度为 25%
氯化钠 （光谱用）	将 40 g A.R 级氯化钠在加热搅拌下溶解于 120 mL 高纯热水中，再加入 2 ~ 3 mL 的标准溶液（Fe^{3+} 1mg/mL），搅拌均匀后，滴加提纯的氨水，至溶液的 pH = 10 左右时为止。将此溶液在水浴上加热，使生成的氢氧化物沉淀凝聚，冷至室温后过滤，将滤液放入铂皿中，在低温电炉上密闭蒸发至有结晶薄膜出现为止，冷却后进行抽滤，析出的晶体用纯酒精洗涤。在真空干燥箱中于 105℃ 和 20 mmHg 的压力下干燥至无水	用此法提纯的氯化钠，经光谱定性分析，含有微量的硅、铝、镁和痕量的钙。这样提纯的氯化钠可用作光谱分析中的载体，也可作为标准原始物质。用这个方法也可提纯氯化钾
重铬酸钾	将 100 g A.R 级重铬酸钾溶解在 200 ~ 300 mL 热的高纯水中，用 2 号砂芯漏斗抽滤，将滤液在电炉上蒸发至 150 mL，在强烈搅拌下将溶液倒入一个被冰水冷却的大瓷皿中，使之形成一薄层，以制取小粒结晶，再用布氏漏斗进行抽滤，得到的结晶用少量冷水洗涤，并按上述方法重结晶一次。将洗过的二次结晶于 100 ~ 105℃ 下除水干燥 2 ~ 3 h，然后将温度升至 200℃，继续干燥 12 h	用此法提纯的重铬酸钾，可作基准物质使用，在光谱分析中，仅检出有微量的镁、铋和痕量的铝

2. 常见有机试剂的纯化方法

(1) 无水乙醚 沸点 34.51℃，折光率 n_D^{20} 为 1.3526，相对密度 ρ_4^{20} 为 0.71378。普通乙醚常含有 2% 乙醇和 0.5% 水。久藏的乙醚常含有少量的过氧化物。

①过氧化物的检验和除去 在干净的试管中放入 2~3 滴浓硫酸，1 mL 2% 碘化钾溶液（若碘化钾溶液已被空气氧化，可用亚硫酸钠稀溶液滴至黄色消失）和 1~2 滴淀粉溶液，混合均匀后加入乙醚，出现蓝色即表示有过氧化物存在。除去过氧化物可用新配制的硫酸亚铁稀溶液（配制方法是 $FeSO_4 \cdot 7H_2O$ 60 g，100 mL 水和 6 mL 浓硫酸），将 100 mL 乙醚和 10 mL 新配制的硫酸亚铁溶液放在分液漏斗中洗数次，至无过氧化物为止。

②醇和水的检验和除去 在乙醚中加入少许高锰酸钾粉末和一粒氢氧化钠，放置后，氢氧化钠表面附有棕色树脂，即证明有醇存在。水的存在用无水硫酸铜检验。先用无水氯化钙除去大部分水，再经金属钠干燥。其方法是：将 100 mL 乙醚放在干燥锥形瓶中，加入 20~25 g 无水氯化钙，瓶口用软木塞塞紧，放置一天以上，并间断摇动，然后蒸馏，收集 33~37℃ 的馏分。用压钠机将 1 g 金属钠直接压成钠丝并放入盛乙醚的瓶中，用带有氯化钙干燥管的软木塞塞住，或在木塞中插一末端拉成毛细管的玻璃管，这样，既可防止潮气浸入，又能使产生的气体逸出。放置至无气泡发生后即可使用；放置后，若发现钠丝表面已变黄变粗时，须再蒸馏一次，然后再加入钠丝。

(2) 无水乙醇 沸点 78.5℃，折光率 n_D^{20} 为 1.3616，相对密度 ρ_4^{20} 为 0.7893。

制备无水乙醇的方法很多，根据对无水乙醇质量要求的不同而选择不同的方法。

若要求 98%~99% 的乙醇，可采用下列方法。

①利用苯、水和乙醇形成低共沸混合物的性质，将苯加入乙醇中再分馏，在 64.9℃ 时蒸出苯、水、乙醇的三元恒沸混合物，多余的苯在 68.3℃ 与乙醇形成二元恒沸混合物，最后蒸出。工业上多采用此法。

②用生石灰脱水。于 100 mL 95% 乙醇中加入新鲜的块状生石灰 20 g，回流 3~5 h，然后进行蒸馏。

若要 99% 以上的乙醇，可采用下列方法。

①在 100 mL 99% 乙醇中，加入 7 g 金属钠，待反应完毕后，再加入 27.5 g 邻苯二甲酸二乙酯或 25 g 草酸二乙酯，回流 2~3 h，然后进行蒸馏。

金属钠虽能与乙醇中的水作用，生成氢气和氢氧化钠，但所生成的氢氧化钠又与乙醇发生如下的平衡反应：

$$NaOH + C_2H_5OH \Longrightarrow C_2H_5ONa + H_2O$$

因此单独使用金属钠不能完全除去乙醇中的水，须加入过量的高沸点酯，如邻苯二甲酸二乙酯与生成的氢氧化钠作用，抑制上述反应，从而达到进一步脱水的目的。反应如下：

$$C_6H_4(COOC_2H_5)_2 + 2NaOH \Longrightarrow C_6H_4(COONa)_2 + 2C_2H_5OH$$

②在 60 mL 99% 乙醇中，加入 5 g 镁和 0.5 g 碘，等镁溶解生成醇镁后，再加入 900 mL 99% 乙醇，回流 5 h 后，蒸馏，可得到 99.9% 乙醇。反应如下：

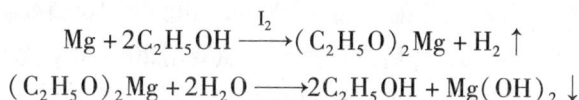

$$Mg + 2C_2H_5OH \xrightarrow{I_2} (C_2H_5O)_2Mg + H_2\uparrow$$

$$(C_2H_5O)_2Mg + 2H_2O \longrightarrow 2C_2H_5OH + Mg(OH)_2\downarrow$$

由于乙醇具有非常强的吸湿性，所以所用仪器必须事前干燥好，在操作时，动作要迅速，尽量减少转移次数以防空气中的水分进入。

（3）丙酮 沸点 56.2℃，折光率 $n_D^{20}1.3588$，相对密度 $\rho_4^{20}0.7899$。

普通丙酮常含有少量的水及甲醇、乙醛等还原性杂质。其纯化方法如下。

①于 250 mL 丙酮中加入 2.5 g 高锰酸钾回流，若高锰酸钾紫色很快消失，再补充少量高锰酸钾继续回流，至紫色不褪为止。然后将丙酮蒸出，用无水碳酸钾或无水硫酸钙干燥，过滤后蒸馏，收集 55～56.5℃ 的馏分。用此法纯化丙酮时，须注意丙酮中含还原性物质不能太多，否则会过多地消耗高锰酸钾和丙酮，使处理时间增长。

②将 100 mL 丙酮装入分液漏斗中，先加 4 mL 10% 硝酸银溶液，再加入 3.6 mL 1 mol/L 氢氧化钠溶液，振摇 10 min，分出丙酮层，再加入无水硫酸钾或无水硫酸钙进行干燥。最后蒸馏收集 55～56.5℃ 馏分，此法比方法 ① 要快，但硝酸银较贵，只适宜小量试剂的纯化。

（4）氯仿 沸点 61.7℃，折光率 $n_D^{20}1.4459$，相对密度 $\rho_4^{20}1.4832$。

氯仿在日光下易氧化成氯气、氯化氢和光气（剧毒），故氯仿应贮于棕色瓶中。市场上供应的氯仿多用 1% 酒精做稳定剂，以消除产生的光气。氯仿中乙醇的检验可用碘仿反应；游离氯化氢的检验可用硝酸银的醇溶液。

除去乙醇的一种方法为：在氯仿中加入其二分之一体积的水振摇数次，分离下层的氯仿，用氯化钙干燥 24 h，然后蒸馏。

另一种纯化方法为：将氯仿与少量浓硫酸一起振动两三次。每 200 mL 氯仿用 10 mL 浓硫酸，分去酸层以后的氯仿用水洗涤，干燥，然后蒸馏。

除去乙醇后的无水氯仿应保存在棕色瓶中并避光存放，以免光化作用产生光气。

(5)二氯甲烷　沸点40℃，折光率$n_D^{20}1.4242$，相对密度$\rho_4^{20}1.3266$。

使用二氯甲烷比氯仿安全，因此常常用它来代替氯仿作萃取剂。普通的二氯甲烷一般都能直接做萃取剂用。如需纯化，可用5%碳酸钠溶液洗涤，再用水洗涤，然后用无水氯化钙干燥，蒸馏收集40~41℃的馏分，保存在棕色瓶中。

(6)乙酸乙酯　沸点77.06℃，折光率$n_D^{20}1.3723$，相对密度$\rho_4^{20}0.9003$。

乙酸乙酯一般含量在95%~98%，含有少量的水、乙醇和醋酸，可用下述方法纯化：于1000 mL乙酸乙酯中加入100 mL醋酸酐、10滴浓硫酸，加热回流4 h，除去乙醇和水等杂质，然后进行蒸馏。馏液用20~30 g无水碳酸钾振荡，再蒸馏。产物沸点为77℃，纯度可达99%以上。

(7)甲醇　沸点64.96℃，折光率$n_D^{20}1.3288$，相对密度$\rho_4^{20}0.7914$。

普通未精制的甲醇约含有0.02%的丙酮和0.1%的水。而工业甲醇中这些杂质的含量达0.5%~1%。为了制得纯度达99.9%以上的甲醇，可将甲醇用分馏柱分馏，收集64℃的馏分，再用金属镁去水(与制备无水乙醇相同)。甲醇有毒，处理时应防止吸入其蒸气。

(8)石油醚　石油醚为轻质石油产品，是相对分子质量低的烷烃类的混合物。其沸程为30~60℃，60~90℃，90~120℃等沸程规格的石油醚，其中含有少量不饱和烃，沸点与烷烃相近，用蒸馏法无法分离。

石油醚的精制通常将石油醚用其1/10体积的浓硫酸洗涤2~3次，再用10%的硫酸加入高锰酸钾配成的饱和溶液洗涤，直至水层中的紫色不再消失为止。然后再用水洗，经无水氯化钙干燥后蒸馏。若需绝对干燥的石油醚，可加入钠丝(与纯化无水乙醚相同)。

(9)四氯化碳　沸点76.8℃，折光率$n_D^{20}1.4603$，相对密度$\rho_4^{20}1.595$。将60 g氢氧化钾溶于60 mL水中然后和100 mL乙醇溶液混在一起，在50~60℃时振摇30 min，然后水洗，再将四氯化碳按上述方法重复操作一次(氢氧化钾的用量减半)。四氯化碳中残余的乙醇可以用氯化钙除掉。最后四氯化碳用氯化钙干燥，过滤，蒸馏收集76.7℃馏分。四氯化碳不能用金属钠干燥，因有爆炸危险。

(10)吡啶　沸点6115.5℃，折光率$n_D^{20}1.5095$，相对密度$\rho_4^{20}0.9819$。

分析纯吡啶含有少量水分，可供一般实验用。如要制得无水吡啶，可将吡啶与粒状氢氧化钾(钠)一同回流，然后隔绝潮气蒸出备用。干燥的吡啶吸水性很强，保存时应将容器口用石蜡封好。

（11）N，N－二甲基甲酰胺（DMF）　沸点 149～156℃，折光率 n_D^{20} 1.4305，相对密度 ρ_4^{20} 0.9487。无色液体，可与多种有机溶剂和水任意混合，对有机和无机化合物的溶解性能较好。N，N－二甲基甲酰胺含有少量水分。常压蒸馏时有些分解，产生二甲胺和一氧化碳。在有酸或碱存在时，分解加快。所以加入固体氢氧化钾（钠）在室温放置数小时后，会有部分分解。因此，常用硫酸钙、硫酸镁、氧化钡、硅胶或分子筛干燥，然后减压蒸馏，收集 76℃/4800Pa（36 mmHg）的馏分。如含水较多时，可加入其 1/10 体积的苯，在常压及 80℃ 以下蒸去水和苯，然后再用无水硫酸镁或氧化钡干燥，最后进行减压蒸馏。纯化后的 N，N－二甲基甲酰胺要避光贮存。

N，N－二甲基甲酰胺中如有游离胺存在，可用 2，4－二硝基氟苯是否产生颜色来检查。

（12）二甲基亚砜（DMSO）　沸点 189℃，熔点 18.5℃，折光率 n_D^{20} 1.4783，相对密度 ρ_4^{20} 1.100。二甲基亚砜能与水混合，可用分子筛长期放置加以干燥。然后减压蒸馏，收集 76℃/1600Pa（12 mmHg）馏分。蒸馏时，温度不可高于 90℃，否则会发生歧化反应生成二甲砜和二甲硫醚。也可用氧化钙、氢氧化钙、氧化钡或无水硫酸钡来干燥，然后减压蒸馏。也可用部分结晶的方法纯化。

二甲基亚砜与某些物质混合时可能发生爆炸，例如氢化钠、高碘酸或高氯酸镁等，应予注意。

（13）四氢呋喃　沸点 67℃（64.5℃），折光率 n_D^{20} 1.4050，相对密度 ρ_4^{20} 0.8892。

四氢呋喃与水混溶，故常含有少量水分及过氧化物，如要制得无水四氢呋喃，可用氢化铝锂在隔绝潮气下回流（通常 1000 mL 需 2～4 g 氢化铝锂）除去其中的水和过氧化物，然后蒸馏，收集 66℃ 的馏分（蒸馏时不要蒸干，剩余少量残液时即倒出）。精制后的液体加入钠丝并应在氮气氛中保存。如需较久放置，应加 0.025% 4－甲基－2，6－二叔丁基苯酚作抗氧剂。

处理四氢呋喃时，应先用少量液体进行试验，在确定其中只含有少量水和过氧化物，作用不致过于激烈时，方可进行纯化。四氢呋喃中的过氧化物可用酸化的碘化钾溶液来检验。如过氧化物较多，应另行处理为宜。

（14）1，4－二氧环己烷（二氧六环）　沸点 101.5℃，熔点 12℃，折光率 n_D^{20} 1.4224，相对密度 ρ_4^{20} 1.0336。

二氧六环能与水任意混合，常含有少量乙二醇缩醛和水，久贮的二氧六环可能含有过氧化物(鉴定和除去方法参阅乙醚)。二氧六环的纯化方法是在500 mL 二氧六环中加入 8 mL 浓盐酸和 50 mL 水的溶液,回流 6 ~ 10 h,在回流过程中,慢慢通入氮气以除去生成的乙醛。冷却后,加入固体氢氧化钾,直到不能再溶解为止,分去水层,再用固体氢氧化钾干燥 24 h。然后在金属钠存在下加热回流 8 ~ 12 h,最后在金属钠存在下蒸馏,压入钠丝密封保存。装入精制过的 1,4 - 二氧环己烷时应当避免与空气接触。

2.4.2.2　试剂的干燥

除去固体、气体或液体试剂中的少量水分的过程称为干燥。不同的试剂干燥的方法也不同,如加热烘干、用干燥剂脱水等。下面分别介绍几种基本的干燥方法和有关技术。

1. 液体的干燥

(1)干燥剂的选择

液体有机化合物的干燥,通常是用干燥剂直接与其接触。因而所用的干燥剂必须不与该物质发生化学反应或催化作用,不溶解于该液体中。例如酸性物质不能用碱性干燥剂;而碱性物质则不能用酸性干燥剂。有的干燥剂能与某些干燥的物质生成配合物,如氯化钙易与醇类、胺类形成配合物,因而不能用来干燥这些液体。强碱性干燥剂如氧化钙、氢氧化钠能催化某些醛类或酮类发生缩合、自动氧化等反应;也能使酯类或酰胺类发生水解反应。氢氧化钾(钠)还能溶解于低级醇中。

在使用干燥剂时,还要考虑干燥剂的吸水容量和干燥效能。吸水容量是指单位重量干燥剂所吸收的水量;干燥效能是指达到平衡时液体干燥的程度。对于形成水合物的无机盐干燥剂,常用吸水后结晶水的蒸气压表示。例如,硫酸钠形成 10 个结晶水的水合物,其吸水容量达 1.25。氯化钙最多能形成 6 个结晶水的水合物,其吸水容量为 0.97。两者 25℃ 时的水蒸气压分别为 0.26 kPa 及 0.04 kPa,因此,硫酸钠的吸水量较大,但干燥效能弱;而氧化钙的吸水量较小但干燥效能强。所以在干燥含水量较多而又不易干燥的(含有亲水性基团)化合物时,常先用吸水量较大的干燥剂除去大部分水分,然后再用干燥性能强的干燥剂干燥。通常第二类干燥剂的干燥效能较第一类为高,但吸水量较小,所以都是用第一类干燥剂干燥后,再用第二类干燥剂除去残留的微量水分。而且只是在需要彻底干燥的情况下才使用第二类干燥剂。

此外选择干燥剂时还要考虑干燥剂的干燥速度和价格,常用干燥剂的性能见表 2 - 5 和表 2 - 6。

表 2−5　常用干燥剂的性能与应用范围

干燥剂	吸水作用	吸水容量	干燥性能	干燥速度	应用范围
氯化钙	形成 $CaCl_2 \cdot nH_2O$ $n = 1,2,4,6$	0.97 按 $CaCl_2 \cdot 6H_2O$ 计	中等	较快，但吸水后表面为薄层液体所盖，故放置时间要长些为宜	能与醇、酚、胺、酰胺及某些醛酮形成配合物，因而不能用来干燥这些化合物。工业品中可能含氢氧化钙等碱或氧化钙，故不能用来干燥酸类
硫酸镁	形成 $MgSO_4 \cdot nH_2O$ $n = 1,2,4,$ $5,6,7$	1.05 按 $MgSO_4 \cdot 7H_2O$ 计	较弱	较快	中性，应用范围广，可代替 $CaCl_2$，并可用以干燥酯、醛、酮、腈、酰胺等不能用 $CaCl_2$ 干燥的化合物
硫酸钠	$Na_2SO_4 \cdot 10H_2O$	1.25	弱	缓慢	中性，一般用于有机液体的初步干燥
硫酸钙	$2CaSO_4 \cdot H_2O$	0.06	强	快	中性，常与硫酸镁(钠)配合，作最后干燥之用。
碳酸钾	$K_2CO_3 \cdot H_2O$	0.2	较弱	慢	弱碱性，用于干燥醇、酮、酯、胺及杂环等碱性化合物，不适于酸、酚及其他酸性化合物的干燥。
氢氧化钾(钠)	溶于水	—	中等	快	强碱性，用于干燥胺、杂环等碱性化合物，不能用于干燥醇、酯、醛、酮、酸、酚等
金属钠	$Na + H_2O \rightarrow$ $NaOH + 1/2H_2$	—	强	快	限于干燥醚、烃类中痕量水分。用时切成小块或压成钠丝
氧化钙	$CaO + H_2O \rightarrow$ $Ca(OH)_2$	—	强	较快	适于干燥低级醇类
五氧化二磷	$P_2O_5 + 3H_2O \rightarrow$ $2H_3PO_4$	—	强	快，但吸水后表面为黏浆液覆盖，操作不便	适于干燥醚、烃、卤代烃、腈等中的痕量水分，不适用于醇、酸、胺、酮等
分子筛	物理吸附	约 0.25	强	快	适用于各类有机化合物的干燥

表 2 - 6　各类有机物常用的干燥剂

化合物类型	干　燥　剂
烃	$CaCl_2$，Na，P_2O_5
卤代烃	$CaCl_2$，$MgSO_4$，Na_2SO_4，P_2O_5
醇	K_2CO_3，$MgSO_4$，CaO，Na_2SO_4
醚	$CaCl_2$，Na，P_2O_5
醛	$MgSO_4$，Na_2SO_4
酮	K_2CO_3，$CaCl_2$，$MgSO_4$，Na_2SO_4
酸、酚	$MgSO_4$，Na_2SO_4
酯	$MgSO_4$，Na_2SO_4，K_2CO_3
胺	KOH，$NaOH$，K_2CO_3，CaO
硝基化合物	$CaCl_2$，$MgSO_4$，Na_2SO_4

（2）干燥剂的用量

以最常用的乙醚和苯两种溶液为例。水在乙醚中的溶解度室温时为 1% ~ 1.5%，如用无水氯化钙来干燥 100 mL 含水的乙醚时，假定无水氯化钙全部转变成六水化合物，这时的吸水量是 0.97，即 1 g 无水氯化钙大约可吸去 0.97 g 水，因此无水氯化钙的理论用量至少要 1 g。但实际上则远较 1 g 多，例如，100 mL 含水乙醚常需用 7 ~ 10 g 无水氯化钙。这是因为萃取时，乙醚层中的水分不可能完全干净，其中还有悬浮的微细水滴。另外达到高水合物需要的时间很长，往往不能达到它应有的吸水量，因而干燥剂的实际用量是大大过量的。水在苯中的溶解度极小（约 0.05%），理论上讲只需要很少量的干燥剂。由于上面的一些原因，实际用量还是比较多的，但可少于干燥乙醚时的用量。干燥其他液体有机物时，可从溶解度手册查出水在其中的溶解度（若查不到水的溶解度，则可从它在水中的溶解度来推测，难溶于水者，水在它里面的溶解度也不会大），或根据它的结构来估计干燥剂的用量（在极性有机物中水的溶解度较大，有机分子中若含有能与氧原子配位的基团时，水的溶解度亦大）。一般对于含有亲水性基团的（加醇、醚、胺等）化合物，所用的干燥剂要过量多些。由于干燥剂也能吸附一部分有机物，所以干燥剂的用量应控制得严些。必要时，可先用一般干燥剂干燥，过滤后再使用干燥效能较强的干燥剂。一般干燥剂的用量为每 10 mL 液体需 0.5 ~ 1 g，但由于液体中的水分含量不等，干燥剂的质量、颗粒大小和干燥时的温度等不同，以及干燥剂对副产物的吸收（如氧化钙吸收醇)等诸多因素不同，因此很难规定具体的数量，上述数据仅供参考。操作者应细心积累这方面的经验，在实际操作中，干燥一定时间后，要观察干燥剂的形态，若它的大部分棱角还清楚可辨，则表明干燥剂的量已足够了。

(3)实验操作

在干燥前应将被干燥的液体中的水分尽可能分离干净[宁可损失一些有机物，也不能存在可见的水层。如果有机液体中存在较多的水分，实验过程中还有可能出现少量的水层(例如在用氧化钙干燥时)，必须将此水层分去或用吸管将水吸去]。将该液体置于锥形瓶中，用骨勺取适量的干燥剂直接放入液体中(干燥剂颗粒大小要适宜，太大时因表面积小吸水很慢，且内层干燥剂不起作用；太小时则不易过滤，吸附有机物太多)。用软木塞塞紧，振摇片刻。如果发现干燥剂附着瓶壁，互相黏结，通常是表示干燥剂不够，应继续添加；放置一段时间(至少半小时，最好放置过夜)，并不时加以振摇。干燥过程中，浑浊的液体会变为澄清，这并不一定说明它已被干燥，澄清与否和水在该化合物中的溶解度有关。滤去干燥剂，再进一步蒸馏处理，由于金属钠、生石灰、五氧化二磷等和水反应后生成比较稳定的产物，有时可不必过滤就能直接进行蒸馏。

此外还可以利用分馏或二元、三元共沸混合物来除去水分，该方法属于物理方法。对于不与水生成共沸混合物的液体有机物，例如甲醇和水的混合物，由于沸点相差较大，用精密分馏柱即可完全分开。有时利用某些有机物可与水形成共沸混合物的特性，向待干燥的有机物中加入另一有机物，利用此有机物与水形成最低共沸点的性质，在蒸馏时逐渐将水带出，从而达到干燥的目的。例如，工业上制备无水乙醇的方法之一就是将苯加到95%乙醇中进行共沸蒸馏。近年来工业生产上常应用离子交换树脂脱水以制备无水乙醇。

2. 固体化合物的干燥

固体化合物常用干燥器进行干燥。

(1)普通干燥器(图2-22)　其盖与缸身之间的平面经过磨砂，在磨砂处涂以润滑脂，使之密闭。缸中有多孔瓷板，瓷板下面放置干燥剂，上面放置表面皿，待干燥样品放在表面皿中。

(2)真空干燥器(图2-23)　它的干燥效率较普通干燥器高。真空干燥器上的玻璃活塞，用以抽真空，活塞下端呈弯钩状，口向上，防止在通入空气时，因为气流太猛将固体冲散。最好用另一表面皿覆盖盛有样品的表面皿。在用水泵抽气过程中，最好能用金属丝(或用布)围在干燥器外围，以保证安全。

应按样品所含的溶剂来选择使用的干燥剂类型。例如，五氧化二磷可吸水；生石灰可吸水或酸；无水氯化钙可吸水或醇；氢氧化钠吸收水或酸；石蜡片可吸收乙醚、氯仿、四氯化碳和苯等。有时在干燥器中同时放有几种干燥

剂，如在瓷板上用浅器皿盛放氢氧化钠，在底部放硫酸（在 1 L 浓硫酸中溶有
18 g 硫酸钡的溶液，如已吸收了大量水分，则硫酸钡就沉淀出来，表明不再适
用于干燥而需要重新更换），这样可吸收水和酸，效率更高。

（3）真空恒温干燥器（图 2 - 24）　此设备适用于少量物质的干燥（若所需
干燥物质的数量较大时，可用真空恒温干燥箱），如图 2 - 24 所示，在 2 中放置
五氧化二磷。将待干燥的样品置于 3 中，烧瓶 A 中放置有机液体，通过活塞 1
将仪器抽真空，加热回流烧瓶 A 中的液体，利用蒸汽加热外套 4，从而使样品
在恒定温度下得到干燥，干燥温度应与被处理物沸点接近。

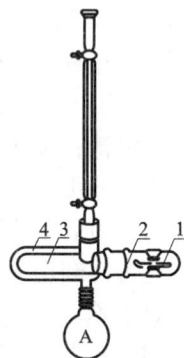

图 2 - 22　普通干燥器　　图 2 - 23　真空干燥器　　图 2 - 24　真空恒温干燥器

2.4.2.3　试剂的取用方法

1. 液体试剂的取用规则

（1）从滴瓶中取用液体试剂时，应垂直滴入试管或烧杯中。滴管决不能触
及所使用的容器器壁以免玷污（图 2 - 25）。滴管放回原瓶时不要放错，不准用
其他滴管或自己的滴管到瓶中吸取试剂。

（2）从小口瓶中取用试剂时，先将瓶塞反放在桌面上，不要弄脏，把试剂
瓶上贴有标签的一面握在手心，逐渐倾斜瓶子，倒出试剂，让试剂沿管壁慢慢
流入试管或沿着洁净的玻璃棒注入烧杯中（图 2 - 26）。取出所需用量后，将小
口瓶竖起，用干净的滤纸吸干瓶口的试液，注意瓶塞不能盖错。

（3）用量筒或移液管定量取用试剂时，多取的试剂不能倒回原瓶中，可倒
入指定的容器内。

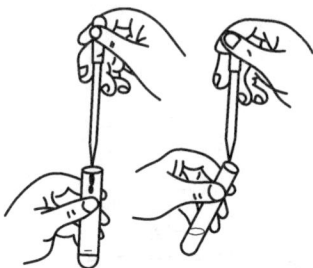

图 2 - 25　用滴管将试剂加入试管中　　图 2 - 26　液体试剂倒入烧杯

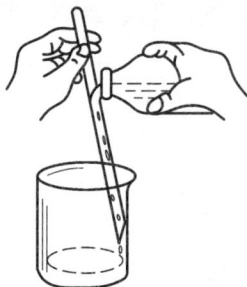

2．固体试剂的取用规则

（1）要用洁净的药勺取试剂。用过的药勺必须洗净和擦干后才能再使用，以免玷污试剂。

（2）称量药品时，注意不要多取，取多的药品，不能倒回原瓶，可放在指定的容器中。

（3）取出试剂后，应立即盖紧瓶盖，不要盖错盖子。

（4）一般的药品可以在干净的纸或表面皿上称量。具有腐蚀性、强氧化性或易潮解的药品，应使用干净的表面皿或称量瓶称量。

（5）有毒药品应在教师指导下取用，不要将药品洒落在实验台上和地面上。

2.5　加热仪器及其使用方法

2.5.1　燃烧加热器及其使用方法

2.5.1.1　燃烧加热器

实验室中常用的燃烧加热器有酒精灯（图 2 - 27）、酒精喷灯（图 2 - 28）、煤气灯（图 2 - 29）。

1．酒精灯的使用方法

酒精灯一般是玻璃制品，配有带磨口的灯罩。不用时，必须将灯罩罩上，以免酒精挥发。酒精易燃，使用时必须注意安全：①灯内酒精不能装得太满，一般不宜超过其总容积的 2/3；②点燃酒精灯之前，应先将灯头提起，吹去聚集在灯内的酒精蒸气；③应该用火柴点燃酒精灯，不要用燃着的其他酒精灯直接去点燃，否则灯内的酒精会洒在外面，引起火灾；④需要添加酒精时，应先将火焰熄灭，然后，将酒精加入灯内；⑧要熄灭灯焰时，可将灯罩盖上（切勿用嘴去吹），然后再提起灯罩，待灯口稍冷，再盖上灯罩；可以防止灯口破裂。

图 2 - 27　酒精灯

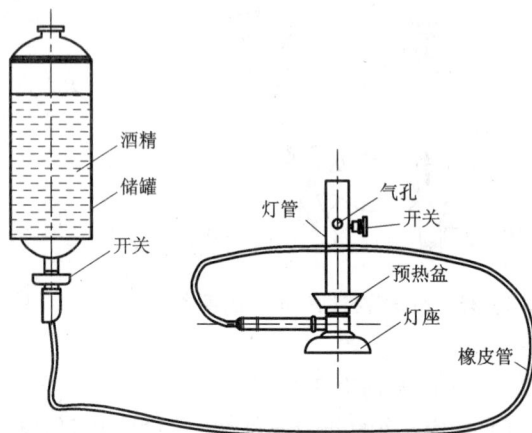

图 2 - 28　酒精喷灯及其构造

2. 酒精喷灯的使用方法

酒精喷灯一般是金属制品。使用前，先在预热盆上注入酒精至满，然后点燃盆内的酒精，以加热铜质灯管。待盆内酒精将燃尽时开启开关，这时在灼热灯管内汽化的酒精，与来自气孔的空气混合被点燃，可得到温度很高的火焰。调节开关的螺丝，可以控制火焰的大小。用毕，向右旋紧开关，即可熄灭酒精喷灯。

应该注意，在开启开关点燃酒精喷灯之前，必须充分灼烧灯管，否则，酒精在灯内不会全部汽化，会有液体酒精由管口喷出，形成"火雨"，甚至引起火灾。遇到这种情况，必须迅速熄灭喷灯。待稍冷后再往预热盘中添满酒精，重新预热灯管。喷灯不用时，必须关好储罐的开关，以免酒精漏失，造成危险。

3. 煤气灯的使用方法

煤气灯的式样有多种，但构造原理是相同的，它由灯管和灯座组成（图 2 - 29）。灯管的下部有

图 2 - 29　煤气灯的构造

螺旋，与灯座相连，灯管下部还有几个圆孔，为空气的入口。旋转灯管，即可完全关闭或不同程度地开启圆孔，以调节空气的进入量。灯座的侧面有煤气的入口，可接上橡皮管把煤气导入孔内。灯座下面（或侧面）有一螺旋针阀，用来调节煤气的进入量。

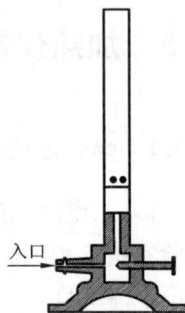

当灯管圆孔完全关闭时，点燃进入煤气灯的煤气，此时的火焰呈黄色（系

碳粒发光所产生的颜色),煤气燃烧不完全,火焰温度不高。逐渐加大空气进入量,煤气燃烧会逐渐完全,火焰分为三层(图2-30)。

焰心(内层)——煤气和空气混合物并未燃烧,温度低,为300℃左右。

还原焰(中层)——煤气不完全燃烧,并分解为含碳的产物,所以这部分火焰具有还原性,称"还原焰"。温度较前者高,火焰呈蓝色。

氧化焰(外层)——煤气完全燃烧。过剩的空气使这部分火焰具有氧化性,故称"氧化焰"。火焰的最高温度处在还原焰顶端上部的氧化焰中,为800~900℃(煤气的组成不同,火焰的温度也有所差异),火焰呈淡紫色。实验时,一般都用氧化焰来加热。

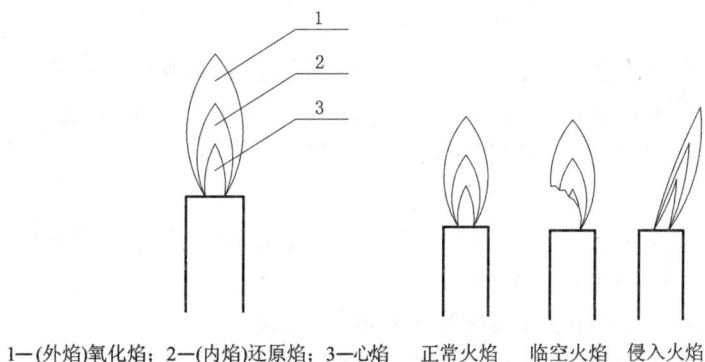

1—(外焰)氧化焰；2—(内焰)还原焰；3—心焰　　正常火焰　　临空火焰　　侵入火焰

图2-30　火焰的结构和种类

当空气或煤气的进入量调节得不合适时,会产生不正常的火焰。当煤气和空气的进入量都很大时,火焰就临空燃烧,称为临空火焰。待引燃用的火柴熄灭时,它也立刻自行熄灭;当煤气进入量很小,而空气进入量很大时,煤气会在灯管内燃烧而不是在灯管口燃烧,这时还能听到特殊的嘶嘶声,并能看到一根细长的火焰,这种火焰叫侵入火焰。它将烧热灯管,一不小心就会烫伤手指。有时在煤气灯使用过程中,因某种原因煤气量会突然减小,这时立即产生侵入火焰,这种现象称为回火。遇到临空火焰或侵入火焰时,应关闭煤气阀,重新调节和点燃。

2.5.1.2　加热方法

实验室中常用的受热容器有烧杯、烧瓶、锥形瓶、蒸发皿、坩埚、试管等。这些仪器能够承受一定的温度,但不能骤热或骤冷,因此在加热前,必须将容器外面的水擦干。加热后不能立即与冷的或潮湿的物体接触。

1. 在试管中加热液体

不易分解的物质，可以放入试管中直接在火焰上加热（图2-31）。

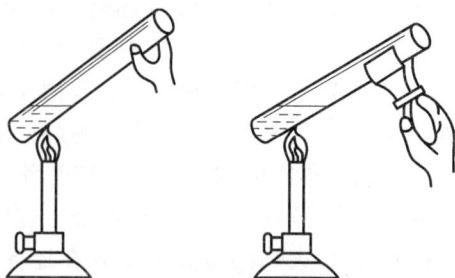

但易分解的物质应在水浴中加热。离心试管由于管底玻璃较薄，不宜直接加热，应在水浴中加热。在火焰上加热试管时，应注意以下几点。

图2-31　加热试管内的液体

（1）应该用试管夹夹持试管的中上部（微热时，可用拇指、食指和中指持试管）。

（2）试管应稍微倾斜，管口向上，以免烧焦试管夹或烤痛手指。

（3）应使液体各部分受热均匀，先加热液体的中上部，再慢慢往下移动，然后不时地上下移动，不要集中加热某一部分，否则将使蒸汽骤然发生，液体冲出管外。

（4）不要将试管口对着别人或自己，以免溶液溅出时把人烫伤（尤其是加热浓酸浓碱时，更要注意）。

2. 在烧杯、烧瓶等玻璃仪器中加热液体

在烧杯、烧瓶等玻璃仪器中加热液体时，玻璃仪器必须放在石棉网上（图2-32），否则容易因受热不均匀而破裂。

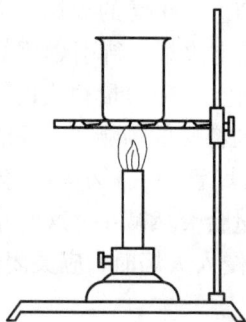

图2-32　烧杯加热

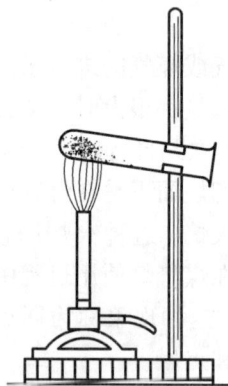

图2-33　加热试管中的固体

3. 在试管中加热固体

在试管中加热固体时，注意不要使凝结在试管上部的水珠流到灼热的管底,否则试管会破裂,因此试管口必须稍微向下倾斜。加热时可用试管夹夹持,也可用铁夹固定试管(图2-33)。

4. 在坩埚中灼烧固体

当需要高温加热固体时,可把固体放在坩埚中用氧化焰灼烧(图2-34)。也可以把盛有固体的坩埚放到马弗炉中灼烧。开始,先用小火烘烧坩埚,使坩埚受热均匀,然后加大火焰灼烧。不要让还原焰接触坩埚底部,以免在坩埚底部结上黑炭,以致坩埚破裂。

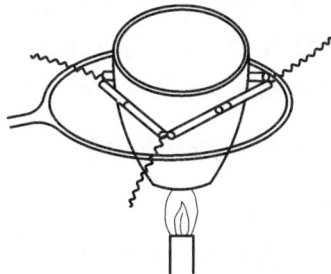

图2-34　灼烧坩埚

要夹取高温坩埚时,必须使用干净的坩埚钳。使用前先在火焰旁预热一下铁钳的尖端,再去夹取。

2.5.2　电热加热器及其使用方法

实验室中经常使用的电热加热器有电炉(图2-35)、管式炉(图2-36)、马弗炉(图2-37)等。

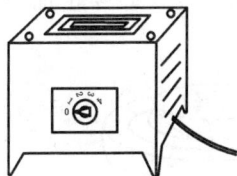

图2-35　电炉

图2-36　管式炉

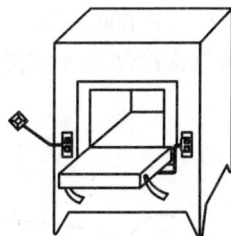

图2-37　马弗炉

电炉可以代替酒精灯或煤气灯加热盛于容器中的液体和固体。容器(烧杯或蒸发皿)和电炉之间要隔一块石棉网,保证其受热均匀。

管式炉有一管状炉膛,利用电阻丝或硅碳棒供热,用电阻丝加热的管式炉最高使用温度为950℃(使用时不应超过900℃),用硅碳棒加热的管式炉最高使用温度可达1300℃(使用时不要超过1200℃)。温度的高低可以通过调节电炉的电阻来控制。炉膛内可插入一根耐高温的瓷管或石英管,瓷管中再放入盛有反应物的瓷舟。反应物可在空气气氛或在其他气氛中受热。

马弗炉是一种用电阻丝或硅碳棒加热的炉子。它的炉膛是长方体,有一炉门,打开炉门就可很容易地放入要加热的坩埚或其他耐高温的器皿。马弗炉的最高使用温度为950～1300℃。测量如此高的温度不能使用普通温度计而是使用高温计，它是由一对热电偶和一只毫伏表所组成。

　　只要把热电偶和温度控制器连接起来，待炉温升到所需温度时，控制器就自动切断电源，炉温停止上升。炉温刚稍低于所需温度时，控制器又自动接通，使炉温上升。如此不断交替，就可以把炉温控制在某一温度附近。

2.5.3　间接加热方法

1. 水浴加热

　　加热温度不超过100℃时，最好采用水浴锅加热，水浴锅一般是铜制的水锅（如图2-38），水浴锅上面可以放置大小不同的铜圈或铝合金圈，以承受各种不同尺寸的器皿（水浴锅也可用盛水的烧杯代替）。

　　采用水浴加热（图2-39），应该注意以下几点：

　　(1)加热温度在90℃以下时，可将盛物料的容器浸在水中但不得接触水浴锅底，调节火焰的大小，把水温控制在需要的范围以内。如果需要加热到100℃时，则可用沸水浴；也可把容器架在水浴环上，利用水蒸气来加热。

　　(2)水浴锅内盛水量不得超过其总容积的2/3，加热过程中随时补充用水，以便保持水位基本不变。

　　(3)水浴锅不能干烧，当不慎将水浴锅中的水烧干时，应立即停止加热，待水浴锅冷却后才恢复使用。

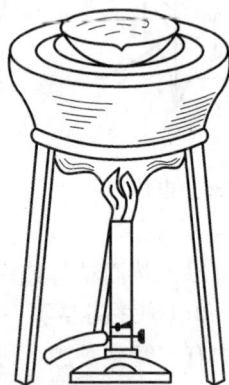

图2-38　水浴锅　　　　　　　　　　图2-39　水浴加热

2. 油浴加热

　　加热温度在100℃以上250℃以下时，可以采用油浴加热。油浴的优点在于温度容易控制在一定范围内，容器内的反应物受热均匀。常用的油浴有液体石蜡豆油、棉子油、硬化油(如氢化棉子油)等。植物油可加热到220℃，药用液体石蜡可加热到220℃，硬化油可加热到250℃。容器内反应物的温度一般要比油浴温度低20℃左右。油浴中应悬挂温度计，以便随时调节灯焰，控制温度。

用油浴加热时，要特别小心，防止着火。当油冒烟情况严重时，应停止加热。万一着火，也不要慌张，可首先熄灭加热灯具(酒精灯、煤气灯)或关闭加热电炉，再移去周围易燃物，然后用石棉板或厚湿布盖住油浴口，火即可熄灭。

加热完毕后，把容器提离油浴液面，仍用铁夹夹住，放置在油浴上面。待附着在容器外壁上的油流完后，用纸和干布把容器擦净。

3. 沙浴加热

用铁盘装沙，将容器的下半部埋在沙中加热的方法称为沙浴加热(图2－40)。

沙浴使用方便，可加热到350℃；沙浴的缺点是沙对热的传导能力较差，沙浴温度分布不均匀，且不容易控制。因此，容器底部

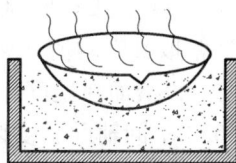

图2－40　沙浴加热

的沙层要薄些，使容器容易受热；而容器周围的沙层要厚些，使容器不易散热。沙浴中应插入温度计，以便控制温度，温度计的水银球应紧靠被加热的容器。使用沙浴时，铁盘下的桌面上要垫隔热石棉板，以防辐射热烤焦桌面。

2.5.4　温度计与测温方法

1. 液体温度计及其校正

实验室常用的测温设备为液体温度计(包括水银温度计和酒精温度计)。液体温度计具有实用、方便的优点，但它的测温范围小(只能测定物质沸点与凝固点之间的温度)。温度计的玻璃毛细管的制作不易均匀，且具有热滞现象，因此在温度要求严格的精密实验中，必须对温度计进行校正。

以水银温度计为例，因为温度计未受热部分的汞柱与玻璃毛细管的热膨胀系数不同。严格说来，只有当温度计上的汞柱全部浸入液体或蒸气中时，它所测得的温度才是准确的，但在实际工作中，温度计的汞柱有一大部分是暴露在所测物之上的，仅有小部分浸入所测物中，这样，实测温度值总是低于真实温度。在要求不太严格的实验中，这种误差可根据温度计的刻度范围用以下差值加以校正。外露汞柱的校正数值差：100℃以下时为1℃；200℃时为3℃；250℃以上可达6～10℃。温度愈高，这个误差的校正数值愈大。如果要获得较准确的温度，则可根据下式进行校正。

$$\Delta t = kn(t_1 - t_2)$$

式中　Δt——汞柱校正值，℃；

n——由液面到温度计读数间的汞柱长度（用读数表示）；

t_1——温度计所指示的读数；

t_2——离开液面的汞柱的平均温度（用另一支辅助温度计进行测定。即将辅助温度计的汞球贴在高出液面的那段汞柱长度的 1/2 处，所测得的温度）。

k——汞与玻璃的膨胀系数之差，此系数与 t_1 有关，如：

$t_1 = 0 \sim 150℃$，$k = 0.000158$；

$t_1 = 150 \sim 250℃$，$k = 0.00016$；

$t_1 = 300℃$，$k = 0.000164$；

通常取其平均值 0.00016。

例如，设液面在温度计的 30℃ 处，实测时温度计的读数为 190℃（t_1），测得 t_2 为 65℃，问辅助温度计的汞球应放在多少度处？并求出该温度计的汞柱校正值 Δt。

由题可知，$n = 190℃ - 30℃ = 160℃$；则辅助温度计应放在 $1/2 \times 160℃ + 30℃ = 110℃$ 处，用 $k = 0.00016$ 来计算，则汞柱校正值

$$\Delta t = 0.00016 \times 160 \times (190 - 65) = 3.2℃$$

校正结果说明准确温度应为 190℃ + 3.2℃ = 193.2℃。

为了获得更谁确的温度数据，可以选择几种不同熔点的标准物质（高纯或基准试剂），通过测定各物质的熔点来校正温度计。根据温度计所示各物质熔点的度数与各对应物质的理论熔点作出曲线图，利用曲线图即可查得准确温度。

温度计不能作搅棒使用。刚测量过高温物体温度的温度计，不能立即用冷水去洗，以免水银球炸裂温度计损坏而洒落水银。温度计也要轻拿轻放，避免打碎。一旦打碎洒出水银，要立即用硫磺粉覆盖。

2. 温度测量仪器

实验室中除使用液体温度计外，也常使用电阻温度计、热电式高温计、光学高温计。

（1）电阻温度计

金属丝（如铂丝等）的电阻随温度的变化而变化，它们之间具有一定的关系。电阻温度计就是利用这一性质，通过测量金属丝的阻值来确定温度的。铂电阻温度计的测量范围在 $200 \sim 1300℃$ 之间，具有较高的精度及较广的测量范围，在测量温度中广泛应用。

（2）热电式高温计

热电式高温计又叫温差电偶温度计。它是利用两种金属丝对温度敏感程度

的不同，在端点接头处所形成的热电势，即利用热电效应制成的一种测温元件，叫做热电偶。

图2－41为两根具有温差电效应不同的金属组成的一个闭合环路，其中的一端焊接在一起，此端称为测量端（或称工作端、热端），这两根不同的金属导体（C，D）称为热电极。由

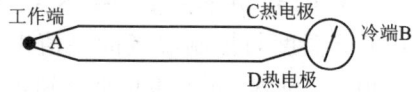

图2－41　热电偶测温原理示意图

于两种金属自由电子的密度不同，因而结点处形成一定的电位差。工作端（A）温度愈高，它与冷端（B）（也称自由端）的温差愈大，回路中产生的电动势就愈大。热电偶就是利用这一原理通过测量电动势来测定温度的。

热电偶一般与测温毫伏计（或温度控制器）和补偿导线组成测温仪，与高温电炉等加热设备配套使用。

3. 恒温装置

在测量理化数据（如化学反应速率、化学平衡常数、配合物稳定常数）、进行水盐体系研究以及无机或有机合成等实验时，都需要在恒温条件下进行。

恒温槽的作用是使被测定物质的温度在一定时间内保持不变。其基本原理为（图2－42）：接通电源后，继电器触点因弹片的作用，C接触，于是加热器开始加热，搅拌器搅拌槽内液体，使各部分的温度均匀。当达到所需温度时，定温器中的水银与其上端的铂丝接通，线圈A产生磁场，吸引弹簧片，使之与C断开，加热停止。当温度低于定值，定温

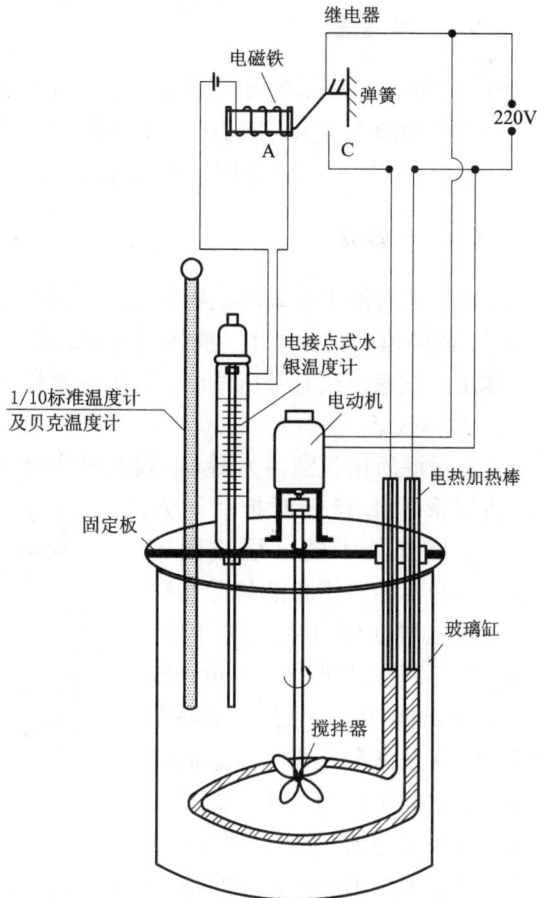

图2－42　恒温槽装置

器中的水银与铂丝脱离接触，线圈 A 失去磁性，弹簧片弹回而与 C 接触，于是又接通了加热器线路，恢复加热。如此反复，从而保持体系恒温。

所谓恒温，并非温度保持不变，而是在一定的温度范围内波动。通常，用槽内实际温度与控制温度的差值来表示恒温槽的灵敏度。一般的恒温槽在 20～40℃范围内，灵敏度可准确到 0.1℃左右，灵敏度高的可精确到 0.05℃。

2.6　溶液与沉淀的分离

有许多化学反应，产物呈固液悬浮态，需要进行固 – 液分离才能得到纯品。常见的固 – 液分离方法有倾泻、过滤（抽滤）、重力沉降、离心沉降等。

2.6.1　倾泻法

倾泻法适用于沉淀物颗粒较大、静置后能很快沉降的体系。倾斜器皿，把上层液体慢慢倾入到另一容器中，留下的沉淀物可再加少量洗液洗涤，搅拌静置沉降分离，再次倾斜处理即可达到分离目的。

2.6.2　过滤法

过滤是从溶液中分离沉淀物最常用的操作方法。当溶液和固体一起通过过滤器时，固体留在过滤器上，液体通过过滤器，达到分离的目的。

常用的过滤方法有常压过滤、减压过滤和热过滤三种。

1. 常压过滤

常压过滤所用过滤器为漏斗，过滤操作前必须准备好滤纸。

将圆形滤纸轻轻对折后再对折，然后展开其中的一层成圆锥形，放入漏斗内，使滤纸与漏斗贴紧，为使三层滤纸处紧贴漏斗壁，可撕去外面两层滤纸的一小角。按住三层滤纸的一侧，用少量水润湿滤纸，使其贴紧漏斗壁，注意滤纸与漏斗壁之间不应有气泡，如有气泡可轻轻按压滤纸，赶出气泡。

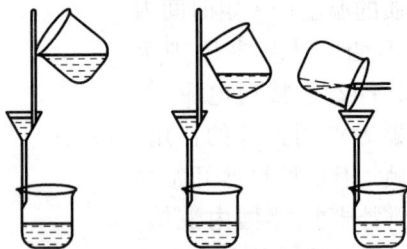

图 2 – 43　沉淀的过滤

一般采用倾泻法注入过滤物，一手持烧杯，另一只手持玻棒垂直紧靠烧杯嘴，让溶液沿玻棒流入漏斗，但不要碰到滤纸。大部分清液过滤后，用玻棒轻轻搅起沉淀，转移至漏斗中，用蒸馏水清洗烧杯和沉淀，将洗液和沉淀物转入漏斗中,如此反复,直至沉淀物全部转移到漏斗中,见图 2 – 43。

2. 减压过滤(抽滤)

抽滤的特点是过滤速度快,沉淀干燥效果好,但胶状沉淀和细颗粒沉淀不宜用此法。

抽滤装置由三部分组成:滤器与接收器、减压系统和安全瓶。滤器为布氏漏斗,接收器为抽滤瓶,抽滤瓶支管连接安全瓶,安全瓶再连接减压系统。

减压过滤时滤纸应剪得比布氏漏斗内径略小些,但又要盖住全部瓷孔。将滤纸放入布氏漏斗中,用水或溶液湿润,减压使滤纸吸紧在漏斗上,然后再进行过滤操作,见图 2 – 44。

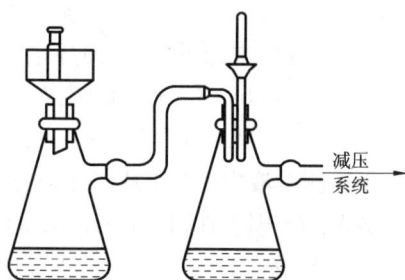

图 2 – 44 减压抽滤装置

3. 热过滤

在过滤过程中为了防止溶质结晶析出,可采用热过滤。热过滤一般要求用无颈或短颈漏斗,漏斗需预热以利保温。减压过滤方法也可采用热过滤,热过滤前布氏漏斗用热水或烘箱预热即可。

2.6.3 离心分离法

少量的沉淀和溶液分离时不能用常规的过滤法,因为沉淀物太少,粘在滤纸上难以取下,此时可选择离心分离法(图 2 – 45)。

将盛有溶液和沉淀的离心试管放在离心机中离心分离沉降后用滴管把清液吸出和沉淀分开,滴管末端接近沉淀时要特别小心,勿使滴管触及和扬起沉淀物。

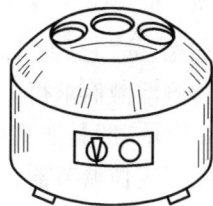

图 2 – 45 电动离心机

2.7 常用测量仪器的使用方法

2.7.1 酸度计的使用方法

pH 计的工作原理是由甘汞电极(参比电极)与玻璃电极与被测溶液组成原电池,该电池的电动势仅仅随溶液中 H^+ 离子浓度而变化,通过标准缓冲溶液校正后,可直接测出被测溶液的 pH。

1. 玻璃电极

pH 计是测定溶液 pH 最常用的仪器之一。它主要是利用一对电极，在不同的 pH 溶液中能产生不同的电动势。这对电极中一支称为指示电极（玻璃电极，图 2-46），另一支称为参比电极（甘汞电极，图 2-47）。玻璃电极是用一种导电玻璃（含 72% SiO_2，22% Na_2O，6% CaO）吹制成极薄的空心小球，球内装有 $0.1\ mol\cdot L^{-1}\ HCl$ 和 $Ag-AgCl$ 电极，把它插入一个待测溶液中，便组成一个电极。

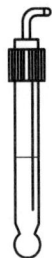

图 2-46　玻璃电极

$$Ag,\ AgCl（固）|0.1\ mol\cdot L^{-1}\ HCl|玻璃|待测溶液$$

这个导电的薄玻璃膜把两种溶液隔开，即有电势产生，小球内氢离子浓度是固定的，所以该电极的电位随待测溶液的 pH 的不同而改变。

$$\varphi_G = \varphi_G^{\ominus} - 0.059\text{pH}$$

式中　φ_G——玻璃电极的电极电位；

图 2-47　甘汞电极

$\varphi_G^{\ominus}$——玻璃电极的标准电极电位。

将玻璃电极和饱和甘汞极组成电池，并与电位差计连接，即可测定电池的电动势。

如果，$\varphi_G^{\ominus}$ 已知，即可从电势求出 pH。$\varphi_G^{\ominus}$ 可以用一个已知 pH 的缓冲溶液代替待测溶液而求得。

pH 计一般是把测得的电池电动势直接用 pH 表示出来。为了方便起见，仪器设置了定位调节器，当测量标准缓冲溶液的时候，利用这一调节器，把读数直接调节在标准缓冲液的 pH 上面校正仪器，这样在测未知溶液的时候，指针就可以直接指出溶液的 pH，一般都把前一步称为"校准"，后一步称为"测量"。一台已经校准过的仪器在一定时间内可以连续测量许多份未知液。如果电极的稳定性还没有完全建立，经常校准还是必要的。

玻璃电极的使用和维护：

（1）玻璃电极的主要部分是下端的一个玻璃泡，此球泡由一层极薄的特种玻璃制成，切忌与硬物接触，一旦发生破裂，则完全失效。

（2）在初次使用时，应先把玻璃电极在蒸馏水中浸泡数小时，最好一昼夜。不用时也最好把电极的玻璃球浸在蒸馏水中，以便在下次使用时可简化浸泡和校正手续。

（3）电极插头上的有机玻璃管具有优良的绝缘性能，切忌与化学药品或油污接触。

（4）玻璃膜不可沾有油污，如果发生这种情况，则应先将其浸入酒精中，再放入乙醚或四氯化碳中，然后再移到酒精中，最后用水冲洗干净，并浸于水中待用。

2. 甘汞电极

常用的参比电极是饱和甘汞电极，它是将汞（Hg）和甘汞（Hg_2Cl_2）的糊状物装入饱和 KCl 溶液中。汞上面插入铂丝，其上再连导线。饱和 KCl 溶液的下部有 KCl 晶体，管的下端开口用多孔陶瓷塞住，溶液能通过陶瓷内的毛细管向外渗漏。

饱和甘汞电极电位（相对于标准氢电极）与温度的关系为

$$\varphi = 0.2420 - 0.00076(t - 25)$$

式中，t 表示实验条件下的温度（℃）。

在测定与 Cl^- 无关的电极电位时，甘汞电极中 KCl 饱和溶液就可以直接作"盐桥"而无须再加盐桥。若对 Cl^- 测定有影响，就得附加盐桥，附加盐桥可用不同的形式。

3. PHS－10B 数字酸度计使用说明

（1）面板控制调节器件说明，见图 2－48。

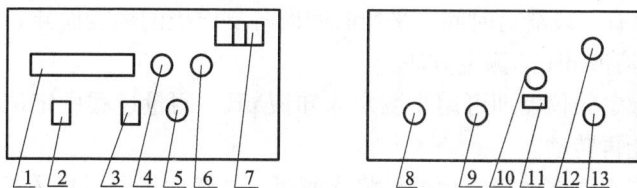

图 2－48　pH 计仪器面板示意图

1—发光二极管（LED）数字显示屏；2—电源开关，按下则电源接通，放开则电源断开；3—工作选择开关；4—定位调节；5—零点调节；6—斜率补偿；7—温度补偿；8—参比电极接线柱；9—测量电极输入插座，插入时插座外套向前按，同时将电极插头插入，然后放开外套；用手拉动电极连线，插头不能拉则表示已插好；如果需拔出插头，只需将外套向前按动，电极插头即可自行跳出；10—输出选择开关，分两挡，分别使输出信号插座得到满量程为 2 V 及 10 mV 信号；11—输出信号插座，与输出选择开关（10）配套使用；12—保险丝；13—电源插座，三芯中顶端一芯为接地端，电源为单相交流 220 V 50 Hz

（2）使用方法

Ⅰ. 准备工作

①仪器插入 220 V 交流电源，按下电源开关，LED 显示屏上即有数字显示，仪器预热十分钟，工作选择（4）在 mV 挡位置时，调节零点读数为 000。

②将已配制好的标准缓冲液(B_4、B_9两种溶液)分别倒入烧杯中,准备蒸馏水和滤纸少许。

③甘汞电极上下端的橡皮塞套应拔去,请注意电极内氯化钾溶液中不能有气泡,溶液不能过少,以防断路。并且溶液中应有氯化钾结晶析出,以确保溶液饱和。

④将甘汞电极和预先经蒸馏水浸泡的玻璃电极稳妥地装置在电极夹具内,甘汞电极应比玻璃电极下放 $2\sim3$ mm,以保护玻璃电极球泡,防止破碎。

⑤将甘汞电极引线旋紧于参比接线柱 8 上,将玻璃电极插头插入插座 9 上(也可使用复合电极,即甘汞电极和玻璃电极同置于一个塑料套管内)。

Ⅱ.用标准缓冲液标定仪器

①根据溶液温度,将温度补偿器置于相应位置,按下工作选择 3 使之位于 pH 位置。

②将电极浸入蒸馏水中轻轻摇动清洗,然后提起电极用滤纸吸去水滴,将电极浸入 B_4 缓冲液,待示值稳定后(为 20 s~3 min)调节定位钮使示值为 4.00(20℃)。

③将电极用蒸馏水清洗并吸干后浸入 B_9 缓冲液,待读数稳定后调节斜率补偿,使仪器读数为 B_9 溶液在 20℃时标准值 9.230。

④重复②、③两步操作,直至两次均相符合。

⑤电极在充分建立平衡后,才能得到稳定读数。电极插入溶液后,要得到稳定读数,总有一段滞后时间。平衡时间因各支电极的性能而异。

Ⅲ.未知溶液 pH 的测定方法

①经过标定的仪器即可用来测定未知液 pH。测量过程中定位调节钮和斜率调节钮不能再转动。

②用蒸馏水清洗电极并用滤纸将水滴吸干,然后将电极插入被测溶液,仪器所显示的稳定读数即为该溶液的 pH。

电极电位的测定方法

①取出电极插头,工作选择 3 位于 mV 位置,调节仪器零点,使读数为 000。

②将电极插入仪器,将电极浸入溶液中,仪器的稳定读数即为电极在该溶液中产生的实际电位,单位为 mV。工作电极的电位高于参比电极的电位时,读数会显示负值。

2.7.2 分光光度计的使用方法

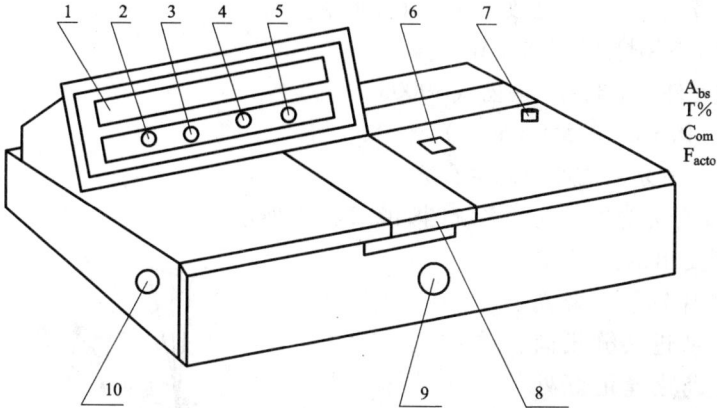

图 2 – 49   722 型分光光度计结构简图

1—数字显示器；2—A/T/C/F；3—SD；4—0%；5—100%

6—波长刻度窗；7—波长手轮；8—比色皿暗室盖；9—试样架拉手；10—干燥器

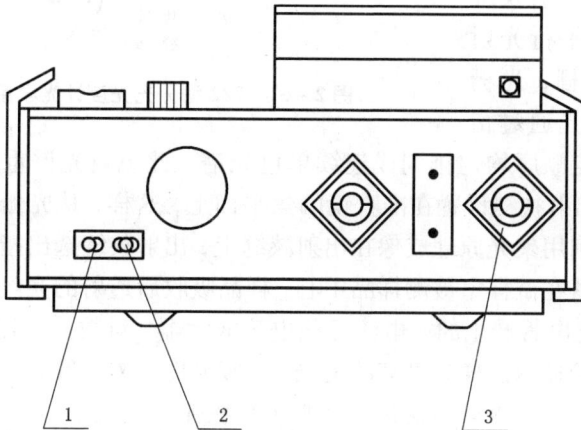

图 2 – 50   722 型分光光度计仪器后视图

1—1.5 A 保险丝；2—电源插头；3—外接插头

1. 722 型分光光度计外部旋钮、开关介绍(图 2 – 49，图 2 – 50)

主要技术指标是：

波长范围：330 ~ 800 nm；波长精度 ±2 nm

电源电压：220 V ±10% , 49.5 ~ 50 Hz

浓度直读范围：0 ~ 2000

吸光度测量范围：0~1.999

透光率测量范围：0~100%

光谱带宽：6 nm；色散元件：衍射光栅

光源：卤钨灯12 V,30 W

接收元件：光电管，端窗式19008

噪声：0.5%（在550 nm处）

2. 722型分光光度计光学系统

722型分光光度计光学系统图如图2-51所示。

钨灯发出的连续辐射经滤光片选择，聚光镜聚光后从进狭缝投向单色器，进狭缝正好处在聚光镜及单色器内准直镜的焦平面上，因此进入单色器的复合光通过平面反射镜反射及准直镜准直变成平行光射向色散元件光栅，光栅将入射的复合光通过衍

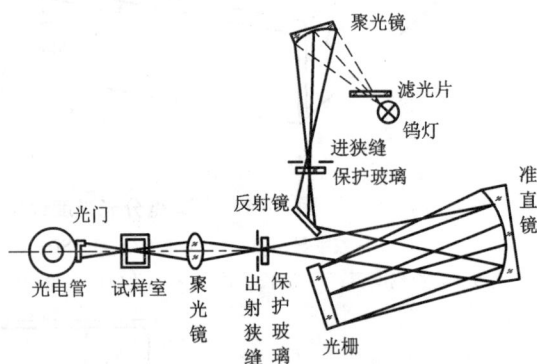

图2-51 722型分光光度计光学系统图

射作用按照一定顺序均匀排列成连续单色光谱。此单色光谱重新回到准直镜上，由于仪器出射狭缝设置在准直镜的焦平面上，这样，从光栅色散出来的光谱经准直镜后利用聚光原理成像在出射狭缝上，出射狭缝选出指定带宽的单色光通过聚光镜落在试样室被测样品中心，样品吸收后透射的光经光门射向光电管阴极面，由光电管产生的光电流经微电流放大器、对数放大器放大后，在数字显示器上直接显示出样品溶液的透光率、吸光度或浓度数值。

3. 722型分光光度计的使用方法及注意事项

(1)将灵敏度旋钮置"1"挡(放大倍率最小)。

(2)开启电源，指示灯亮，仪器预热20 min，选择开关置于"T"。

(3)打开试样室(光门自动关闭)，调节透光率零点旋钮，使数字显示"000.0"。

(4)将装有溶液的比色皿置于比色架中。

(5)旋动仪器波长手轮，把测试所需的波长调节至刻度线。

(6)盖上样品室盖，将参比溶液比色皿置于光路上，调节透光率"100"旋钮，使数字显示T为100.0(若显示不到100.0,则可适当增加灵敏度的挡数，

同时应重复③，调整仪器至"000.0"）。

（7）将被测溶液置于光路中，数字表上直接读出被测溶液的透光率（T）值。

（8）吸光度（A）的测量，参照③、⑥，调整仪器至"000.0"和"100.0"将选择开关置于 A，旋动吸光度调零旋钮，便得数字为"000.0"，然后移入被测溶液，显示值即为试样的吸光度（A）值。

（9）浓度（c）的测量，选择开关由 A 旋至 c，将已标定浓度的溶液移入光路，调节浓度旋钮，使得数字显示为标定值，将被测溶液移入光路，即可读出相应的浓度值。

（10）仪器使用时，应经常参照本操作方法中③、⑥进行调"000.0"和"100.0"的工作。

（11）每台仪器所配套的比色皿不能与其他仪器上的比色皿单个调换。

（12）本仪器数字显示后背部带的外接插座，可输出模拟信号。插座1脚为正，2脚为负接地线。

（13）若大幅度改变测试波长，需等数分钟后才能正常工作（因波长由长波向短波或由短波向长波移动时，光能量急剧变化，光电管受光后响应迟缓，需一段光响应平衡时间）。

（14）仪器使用完毕后应用套子罩住，并放入硅胶保持干燥。

（15）比色皿用完后及时用蒸馏水洗净，用细软的纸或布擦干，存于比色皿盒内。

（16）吸光度测量步骤：

序号	步骤	操作方法	显示
1	通电预热	打开试样室盖，开启电源，预热仪器20 min	
2	选择"T"	调"A/T/C"键，选择"T"。灵敏度旋钮置"1"挡（倍率最小）	"T%"灯亮
3	调 $T=0$	打开试样室盖，调"0%"键	000.0
4	装样品溶液	将待测溶液分别装入比色皿，依次放入样品室	000.0
5	选择波长	旋转波长手轮，选择508 nm	000.0
6	调 $T=100$	盖好样品室盖，参比溶液置于光路中，调"100%"键	100.0
7	调 $A=0$	在上述条件下，按"A/T/C"键，选择"A"，调吸光度调零键	000.0
8	测量样品 A	盖好样品室，将试样架拉手逐格拉出，依次读出各样品的 A 值	
9	结束	取出比色皿洗净、倒置、晾干，灵敏度调至最小，关闭电源	

2.7.3 荧光光度计的使用方法

荧光光度计(图2-52)是用于扫描荧光标记物所发出的荧光光谱的一种仪器,它能提供包括激发光谱、发射光谱、荧光强度、量子产率、荧光寿命、荧光偏振等多种物理参数,可从各种角度反映分子的成键和结构情况。通过对这些参数的测定,不但可以做一般的定量分

图2-52 LS-55荧光光度计

析,而且还可以推断分子的不同环境下的构象变化,从而阐明分子结构与功能之间的关系。荧光分光光度计的激发波长扫描范围一般是190~650 nm,发射波长扫描范围是200~800 nm。荧光光谱法可用于液体、固体(如凝胶条)的光谱扫描,而且,荧光光谱法具有灵敏度高、选择性强、用样量少、方法简便、工作曲线的线性范围宽等优点,因此被广泛应用于生命科学、医学、药学、有机化学和无机化学等领域。

1. 基本原理

当物质的基态分子受到一激发光源照射,被激发到激发态后,在返回基态时,可以产生波长与入射光相同或波长更长的荧光。由于物质结构不同,所能吸收的紫外光波长不同,在返回基态时,所发射的荧光波长也不同,利用这一性质可以定性鉴别物质。对于同种物质,其产生的荧光强度在一定范围内与该物质的浓度呈线性关系,利用这一性质可以进行物质的定量测定。利用物质被紫外光照射后所产生的、能够反映出该物质特性的荧光,以进行该物质的定性分析和定量分享的方法称为荧光分析。

荧光分析的主要特点是灵敏度高,检出限为$10^{-7} \sim 10^{-9}$ g·mL^{-1},比紫外可见光分光光度法高$10 \sim 10^3$倍。荧光分析法的另一特点是选择性高,因为能吸收光的物质并不一定产生荧光,而且不同物质由于结构不同,其激发态能级的分布具有各自不同的特征,这种特征反映在荧光上表现为各种物质都有其特征荧光激发和发射光谱,因此,虽吸收同一波长的光,不同的物质产生的荧光波长也不同。荧光分析法还有方法快捷,重现性好,取样容易,试样需要少等优点。荧光分析法的不足之处主要是和其他方法相比其应用范围还不够广泛,因为有许多物质不产生荧光。

2. 仪器基本结构

由高压汞灯或氙灯发出的紫外光和蓝紫光经滤光片照射到样品池中,激发

样品中的荧光物质发出荧光,荧光经过滤过和反射后,被光电倍增管所接受,然后以图像或数字的形式显示出来。荧光光度计的基本结构如图2-53所示,其中光源与检测器成直角方式排列。

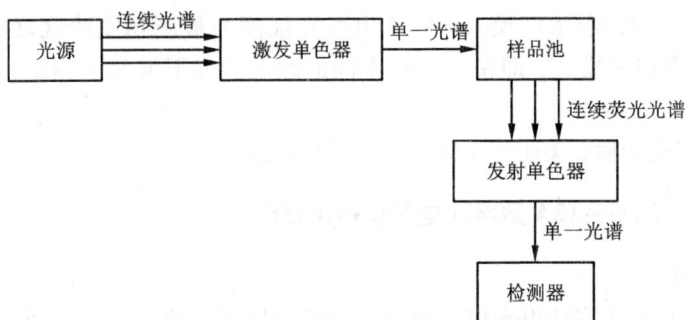

图 2-53　荧光光度计基本结构图

①光源:为高压汞蒸气灯或氙弧灯,后者能发射出强度较大的连续光谱,且在 300~400 nm 范围内强度几乎相等,故较常用。

②激发单色器:置于光源和样品室之间的为激发单色器,筛选出特定的激发光谱。

③发射单色器:置于样品室和检测器之间的为发射单色器,常采用光栅为单色器。筛选出特定的发射光谱。

④样品室:通常由石英池(液体样品用)或固体样品架(粉末或片状样品)组成。测量液体时,光源与检测器成直角安排;测量固体时,光源与检测器成锐角安排。

⑤检测器:一般用光电管或光电倍增管作检测器。可将光信号放大并转为电信号。

3. 操作方法

型号不同的荧光光度计操作方法略有不同,具体参见相应仪器的使用说明书。LS-55 型荧光光度计的操作方法简述如下:

①打开仪器、电脑及相应软件。

②在红色的"Luminescence Mode"中选择"Fluor(荧光)",选定荧光测定模式。点击右上角的红色数字"1(氙灯开)"或"0(氙灯关)",控制光源开关。

③设定仪器参数,包括激发光和发射光的起止波长、狭缝宽度等。

④开始扫描。扫描有三种方式:a. 预扫描,用以找出未知样品的最适宜激发波长;b. 单个单色器扫描,可单独进行激发光或发射光的光谱扫描;c. 同步

扫描，两个单色器以一定的波长间距同步转动扫描，同步扫描又可分为固定波长差和固定能量差两种方式。

⑤其他模式。LS－55 型荧光光度计提供了几种检测模式，可根据测定情况进行选择。a. 时间驱动模式，可在固定波长上记录一定时间内的样品荧光强度变化；b. 强度/浓度模式，可给出荧光强度与被测物质浓度之间的关系；c. 波长编程扫描模式，即在一系列不同的激发和发射波长下测定样品的荧光强度。

⑥测定完毕后，关闭应用软件、仪器和电脑。

2.7.4　DDS－12A 数字式电导仪使用说明

1. 概述

DDS－12A 数字式电导仪是电化学分析常用仪器之一。它广泛应用于化学研究、化学工业、电子工业、医药工业、环境监测、锅炉用水监测等领域。

2. 测量原理

溶液的电导是溶液的一个重要参数，它反映了溶液的导电能力。电导等于电阻的倒数，即 $S_x = 1/R_2$，式中 S_x 为待测电导（单位：西门子，用 S 表示），R_x 为电阻（单位：欧姆，用 Ω 表示）。实际上，物体的电导值不仅取决于组成该物质的导电能力的强弱，而且取决于该物体的几何尺寸，即沿电流流动方向的长度和截面积，因此，为了准确反映物质的导电能力，需要引入电导率这个概念。电导率（以 k 表示）是单位长度和单位截面积的电导值。

$$电导率 = \frac{电导值 \times 长度}{截面积}$$

其单位一般有 $S \cdot cm^{-1}$、$mS \cdot cm^{-1}$、$\mu S \cdot cm^{-1}$ 三种。

最常用的 DJS－Ⅰ型电导仪的电极由两块平行放置并被固定的玻璃支架上的铂片组成，每块铂片上分别引出一根导线。测定时将它们连接到仪器的输入端，介于两块铂片之间的这一部分溶液的电导值就是被测溶液的电导值 S_x。见图 2－54 和图 2－55。

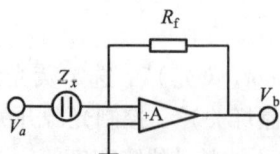

图 2－54　测量原理示意图　　　　图 2－55　电导池等效电路图

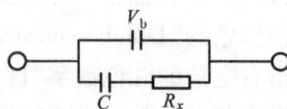

图中：Z_x——电导池模拟阻抗，$Z_z = j_z C_0 / (R_x + C_x)$；

$\quad\quad R_f$——标准比较电阻；

$\quad\quad C_0$——电导电极引线电容；

$\quad\quad C$——电导池等效电容。

设两块铂片的面积都为 A，铂片间的距离为 L，则 L/A 称为电极常数，以 Q 表示。

这样，被测电导率 $k_x = QS_x$。知道电极常数 Q 后，测得电导 S_x，即可算出电导率 K_x。

测量原理见附图 2-54。图 2-55 为电导池即 Z_x 的等效电路。由图 2-54 可知：

$V_b = -R_f \cdot V_a / Z_x$，由 $K_x = QS_x$，得 $K_x = Q/R_x$，所以当 V_a、R_f、Q 为常数时电导率 K_x 的变化必然引起 R_x 的变化，R_x 的变化又引起 Z_x 的相应变化，而 Z_x 的变化又引起 V_b 的变化，所以通过测量 V_b 的大小，就能测量电导率 K_x 值。

3. 仪器的工作过程

仪器测量方框图如图 2-56 所示。

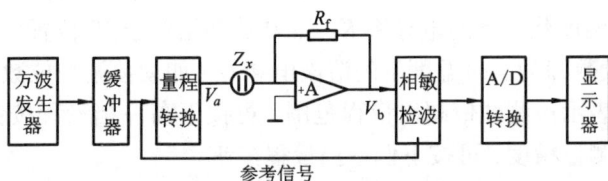

图 2-56　仪器测量方框示意图

由方波发生器输出方波测量电压信号，通过缓冲级后信号经不同程度的反馈得到不同的固定值 V_a，即得到了量程转移。信号 V_a 加在 Z_x、R_f 及反相输入的比例运算放大器 A 上，经比例运算得 Z_x 的大小转移成的相应的电压信号 V_b。V_b 中具有正比于 $1/S_x$ 的电阻性分量和电导电极引线电容 C_0，及电导池等效电容 C 的电抗性分量。V_b 经相敏检测器可将其中电抗性分量除去，同时将交流信号转移成直流信号输出。这样加在 A/D 转换电路输入回路的 V_b 就与被测溶液的电导的 $1/S_x$ 成正比了，从而便可读出被测电导率。

4. 操作方法

仪器面板示意图如图 2-57 所示。

（1）接通电源，仪器预热 10 min。如使用高周则按下 20 mS/cm 按钮，使用低周则放开所有量程，在没有接入电极时使用调零旋钮使仪器读数为 000。

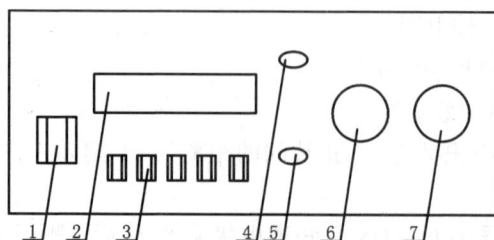

图 2 - 57 DDS - 12A 数字电导仪面板示意图

(2)将电极插座外套向前按动，并插入电极连接插头，然后放开外套，可拉一下电极连线(应不能拉出)，以检查插头是否可靠接好。如电极连接使用接线叉，则固定于仪器接线柱上，连线屏蔽引出端应接黑色接线柱。电极应用干净布拭干，并悬空放置，电极头部不能接触任何物体。按下 2 μS/cm 量程，调节电容补偿，使仪器的读数为 000。

(3)将仪器温度补偿旋钮置于 25℃ 位置，常数补偿置于电极的实际常数相应位置，将电极浸入被测溶液，并按下相应的量程按钮，所得稳定读数即为被测溶液在测量温度状态下的电导率 K。如常数补偿置于 50 位置(即常数为 1 位置)，则仪器读数即为当时温度下的溶液电导 G。如读数第一位数字为 1，后三位数字熄灭，表示被测数值超出量程范围，可按下高一挡量程按钮，如读数很小，为了提高测量精度，可按下低一挡量程按钮。

(4)使用 DJS - 10 电导电极，可以扩展测量范围，即测量结果乘以 10。20 μS/cm 挡可测至 2002 μS/cm，2 μS/cm 可测至 20 μS/cm，此时常数补偿的表示也扩大 10 倍，即补偿 5 ~ 15 的常数范围。

(5)上述测量中温度补偿旋钮在 25℃ 时的电导率。但是由于仪器设置的温度系数为 2%/10℃，与这个温度系数不相符的溶液使用温度补偿器将会产生补偿误差。

(6)仪器可长时连续使用，也可利用输出信号外接记录仪进行连接监测。当需要拔除电极插头时，按一下插座外套，即可自行脱出。

(7)仪器使用完毕后，断开电源开关。

(编者：钱 频 王一凡 刘绍乾)

第三章 基础化学实验项目

3.1 基本技能与操作实验

实验1 渗透压的测定及红细胞形态的观察
Determination of Osmotic Pressure
and Observation of Red Blood Cell Shape

一、实验目的

(1)了解渗透压计测定溶液渗透压的原理和方法。

(2)配制低渗、等渗和高渗溶液,并用渗透压计测量其渗透浓度。

(3)用渗透压计测量体液(尿液)的渗透浓度。

(4)在显微镜下观察红细胞在低渗、等渗和高渗溶液中的不同形态。

二、预习要点

(1)渗透浓度的定义是什么?它的单位与 $mmol \cdot kg^{-1}H_2O$ 有何区别?

(2)为什么渗透压的大小可以用渗透浓度的大小来衡量?

(3)实验测定的凝固点降低值与理论值产生偏差的主要原因是什么?

(4)解释在显微镜下观察到的红细胞在等渗、低渗和高渗溶液中的形态为何不同?

三、实验基本原理

稀溶液的依数性包括溶液的蒸气压下降、溶液的沸点上升、溶液的凝固点降低和溶液的渗透压,其中以溶液的渗透压测定在临床上最为重要,它对纠正体内水、电解质和酸碱平衡失调起着十分重要的作用。

渗透压的测定方法较多,以凝固点降低法的操作最为方便,且精度高、测定迅速,该方法样品用量少,而且对生物样品无变性作用,特别适合于人体各种体液(如尿、血清、胃液、脑脊液、唾液等)的测定。

根据 Raoult F. M. 的凝固点降低原理, 任何 1 mol 的难挥发性非电解质溶于 1 kg 水中, 都可以使水的凝固点由 0℃降低至 −1.857℃。而任何 1 mol 的电解质溶于 1 kg 水中, 其凝固点降低值在理论上应是电解质解离的离子数与 1.857℃的乘积。由于一定浓度的电解质溶液中存在离子间的相互作用力, 故使测得的凝固点降低值往往小于理论值。在计算电解质溶液渗透压时, 引进渗透系数(Φ)加以校正。在临床上, 血浆中单价离子的 Φ 值在 0.91 ~ 0.93 之间。当温度一定时, 溶液的渗透压与溶液的渗透浓度成正比关系。因此, 可用渗透浓度来衡量溶液的渗透压大小, 临床上规定溶液的渗透浓度在 280 ~ 320 mmol·L^{-1}范围内为等渗溶液, 低于或高于此范围的为低渗或高渗溶液。对稀水溶液而言, 溶液的渗透浓度也可用毫摩尔·千克$^{-1}$水(mmol·kg^{-1}H$_2$O)来表示, 由于溶液渗透压值与凝固点降低值呈线性关系, 溶液的 mmol·kg^{-1}H$_2$O 值可通过凝固点降低而测得。渗透压计已将凝固点降低值换算成渗透浓度单位(mmol·kg^{-1}H$_2$O)显示读数。

本实验用渗透压计测定所配制的低渗、等渗、高渗溶液和尿液的 mmol·kg^{-1}H$_2$O值, 并在显微镜下观察红细胞在低渗、等渗和高渗溶液中的形态变化。

四、仪器和试剂

仪器:烧杯(50 mL), 容量瓶(50 mL), 刻度吸管(1 mL), 洗耳球, 滴管, 玻棒, 洗瓶, 小试管, 一次性采血吸管(10 μL), 一次性采血针, 分析天平, 载玻片, 盖玻片, 光学显微镜, FM −5J 简易型冰点渗透压计。

试剂:NaCl(固体, A. R), 70%(V/V)酒精棉球, 消毒干棉球, 擦镜纸, 尿液(新鲜)。

五、实验内容

1. 溶液配制

(1)低渗溶液 准确称取 NaCl 0.31 ~ 0.32 g 置于 50 mL 烧杯中, 加少量蒸馏水, 使之溶解, 转入到 50 mL 容量瓶中, 再用少量蒸馏水淋洗烧杯壁 3 次, 每次淋洗液全部转移入容量瓶中, 加水稀释全刻度, 摇匀, 配成低渗溶液备用。

(2)等渗溶液 准确称取 NaCl 0.44 ~ 0.46 g 于 50 mL 烧杯中, 按上法配成 50 mL 等渗溶液备用。

(3)高渗溶液 准确称取 NaCl 1.25 ~ 1.30 g 于 50 mL 烧杯中, 按上法配成 50 mL 高渗溶液备用。

2. 溶液和体液的渗透压测定

按照《FM-5J 简易型冰点渗透压计》操作步骤测定所配制的低渗、等渗和高渗溶液以及尿液的 $mmol \cdot kg^{-1} H_2O$ 值。每个样品测定 3 次，取其平均值。

3. 红细胞在低渗、等渗和高渗溶液中的形态观察

用 70%（V/V）酒精浸湿的棉花球消毒手指尖部皮肤，待干后，用一次性采血针快速刺入皮肤并立即拔出，使血液自然流出，形成血滴（切勿用手挤压手指，以免组织液稀释血液）。用棉花球擦去第 1 滴血液，然后用一次性采血吸管分别吸取血液 10 μL，加入各装有 1 mL 低渗、等渗和高渗溶液的三支小试管中，摇匀即得红细胞悬液。

从上述三支小试管中各取 1 滴红细胞悬液滴于载玻片上，盖上盖玻片；在显微镜下用高倍镜（40 倍或 45 倍）观察它们的形态变化。

六、实验记录与结果

内容　　　　　　项目	低渗溶液	等渗溶液	高渗溶液	尿液
测定值 1 （$mmol \cdot kg^{-1} H_2O$）				
测定值 2 （$mmol \cdot kg^{-1} H_2O$）				
测定值 3 （$mmol \cdot kg^{-1} H_2O$）				
平均值 （$mmol \cdot kg^{-1} H_2O$）				
相对平均偏差 （d_r）				

（编者：何跃武）

实验 2　凝固点降低法测定相对分子质量
Determination of Molar Mass
Using Freezing Point Depression

一、实验目的

(1)掌握用溶液的凝固点降低法测定溶质的相对分子质量。

(2)掌握 0.1℃分度温度计的使用。

(3)学会分析天平和容量吸管的使用。

二、预习要点

(1)每毫升水的质量以 1 g 计算。

(2)水或溶液开始凝固前,往往有过冷现象发生,即温度已降至冰点以下,仍未结冰(刚刚结出冰屑的温度并非冰点)。结冰后放出热量,使温度回升,至恒定后才是其冰点。

(3)思考并回答下列问题:

①冰水中加入食盐为什么可以作为致冷剂?

②能否应用沸点升高法来计算溶质的相对分子质量?为什么测定溶质(非电解质)的相对分子质量,常用凝固点降低法?

③0.256 g 苯(相对分子质量:128)溶于 25 g 水中,试计算所得溶液的凝固点(已知苯的凝固点为 5.5℃,苯的 K_f 值为 5.10)。

三、实验基本原理

稀溶液具有依数性,凝固点降低是依数性的一种表现。对理想溶液,凝固点的降低与溶质的质量摩尔浓度成正比:

$$\Delta T_f = T_0 - T = K_f \cdot b_B \tag{1}$$

式中:ΔT_f 为凝固点降低值,T_0 和 T 分别为纯溶剂和溶液的凝固点(K),K_f 为凝固点降低常数。

而

$$b_B = \frac{n_B}{m_A} = \frac{m_B/M_B}{m_A} \tag{2}$$

代入(1)式得:

$$M_B = \frac{K_f m_B}{\Delta T_f m_A} \tag{3}$$

因此，只要称取一定量的溶质 $m_B(g)$ 和溶剂 $m_A(g)$ 配成稀溶液，分别测定纯溶剂和稀溶液的凝固点，求得 ΔT_f，再查得溶剂的凝固点降低常数 K_f，即可计算溶质的相对分子质量 M_r。

四、仪器和试剂

仪器： 温度计（0.1℃ 分度），烧杯（400 mL），容量吸管（50 mL），洗耳球，大试管（40×150），放大镜，搅拌器，分析天平，量筒（100 mL）。

试剂： 食盐，冰，尿素（A. R，固体），蒸馏水。

五、实验内容

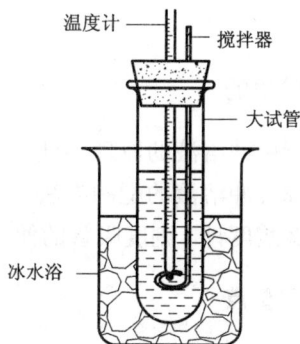

图实 3－1　凝固点降低法测定装置示意图

1. 调节冰盐水的温度

在大烧杯中加大约 250 mL 的水、足量的冰块与食盐，搅拌（勿碰到温度计水银球），使冰盐水的温度为 $-2 \sim -3$℃，在实验过程中，应不断搅拌并间断地补充少量水与食盐，以保持此温度。

2. 测定溶剂（水）的凝固点

用容量吸管准确吸取蒸馏水 50.00 mL，注入干燥大试管中，按图装置，将试管插入冰水浴中。均匀地搅动大试管内的水，直至有冰屑析出。待过冷破坏后，温度略有回升，并恒定几分钟，记下此时温度。然后取出大试管，用手温使冰屑融化，重新插入冰水浴中，重复操作一次。取两次温度平均值（两次所测得的凝固点相差不得超过 0.05℃，否则应重做）即为溶剂的凝固点 T_o。

3. 测定尿素溶液的凝固点

用分析天平准确称取 1.4~1.7 g 尿素，直接倒入上述大试管中，加蒸馏水 50.00 mL，搅拌使其完全溶解（勿使纯水或尿素损失）；按步骤 2 方法，测定尿素溶液的平均凝固点 T。

4. 尿素的相对分子质量的计算

从手册中查得水的 K_f 值为 1.86 K·mol⁻¹kg，代入（3）式，算出尿素的相对分子质量（M_B）以及实验值的相对误差。（尿素相对分子质量的理论值为 60.05）

（编者：钱　频）

实验 3　缓冲溶液的配制及其性质
Preparation of Buffer Solution and Its Property

一、实验目的

(1)掌握缓冲溶液的原理和性质。

(2)掌握缓冲溶液的配制方法。

(3)掌握酸度计、刻度吸管的使用方法。

二、预习要点

(1)缓冲溶液抗酸抗碱及抗稀释的原理是什么?

(2)缓冲溶液稀释前后缓冲容量是否相同?

(3)配制的缓冲溶液,其 pH 的计算值与测量值为何不同? 有哪些因素造成差异?

(4)简述如何配制 pH = 7.40 的缓冲溶液。

三、实验基本原理

缓冲溶液具有抵抗少量强酸或强碱而保持 pH 几乎不变的能力,它一般由弱酸和其对应的共轭碱组成,其 pH 可由下式计算:

$$pH = pK_a + lg \frac{[共轭碱]}{[共轭酸]}$$

其中 K_a 为酸的离解常数。

注意,经上式计算的 pH 是近似的,若要精确计算溶液的 pH,还需考虑离子活度等因素的影响。

四、仪器和试剂

仪器:酸式滴定管(50 mL),刻度吸管(10 mL, 1 mL),酸度计,烧杯,洗耳球。

试剂:Na_2HPO_4(0.2 mol·L^{-1}), KH_2PO_4(0.2 mol·L^{-1}), $NaOH$(2 mol·L^{-1}); KH_2PO_4(2 mol·L^{-1}), HCl(1 mol·L^{-1}), 蒸馏水, $NaOH$(1 mol·L^{-1}), $NaCl$(0.9%, W/V)。

五、实验内容

1. 缓冲溶液的配制

(1)计算配制 pH 为 7.40 的缓冲液 100 mL 所需 0.2 $mol \cdot L^{-1}$ Na_2HPO_4 和 0.2 $mol \cdot L^{-1}$ KH_2PO_4 溶液的用量。($pK'_a = 6.86$)

(2)分别用量筒量取上式所得的 Na_2HPO_4 溶液和 KH_2PO_4 溶液，一并放入 150 mL 烧杯中并混匀，用酸度计测量其 pH（酸度计的使用方法参见本书前文介绍）。

(3)用 2 $mol \cdot L^{-1}$ NaOH 或 2 $mol \cdot L^{-1}$ KH_2PO_4 溶液调节上述溶液的 pH 为 7.40，保留备用。

2. 缓冲溶液的性质

用移液管按下表量取溶液体积，并测定其 pH，然后用刻度吸管按下表量取加入的酸或碱的体积，并测其 pH；根据加入酸、碱前后 pH 的变化，说明缓冲溶液具有哪些性质。

加入酸、碱对各溶液 pH 的影响

编号	溶液体积/mL	pH	加入酸或碱体积/mL	pH	ΔpH	β
1	0.9%（W/V）NaCl 25.00 mL		1 $mol \cdot L^{-1}$ HCl 0.25 mL			
2	0.9%（W/V）NaCl 25.00 mL		1 $mol \cdot L^{-1}$ NaOH 0.25 mL			
3	pH = 7.40 的缓冲液 25.00 mL		1 $mol \cdot L^{-1}$ HCl 0.25 mL			
4	pH = 7.40 的缓冲液 25.00 mL		1 $mol \cdot L^{-1}$ NaOH 0.25 mL			

（编者：刘绍乾　王曼娟）

实验 4　分析天平称量练习
Use of Analytical Balance

一、实验目的

(1)熟悉分析天平的构造,掌握分析天平的正确使用方法。

(2)掌握常用的称样方法——直接称量法和减重称量法。

(3)熟悉干燥器和称量瓶的使用。

二、预习要点

(1)称量前为什么要调节天平处于水平位置?

(2)为什么样品放入分析天平称量前,要先用台秤粗称?

(3)称量时,能否将需称量的试剂直接放在天平托盘上?

(4)为了尽快达到平衡,选取砝码时,是否要遵循"由大到小,中间截取,逐级试验"的原则?

(5)在称量过程中,如何运用优选法较快地确定物体的质量?

(6)称量时,取放物体、加减砝码或圈码时,为什么一定要关闭升降枢?

(7)能否根据光标移动的方向,迅速判断天平的左右两盘哪边轻哪边重?

(8)为什么在试加砝码时,应半开天平升降枢试称,而读数时,应全开天平升降枢?

(9)记录砝码数值时,是否需要再次核对? 记录和计算中,如何正确运用有效数字?

(10)减重法称量中,零点为什么可以不参加计算?

三、实验基本原理

见《分析天平的构造和使用》(2.3.2, 2)。

四、仪器和试剂

仪器:分析天平(附砝码),干燥器,称量瓶,烧杯(50 mL),表面皿(4.5 cm),台秤。

试剂:Na_2CO_3(无水)或细沙。

五、实验内容

1. 称量前的准备工作

详细认真地阅读《分析天平的构造和使用》，并观看台秤、分析天平、称量瓶的使用与操作方法的录像。

2. 称量前的检查工作

取下天平罩，折叠好放在天平箱的上面。逐项检查：

①称量物的温度与天平箱内温度是否一致，称量物的外部是否清洁和干燥。

②天平箱内、秤盘上是否清洁。如有灰尘，用毛刷刷干净。

③天平各部件是否都正常，特别要注意吊耳和圈码。

④天平是否处于水平。

⑤测定或调节天平零点。

3. 直接法称量练习

用直接法准确称量一块表面皿的重量。从干燥器中，用洁净纸条夹取一块洁净、干燥且已知重量的表面皿(有编号并经教师事先称量)，先用台秤粗称其质量 $W_{粗}(g)$，然后放在分析天平上称量，准确至 0.0001 g，记录表面皿质量 $W(g)$，称量完毕，关好升降枢。由教师核对称量结果，误差不能超过 0.0005 g。

4. 减重法称量练习

准确称取 Na_2CO_3(或细沙)的质量。从干燥器中用洁净纸条夹取盛有 Na_2CO_3 粉末(或细沙)的称量瓶，先放在台秤上粗称，记录重量 $W_{粗}(g)$。然后放入分析天平准确称量，记录质量为 $W_1(g)$，关好升降枢。

左手用一洁净纸条从天平中夹取出称量瓶，再用右手取一小片洁净纸包住瓶盖把手，揭开瓶盖，倾斜瓶身，用盖轻轻叩击瓶口，叩出 0.5~0.6 g 样品落入洁净的 50 mL 烧杯中，盖好称量瓶，用纸条夹好放入天平中；拿出纸条，准确称量其质量，记为 $W_2(g)$。两次质量之差，即为试样的质量。按上述方法连续递减，可称取多份试样。

5. 称量结束　应关闭天平，取出称量物、砝码，圈码的指数盘恢复到"0.00"位，关好天平门，罩好天平罩，填写使用登记卡，经老师同意后，方可离开天平室。

六、实验记录与结果

1. 直接法称量记录

天平编号：　　　　　　　　表面皿编号：

$W_{粗}(g) =$ 　　　　　　　准确质量 $W(g) =$

2. 减重法称量记录

天平零点：		$W_{粗}(g) =$		
称量记录	$W_1(g)$	$W_2(g)$	$W_3(g)$	$W_4(g)$
样品质量 $\Delta W(g)$		$W_1 - W_2 =$	$W_2 - W_3 =$	$W_3 - W_4 =$

（编者：何跃武）

实验 5　酸碱标准溶液的配制及标定
Preparation of Acid-Base Standard Solution
and Determination of Its Concentration

一、目的要求

（1）掌握容量仪器的洗涤方法；

（2）学会量筒、容量瓶、移液管、滴定管的正确使用方法，掌握滴定操作的基本方法；

（3）掌握用基准物质标定 HCl 溶液的方法。

二、预习要点

（1）量筒、容量瓶、移液管、滴定管的用途及使用方法。

（2）滴定操作的基本方法，特别是滴定终点的操作要点要求。

（3）思考下列情况对标定 HCl 溶液的浓度是否有影响。

①装 HCl 溶液的滴定管没有用 HCl 溶液润洗。

②滴定管中 HCl 溶液的初读数应为 0.01 mL，而记录数据时误记为 0.10 mL。

③锥形瓶用 Na_2CO_3 标准溶液润洗。

④滴定完后，尖嘴内留有气泡。

⑤滴定过程中，往锥形瓶中加入少量蒸馏水。

三、实验基本原理

（1）一般的酸碱物质因含有杂质或稳定性较差，不可能直接配制成准确浓度的溶液，通常先配成近似浓度的溶液，然后再用适当的基准物质加以标定。

标定 HCl 溶液的基准物质常用分析纯的无水 Na_2CO_3，滴定到达等量点时，溶液呈酸性，可选用甲基橙作指示剂，溶液由黄色变成橙色即为终点。

根据 $c(HCl) \cdot V(HCl) = \dfrac{W(Na_2CO_3)}{M\left(\frac{1}{2}Na_2CO_3\right)} \times 1000$ 即可算出 HCl 溶液的准确浓度。

式中　$W(Na_2CO_3)$——每次滴定所用的 Na_2CO_3 的质量（g）；

$M\left(\dfrac{1}{2}\mathrm{Na_2CO_3}\right)$——$\mathrm{Na_2CO_3}$ 的摩尔质量($\mathrm{g/mol}$)；

$V(\mathrm{HCl})$——每次滴定时所消耗盐酸标准溶液的体积(mL)；

$c(\mathrm{HCl})$——所求盐酸溶液的准确浓度($\mathrm{mol/L}$)。

标定 NaOH 溶液的基准物质常用分析纯的邻苯二甲酸氢钾($\mathrm{KHC_8H_4O_4}$)或草酸($\mathrm{H_2C_2O_4 \cdot 2H_2O}$)。

(2)用已知准确浓度的盐酸标准溶液也可标定未知浓度的碱溶液。

酸碱反应的实质是

$$\mathrm{H_3O^+ + OH^- \Longrightarrow 2H_2O}$$

HCl 与 NaOH 完全反应时

$$n(\mathrm{HCl}) \Longrightarrow n(\mathrm{NaOH})$$

NaOH 溶液的准确浓度可用下式计算：

$$c_{(\mathrm{HCl})} \times V_{(\mathrm{HCl})} = c_{(\mathrm{NaOH})} \times V_{(\mathrm{NaOH})}$$

四、仪器和试剂

仪器：量筒(10 mL、100 mL、500 mL)，试剂瓶(500 mL)，酒精灯，锥形瓶(250 mL)，酸式滴定管(50 mL)，碱式滴定管(50 mL)，电子天平，滴定管架，烧杯(50 mL)，滴管，容量瓶(100 mL)，移液管(25 mL)。

试剂：浓 HCl(相对密度 1.19，A. R)，NaOH(10 $\mathrm{mol \cdot L^{-1}}$)，无水 $\mathrm{Na_2CO_3}$(A. R)，甲基橙指示剂、酚酞指示剂。

五、实验内容

1. 近似 0.1 $\mathrm{mol \cdot L^{-1}}$ HCl 溶液及 0.1 $\mathrm{mol \cdot L^{-1}}$ NaOH 溶液的配制

(1)0.1 $\mathrm{mol \cdot L^{-1}}$ HCl 溶液的配制。

①计算配制 0.1 $\mathrm{mol \cdot L^{-1}}$ HCl 溶液 400 mL 所需浓盐酸的体积(mL)。

②用 10 mL 量筒量取算出的浓盐酸的体积，倒入具有玻塞、洁净的 500 mL 试剂瓶内，加蒸馏水至 400 mL，塞好玻塞，充分摇匀，贴上标签(写明试剂名称、班级、姓名及配制日期)。

(2)0.1 $\mathrm{mol \cdot L^{-1}}$ NaOH 溶液的配制。

用 10 mL 量筒量取 10 $\mathrm{mol \cdot L^{-1}}$ NaOH 溶液 4 mL，倒入具有橡皮塞、洁净的 500 mL 试剂瓶内，加蒸馏水至 400 mL 塞好瓶塞，充分摇匀，贴上标签(写明试剂名称、班级、姓名及配制日期)。

2. $\mathrm{Na_2CO_3}$ 标准溶液的配制

在电子天平上精确称取经 105℃ 干燥至恒重的无水 $\mathrm{Na_2CO_3}$ 0.48～0.52 g，

置于洁净的 50 mL 烧杯中，加入蒸馏水 30 mL，用玻棒小心搅拌，使之溶解。然后用玻棒引流将溶液转移到 100 mL 的容量瓶中，再用少量蒸馏水淋洗烧杯 2 ~ 3 次，每次淋洗液均转移到容量瓶中，再加蒸馏水至接近容量瓶刻度标线时，用滴管小心加入蒸馏水至刻度标线，盖紧瓶塞，充分摇匀。

3. HCl 溶液的标定

(1)将洁净的酸式滴定管用少量上述配制好的近似 $0.1\ mol\cdot L^{-1}$ HCl 溶液润洗 2 ~ 3 次，然后装入该 HCl 溶液，使液面恰好在"0"刻度或稍低于"0"刻度处。静置 1 min，准确记录滴定管的初始读数(准确至小数点后第二位)。

(2)用移液管移取 25.00 mL 上述配制好的 Na_2CO_3 标准溶液至锥形瓶中，加甲基橙指示剂 2 滴，溶液呈黄色。

(3)从滴定管中将 HCl 溶液滴入锥形瓶中，不断振摇，滴定接近终点时，用洗瓶冲洗锥形瓶内壁，加热煮沸以除去 CO_2，然后再逐滴加入 HCl 溶液，滴至溶液由黄色恰好变为橙色；且经振摇在半分钟内不再消失为止，静置 1 min，记录终点读数，前后两次读数之差，即为滴定时消耗 HCl 标准溶液的毫升数。

(4)重复上述滴定操作，直到两次滴定消耗 HCl 溶液的体积相差不超过 0.05 mL 为止，计算 HCl 溶液的平均浓度(保留四位有效数字)。

4. 测定 NaOH 溶液的准确浓度

(1)用移液管移取 25.00 mL 已知准确浓度的 HCl 溶液至锥形瓶中，加酚酞指示剂 2 滴，溶液无色。

(2)将洁净的碱式滴定管用少量上述配制好的近似 $0.1\ mol\cdot L^{-1}$ NaOH 溶液润洗 2 ~ 3 次，然后装入该未知准确浓度的 NaOH 溶液，赶出气泡，调节滴定管内溶液的弯月面在"0"刻度或稍低于"0"度处，静置 1 min，准确记录初始读数。

(3)滴定。开始时可以稍快，接近终点时应逐滴加入碱溶液，并用洗瓶吹洗锥形瓶内壁，继续逐滴滴入碱溶液，直到溶液恰至粉红色，且经振摇在半分钟内不再消失，即为终点，静置 1 min，准确记录碱式滴定管的终点读数。

(4)重复上述滴定操作，直到两次滴定消耗的 NaOH 溶液的体积相差不超过 0.05 mL 为止，计算 NaOH 溶液的平均浓度(保留四位有效数字)。

六、实验记录与结果

HCl 溶液的标定

数据记录与计算　　　　　测定序号		1	2	3
Na₂CO₃ 标准溶液净用量/mL		25.00	25.00	25.00
HCl	初始读数/mL			
	终点读数/mL			
	净用量/mL			
HCl 标准溶液的浓度/mol·L⁻¹				
平均值/mol·L⁻¹				
相对平均偏差（d_r）				

测定 NaOH 溶液的准确浓度

数据记录与计算　　　　　测定序号		1	2	3
HCl 标准溶液的净用量/mL		25.00	25.00	25.00
NaOH	初始读数/mL			
	终点读数/mL			
	净用量/mL			
NaOH 标准溶液的浓度/mol·L⁻¹				
平均值/mol·L⁻¹				
相对平均偏差（d_r）				

（编者：李战辉）

实验 6　$KMnO_4$ 标准溶液的配制及标定
Preparation and Determination of Potassium Permanganate Standard Solution

一、实验目的

(1)掌握高锰酸钾标准溶液的配制和标定方法。

(2)掌握滴定管中装入深色溶液时的读数方法。

(3)熟悉使用物质自身指示剂判断滴定终点的方法。

二、预习要点

(1)$KMnO_4$ 标准溶液是否可以直接配制? 为什么?

(2)配制的 $KMnO_4$ 溶液为什么必须煮沸,并保持微沸 1 个小时,然后最好静置 2~3 天才过滤? 能否用滤纸过滤?

(3)用 $Na_2C_2O_4$ 作基准物质标定 $KMnO_4$ 溶液时,应注意哪些条件?

(4)如何在滴定管中读取有色溶液的读数?

(5)何谓自身催化作用? 解释滴定时,第一滴 $KMnO_4$ 溶液褪色很慢的原因。

(6)滴定速度与反应速度相适应,若滴定速度过慢或过快会造成什么影响?

(7)溶液的酸度过低或过高对本实验有何影响?

三、实验基本原理

$KMnO_4$ 试剂中常含有少量的 MnO_2 和其他杂质,蒸馏水中也常含有微量还原性物质,它们与 MnO_4^- 反应会析出 $MnO(OH)_2$ 沉淀,MnO_2 和 $MnO(OH)_2$ 又能进一步促进 $KMnO_4$ 溶液的分解。因此,$KMnO_4$ 标准溶液不能用直接法进行配制,通常先配成近似浓度的溶液,然后再进行标定。

标定 $KMnO_4$ 的基准物质有 $H_2C_2O_4 \cdot 2H_2O$、$FeSO_4 \cdot (NH_4)_2SO_4 \cdot 6H_2O$、$Na_2C_2O_4$,$As_2O_3$ 和纯铁丝等。其中 $Na_2C_2O_4$ 不含结晶水,容易提纯,且没有吸湿性,因此最常用。在酸性溶液中,$KMnO_4$ 与 $Na_2C_2O_4$ 的反应式如下:

$$2MnO_4^- + 5C_2O_4^{2-} + 16H^+ = 10CO_2\uparrow + 2Mn^{2+} + 8H_2O$$

由于这个反应进行得很慢,所以必须将溶液加热到 348～358 K(75～85℃)。但温度也不能太高,否则会引起 $H_2C_2O_4$ 分解:

$$H_2C_2O_4 = CO_2\uparrow + CO\uparrow + H_2O$$

四、仪器和试剂

仪器:500 mL 烧杯,玻璃棒,500 mL 棕色试剂瓶,表面皿,微孔玻璃漏斗,酸式滴定管,锥形瓶(3 只),10 mL 量杯,台秤,分析天平,称量瓶,250 mL 容量瓶,100 mL 小烧杯。

试剂:$KMnO_4(s)$,$Na_2C_2O_4(AR)$,$H_2SO_4(3\ mol\cdot L^{-1})$溶液。

五、实验内容

1. $0.05\ mol\cdot L^{-1}\left(\dfrac{1}{5}KMnO_4\right)$标准溶液的配制

称取约 0.4 g $KMnO_4$ 于洁净的大烧杯中,加蒸馏水 250 mL,加热促进其溶解,盖上表面皿,加热至沸并保持微沸状态 1 h。冷却后,用微孔玻璃漏斗过滤,滤液贮存在清洁带塞的棕色瓶中,最好将溶液于室温下静置 2～3 天后过滤备用。

2. $0.05\ mol\cdot L^{-1}\left(\dfrac{1}{5}KMnO_4\right)$标准溶液的标定

准确称取烘干的 $Na_2C_2O_4$ 0.7～0.9 g,于 100 mL 小烧杯中,加适量蒸馏水溶解后,定量转入 250 mL 容量瓶中,定容。

用移液管取上述 $Na_2C_2O_4$ 溶液 25.00 mL 于锥形瓶中,加蒸馏水 20～30 mL 和 3 mol·L⁻¹ H_2SO_4 10 mL,加热至 75～85℃(瓶口开始冒蒸汽时的温度),趁热用 $KMnO_4$ 溶液滴定。开始反应速度很慢,所以滴入第一滴后,要摇动锥形瓶,等 $KMnO_4$ 的颜色退去后,再继续滴定,由于生成的 Mn^{2+} 对滴定反应有催化作用,所以滴定速度逐渐加快,但仍然必须逐滴加入,如此小心滴定至溶液呈微红色,半分钟内不褪色即为终点,记录所消耗的 $KMnO_4$ 溶液的体积。平行标定 3 份样品。$KMnO_4$ 标准溶液的浓度按下式计算:

$$c_{\frac{1}{5}KMnO_4} = \frac{m_{Na_2C_2O_4}}{M_{\frac{1}{2}Na_2C_2O_4} \times \dfrac{V_{KMnO_4}}{1000}} \times \frac{25.00}{250.0}$$

六、实验记录与结果

项目 内容	1	2	3
$V_{Na_2C_2O_4}$/mL			
V_{KMnO_4}(初)/mL			
V_{KMnO_4}(末)/mL			
V_{KMnO_4}(消耗)/mL			
c_{KMnO_4}/mol·L^{-1}			
c_{KMnO_4}/mol·L^{-1}(平均)			
偏差(d)			
相对平均偏差(d_r)			

（编者：肖旭贤）

实验7　EDTA 标准溶液的配制及标定
Preparation and Determination of EDTA Standard Solution

一、实验目的

(1)掌握 EDTA 标准溶液的配制方法。

(2)熟悉螯合滴定的原理和方法。

二、预习要点

(1)掌握配位滴定的基本原理。

(2)ZnO 用 HCl 溶解后的溶液,为什么要滴加氨试液至溶液显微黄色?(提示:甲基红指示剂的碱色为黄色)

(3)滴定过程中为什么要使用 $NH_3 \cdot H_2O - NH_4Cl$ 缓冲溶液?

三、实验基本原理

由于 EDTA 是难溶的酸性物质,EDTA 标准溶液常用易溶于水的 $EDTA \cdot 2Na \cdot 2H_2O$ 盐(即乙二胺四乙酸二钠盐,$M_r = 392$)配制。$EDTA \cdot 2Na \cdot 2H_2O$ 是一种白色固体,可制成基准物质,但一般不直接用它配制标准溶液,而是先配制成一定浓度的溶液,然后用 ZnO 基准物质标定其浓度。滴定在 pH 为 10 的条件下进行,以铬黑 T 为指示剂,滴定至溶液从紫红色变为纯蓝色即为终点。反应如下:

$$Zn^{2+} + In^{3-} =\!=\!= ZnIn^-$$

$$Zn^{2+} + Y^{4-} =\!=\!= ZnY^{2-}$$

终点时:　　　　　$ZnIn^- + Y^{4-} =\!=\!= ZnY^{2-} + In^{3-} (pH = 10)$

　　　　　　　　　紫红　　　　　　　　　纯蓝

四、仪器和试剂

仪器:台秤,酸式滴定管(50 mL),试剂瓶(500 mL),移液管(25 mL),容量瓶(250 mL),烧杯(50 mL),刻度吸管(10 mL),分析天平,洗瓶,量筒(50 mL、500 mL),滴定管架,洗耳球,锥形瓶(250 mL),玻棒。

试剂:EDTA 二钠(A. R),ZnO(A. R),铬黑 T 指示剂溶液,pH = 10 的 $NH_3 \cdot H_2O - NH_4Cl$ 缓冲溶液,HCl(6 $mol \cdot L^{-1}$),氨试液。

五、实验内容

1. EDTA 标准溶液($0.05\ mol\cdot L^{-1}$)的配制

取 EDTA·2Na·2H$_2$O 约 9.5 g，加蒸馏水 500 mL 使其溶解，摇匀，贮存于硬质玻璃瓶或聚乙烯塑料瓶中。

2. EDTA 标准溶液的标定

准确称取已在 800℃灼烧至恒重的基准物 ZnO 约 0.12 g 于锥形瓶中，加稀 HCl 3 mL 溶解，加蒸馏水 25 mL 和甲基红的乙醇液 1 滴，滴加氨试液至溶液显微黄色，再加蒸馏水 25 mL，NH$_3$·H$_2$O–NH$_4$Cl 缓冲溶液(pH 为 10.0)10 mL 和铬黑 T 指示剂少量(约 2 滴)，用 EDTA 溶液(约 $0.05\ mol\cdot L^{-1}$)滴定至溶液自紫红色转变为纯蓝色即为终点。重复操作 2 次，取其平均值。

$$c_{EDTA} = \frac{W_{ZnO}(克)}{V_{EDTA}(mL) \times \dfrac{M_{ZnO}}{1000}} \qquad (M_{ZnO} = 81.4)$$

六、实验记录与结果

按照下表处理实验数据：

项　　目	测定序号		
	1	2	3
W_{ZnO}/g			
$V_{EDTA}(初)/mL$			
$V_{EDTA}(末)/mL$			
$V_{EDTA}(消耗)/mL$			
$c_{EDTA}/mol\cdot L^{-1}$			
$c_{EDTA}/mol\cdot L^{-1}$(平均值)			
偏差(d)			
相对平均偏差(d_r)			

（编者：王曼娟）

实验 8　水体化学耗氧量测定
Determination of Chemical Consuming of Oxygen by Water

一、目的要求

(1)了解测定水体化学耗氧量(COD)的意义;

(2)掌握 COD 的测定原理及方法;

(3)了解氧化还原滴定法滴定的原理和条件以及在环境分析中的应用。

二、预习要点

(1)氧化还原滴定法的原理及条件。

(2)测定水体化学耗氧量(COD_{Mn})的有关计算。

(3)测定中有关酸度、温度、指示剂以及滴定速度的要求。

三、实验基本原理

化学耗氧量(COD)是指水中发生化学氧化还原反应所消耗的氧量,它是表示水中有机物含量的一项水质指标,是水质监测的重要参数。化学耗氧量越高,表示水中有机物污染越重。化学耗氧量的测定,大多采用 $KMnO_4$ 煮沸消解方法(COD_{Mn})和 $K_2Cr_2O_7$ 加热回流法(COD_{Cr})。本实验采用高锰酸钾法测定,并用 1 L 水中有机物或还原性物质在规定条件下被高锰酸钾氧化消耗的氧的毫克数表示。

在酸性条件下用高锰酸钾($KMnO_4$)将水样中某些有机物及还原性物质氧化,剩余的 $KMnO_4$ 用过量草酸($H_2C_2O_4$)还原,再以 $KMnO_4$ 标准溶液回滴过量的 $H_2C_2O_4$,然后计算水中有机物和还原性物质所消耗的 $KMnO_4$ 的量,并换算成含量。

$$5H_2C_2O_4 + 2MnO_4^- + 16H^+ =\!=\!= 2Mn^{2+} + 10CO_2\uparrow + 8H_2O$$

实验时应注意以下几个问题。

1. 酸度

实验时应使溶液保持足够的酸度,酸度过低 MnO_4^- 会部分还原为 MnO_2 沉淀。

2. 温度

滴定操作应保持 70～85℃为宜。因室温下此反应缓慢,温度高于90℃则草酸分解:

$$H_2C_2O_4 \Longrightarrow CO_2 + CO + H_2O$$

3. 指示剂

溶液中 MnO_4^- 稍微过量（约 10^{-5} $mol \cdot L^{-1}$）即可显示出粉红色，所以滴定时不需另加指示剂。

4. 滴定速度

MnO_4^- 与 $C_2O_4^{2-}$ 的反应是自动催化反应（Mn^{2+} 作催化剂），所以滴定时，加入第一滴 $KMnO_4$ 溶液退色很慢。在 $KMnO_4$ 红色没有褪去以前，不要加第二滴。等几滴 $KMnO_4$ 溶液作用之后，滴定速度才可稍快，但仍应控制速度，否则加入的 $KMnO_4$ 溶液来不及完全与 $C_2O_4^{2-}$ 反应（MnO_4^- 与 $C_2O_4^{2-}$ 反应速度慢），在热的酸性溶液中会部分发生分解，使测定结果偏高。

$$4MnO_4^- + 12H^+ \longrightarrow 4Mn^{2+} + 5O_2 \uparrow + 6H_2O$$

若在滴定前加入几滴 $MnSO_4$ 溶液，滴定一开始反应速度则较快。

四、仪器和试剂

仪器：移液管（50 mL），锥形瓶，酸式滴定管，滴定管架，电炉或电热板，温度计。

试剂：H_2SO_4（1:3），$H_2C_2O_4$（0.005000 mol · L^{-1}），$KMnO_4$（0.002000 mol · L^{-1}），玻璃珠。

五、实验内容

（1）取 50 mL 自来水样 2 份，分别加入 2 只锥形瓶中，以蒸馏水稀释到 100 mL，各加入 5 mL 1:3 H_2SO_4 溶液，再用滴定管加入 10 mL 0.002 mol · L^{-1} $KMnO_4$ 溶液（设加入量为 V_1 mL），并加入用粗砂纸磨过的玻璃珠 5~8 粒。

（2）将锥形瓶用均匀火加热，从开始沸腾计时，准确煮沸 10 min（严格保证，否则 COD 偏大！），如加热过程中红色明显减退，说明水样的耗氧量过高，应减少水样量另行测定。

（3）取下锥形瓶，在 70~80℃ 时用滴定管加入 10 mL 0.005 mol · L^{-1} $H_2C_2O_4$ 溶液（设加入量为 V_2 mL），充分振荡，此时，剩余的高锰酸钾红色应完全消失。

（4）趁热用滴定管滴入 0.002 mol · L^{-1} $KMnO_4$ 溶液至出现微红色（且经振摇在半分钟内红色不再消失）即为终点（设滴入体积为 V_1' mL）。

（5）根据下式计算耗氧量

$$COD_{Mn} = \frac{\{5c(KMnO_4) \times (V_1 + V_1')_{KMnO_4} - 2c(H_2C_2O_4) \times V_2\} \times 8 \times 100}{V_{水样}}$$

式中 $c(KMnO_4)$——$KMnO_4$ 的浓度$(mol \cdot L^{-1})$；

　　　　$c(H_2C_2O_4)$——$H_2C_2O_4$ 的浓度$(mol \cdot L^{-1})$；

　　　　V_1——开始加入 $KMnO_4$ 溶液的体积(mL)；

　　　　V'_1——滴定时消耗 $KMnO_4$ 溶液的体积(mL)；

　　　　V_2——加入 $H_2C_2O_4$ 的体积(mL)；

　　　　$V_{水样}$——所取水样的体积(mL)。

六、实验记录与结果

COD_{Mn}：

测定序号　　　　　　　数据记录与计算		1	2	3
水样量 $V_{水样}$/mL				
加入 $KMnO_4$ 溶液的体积 V_1/mL				
加入 $H_2C_2O_4$ 的体积 V_2/mL				
滴定消耗的 $KMnO_4$ V'_1	初始读数/mL			
	终点读数/mL			
	净用量/mL			
COD_{Mn}/mg·L^{-1}				
平均值/mg·L^{-1}				
相对平均偏差(d_r)				

（编者：李战辉）

实验 9　电泳和电渗
Electrophoresis and Electroosmosis

一、实验目的

(1)了解溶胶的制备方法。

(2)观察并熟悉胶体的电泳现象和电渗现象,从而确定胶粒所带的电荷。

(3)掌握电泳法测量 ζ 电位的技术。

二、预习要点

(1)预习胶体的电泳、电渗现象和 ζ 电位及其两种测定方法;预习渗析膜和溶胶的制备方法、溶胶的渗析方法、电泳仪的操作、电泳速度的测定和电渗仪的操作及电渗现象的观察记录等知识点。

(2)思考并回答下列问题:

①胶粒带电的符号可由哪些方法进行测定?

②为什么要制备电导率与 $Fe(OH)_3$ 溶胶相同的 HCl 溶液?其作用如何?

三、实验基本原理

在外电场的作用下,胶粒在分散介质中定向移动的现象称为电泳。中性粒子在外电场中不可能发生定向移动,所以电泳现象说明胶粒是带电的。处在溶液中的带电固体表面,由于有静电吸引力的存在,它必然要吸引等电量的、与固体表面上带有相反电荷的离子(这种离子可简称为反离子或异电离子)环绕在固体粒子的周围,这样便在固液两相之间形成双电层。

反离子在溶液中同时受到两个方向相反的作用:静电吸引力使其趋于靠近固体表面;热运动所产生的扩散作用又使反离子趋向于均匀分

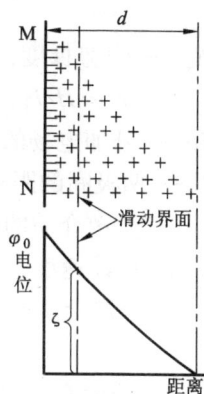

图 3 – 2　扩散层中离子的分布和电位随距离的变化

布。靠近固体表面处反离子浓度大些,随着与表面距离的增大,反离子由多到少。

如图 3 – 2 所示。以 MN 代表胶粒表面。设此表面吸附负离子,正离子(即反离子)扩散分布在胶核周围。带电表面及这些反离子构成的双电层,称为扩散双电层,其厚度 d 随溶液中离子浓度和价数的不同而不同。

　　φ_0 是固体表面与溶液本体之间的电势差,即热力学电势。胶粒在电场作用下与介质发生的相对移动,此分界面不在固液面 MN 处,而是有一液体牢固地附在固体表面,随表面一起运动,一旦固液两相发生相对移动,滑动面便呈现出来。测定电泳速度算出的就是滑动面与溶液本体之间的电势差,称为电动电势或 ζ 电势。

　　由于滑动面内的反离子部分抵消了固体表面的电荷,故 ζ 电势在数值上小于热力学电势,若介质中反离子浓度增大,将压缩扩散层使其变薄,把更多的反离子挤进滑动面内,使 ζ 电势变小,当 ζ 电势为零时,称为等电态,此时胶粒不带电,电泳、电渗的速度为零。

　　在电泳中胶粒的运动方向可判断胶粒所带电性。ζ 电势的大小由电泳(或电渗)速度算出。在外加电场作用下,若分散介质对分散相发生相对移动,称为电渗。

　　ζ 电势与溶胶的稳定性有关。ζ 值越大,溶胶越稳定(不易沉降);反之,ζ 电势趋于零时,溶胶有聚沉现象。因此,无论制备或破坏胶体,都需要研究胶体的 ζ 电势。ζ 电势可根据赫姆霍兹公式计算:

$$\zeta = \frac{\eta \cdot u}{E_e} \tag{1}$$

式中　E_e——电势梯度,$V \cdot m^{-1}$,可用下式计算:

　　　　$E_e = H/l$;

　　　　H——外加电场的电压,V;

　　　　l——两极间的距离,m;

　　　　η——分散介质的黏度,$Pa \cdot s$;

　　　　u——电泳速度,$m \cdot s^{-1}$;

　　　　e——分散介质的介电常数,$F \cdot m^{-1}$;

　　　　　　$e = e_r \cdot e_0$

式中　e_r——分散介质的相对介电常数;

　　　　e_0——真空介电常数。

　　　　$e_0 = 8.854 \times 10^{-12} F \cdot m^{-1} (1F = 1C \cdot V^{-1} = 1\ A \cdot s \cdot V^{-1})$

　　当 e、η、H、l 都是已知常数时,只要用电泳法测定 u、ζ 电势即可求出。

　　ζ 电势还可以通过电渗实验求出,从电渗实验求 ζ 电势公式为:

$$\zeta = \frac{\eta \cdot \kappa \cdot V}{\varepsilon I} \tag{2}$$

式中　V——液体流过毛细管的体积,m^3;

I——通过液体介质的电流，A；

k——液体介质的电导率，$S \cdot m^{-1}$。

当 η、k、e 为已知常数时，只要测定电流 I，并用电渗法测定 V，即可由上式算出 ζ 电势。

四、仪器和试剂

仪器：电泳仪 1 套，电渗仪 1 套，电导仪 1 台，电炉 1 个，秒表 1 个，铂电极 4 支，量筒 100 mL 1 个，移液管 2 mL 1 支，锥形瓶 100 mL 1 个，烧杯（1000 mL，250 mL，100 mL 各 1 个）。

试剂：HCl 溶液（约 0.0001 $mol \cdot L^{-1}$），$FeCl_3$ 溶液（10%），火棉胶溶液（5%，溶剂是乙醇与乙醚混合液，体积比为 1:3）。

五、实验内容

1. 渗析膜的制备

取 10 mL 5% 火棉胶液（它是硝化纤维素的乙醇乙醚混合溶液，要远离火焰！）倒入洗净的 100 mL 锥形瓶中，倾斜瓶子，慢慢转动使火棉胶液刚好达到瓶口，转动几次后，使其在瓶壁形成一层均匀薄层。倾出多余的火棉胶液于回收瓶中，把瓶子倒置在铁圈上，让剩余火棉胶流尽，并使乙醚挥发完（可借助电吹风以冷风吹），大约 15 min 后，火棉胶膜风干（不黏手）。在瓶内加满蒸馏水，溶去剩余的乙醇，放置 3 min，倒去水，在瓶口剥开一部分膜，在膜与壁之间注入蒸馏水，膜就会脱离瓶壁，轻轻取出火棉胶袋。将袋盛满蒸馏水，检查是否漏水，然后泡在蒸馏水中备用。

2. 制备 $Fe(OH)_3$ 溶胶

在 250 mL 烧杯中加入 100 mL 蒸馏水，加热至沸腾，用移液管将 2 mL 10% $FeCl_3$ 溶液边搅拌边一滴滴地加到水中，再煮沸 3 min，看到红棕色溶胶生成，冷却待用。

3. 溶胶的渗析

把 $Fe(OH)_3$ 溶胶倒入火棉胶袋中，用线拴住口袋，将其悬在装满蒸馏水的 1000 mL 大烧杯中，水温保持在 60～70℃ 之间进行热渗析。经常换水，直至蒸馏水中无 Cl^- 时，渗析结束。

4. 制备电导率与 $Fe(OH)_3$ 溶胶相同的 HCl 溶液

用电导率仪测定 $Fe(OH)_3$ 溶胶的电导率。取 50 mL 稀盐酸放入 150 mL 的烧杯中，然后用吸管吸取蒸馏水滴入稀盐酸中，测定稀盐酸的电导率。稀盐酸中不断滴入蒸馏水并测定电导率，直至其电导率与 $Fe(OH)_3$ 溶胶的相同为止，作测定辅助液用。

5. 电泳速度的测定

(1)如图 3-3 把电泳仪的两个大活塞打开，在活塞下部的 U 形管中充满 $Fe(OH)_3$ 溶胶。应注意：管内不能停留有气泡。溶胶装好后关上大活塞，活塞上方多余的溶胶倒掉，并按顺序用蒸馏水、稀 HCl 洗涤 2～3 次。在活塞上方及支管中充满稀 HCl 溶液。固定好电泳仪，在弯管处插入铂电极，电极的位置要固定。

(2)打开电泳仪横梁上的小活塞，使两臂液面达到同一水平，然后关上小活塞。接好线路，轻轻打开两只大活塞，注意维持界面清晰。

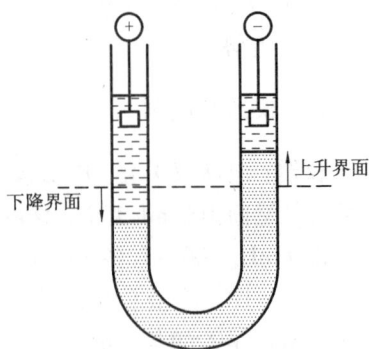

图 3-3　电泳仪

(3)通电(注意用电安全，切勿接触导线外露部分)，开始实验。电压为 100～150 V，保持电压稳定。左或右臂界面从旋钮处移到清晰可读的位置时打开秒表记时，并记录刻度位置，半小时后记录界面移动的距离，计算电泳速度 u，观察界面移动的方向，判断胶粒所带电荷的符号，切断电源。

(4)量出两电极间的准确距离。

(5)重复上述操作数次。

6. 电渗现象

(1)按图 3-4 所示将仪器装置好。把砖板固定，用水(为减少电阻，可加入少许 KCl 溶液)充满电渗仪，以没有气泡为度，在两端插入铂电极，中间插入盐桥，并记录毛细管末端水位的位置。

(2)接通电源，电压为 50 V 直流电，记录电流 I。

(3)从毛细管末端观察水的移动情况，记录一定时间后毛细管中水柱的移动距离，判定胶粒(砖板)带何种电荷。

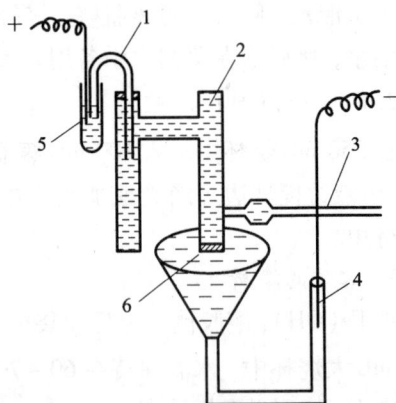

图 3-4　电渗仪装置图

1—盐桥；2—H_2O(或 KCl 溶液)；3—毛细管；

4，5—铂电极；6—砖板

六、数据记录与结果

1. 电泳速度的测定

电泳开始时电泳仪左或右臂界面的刻度位置为_____。

秒表计时半小时后界面的刻度位置为_____。

界面移动的距离为_____。

电泳速度 u _____。

界面移动的方向为_____。

胶粒所带电荷的符号为为_____。

两电极间的准确距离为_____。

2. 电渗现象的观测

电渗前毛细管末端水位的位置为_____。

电渗一定时间后毛细管末端水位的位置为_____。

毛细管中水柱的移动距离为_____。

胶粒(砖板)所带电荷符号为_____。

电压为 50 V 直流电的电流 I _____。

（编者：王一凡）

3.2　理化常数与测定实验

实验 10　氯化铅标准溶解热($\Delta_r H_m^\ominus$)的测定
Determination of Standard Solution Heat of Lead Chloride

一、实验目的

(1)了解难溶电解质标准溶解热($\Delta_r H_m^\ominus$)的测定原理。

(2)学会在不同温度下测定溶度积常数($K_{sp}^\ominus$)。

(3)熟悉测定氯化物水溶液中氯离子含量的莫尔(Mohr)法。

二、预习要点

(1)预习由莫尔(Mohr)滴定法测得一定温度下 $PbCl_2$ 饱和溶液中氯离子浓度与 $PbCl_2$ 的溶解度 S 以及溶度积常数 K_{sp} 的关系式；预习溶度积常数与难溶电解质溶解过程的标准溶解热 $\Delta_r H_m^\ominus$ 之间的关系式的推导和应用，即用不同温度下测得的不同溶度积常数，以 $\lg K_{sp}$ 对 $1/T$ 作图，由得出的直线斜率计算出标准溶解热 $\Delta_r H_m^\ominus$ 值；预习恒温水浴槽的工作原理与水银接点温度计的使用；预习时应注意实验操作中的有关事项：为了使锥形瓶内溶液的温度与所测温度达到平衡，应注意在此过程中不断地转动锥形瓶；静置应达到约 15 min 以上，待瓶内不溶的 $PbCl_2$ 完全聚沉；应及时记下温度计测量瓶内 $PbCl_2$ 饱和溶液的温度，并立即慢慢吸取上清液；滴定时应注意滴定终点的判断。

(2)思考并回答下列问题：

①25℃时 $PbCl_2$ 在水中溶解并离解成 Pb^{2+}、Cl^- 离子时的标准焓变 $\Delta_r H_m^\ominus$ 应为多少($kJ \cdot mol^{-1}$ 理论计算值)？

②用 Mohr 法测得 $PbCl_2$ 的 K_{sp} 值与其实际值有何差异？为什么？

③用本实验的方法测得 $PbCl_2$ 离解成 Pb^{2+}、Cl^- 离子的标准焓变 $\Delta_r H_m^\ominus$，与理论值相比，有何差异？为什么？

④影响实验成败的因素有哪些？

三、实验基本原理

PbCl$_2$ 在水中的溶解度随温度的升高而增大，因此 PbCl$_2$ 的溶解过程应是吸热的，也即其溶解热 $\Delta_r H$ 为正值。热力学中等压热效应 $\Delta_r H$ 的求法有多种，考虑到 PbCl$_2$ 为难溶电解质，故本实验采用在不同温度下测得 PbCl$_2$ 的溶解度，来求溶度积常数；并通过溶度积常数与标准自由能变化的关系，再由 Gibbs 方程式 $\Delta_r G_m^\ominus = \Delta_r H_m^\ominus - T\Delta_r S_m^\ominus$ 最终获得求算 PbCl$_2$ 的标准溶解热 $\Delta_r H_m^\ominus$ 的方法。溶解过程标准自由能变化 $\Delta_r G_m^\ominus$ 与溶度积常数 K_{sp} 的有关公式如下：

$$\Delta_r G_m^\ominus = -RT \ln K_{sp} \tag{1}$$

$$-RT \ln K_{sp} = \Delta_r H_m^\ominus - T\Delta_r S_m^\ominus \tag{2}$$

$$\ln K_{sp} = -\frac{\Delta_r H_m^\ominus}{RT} + \frac{\Delta_r S_m^\ominus}{R} \tag{3}$$

由于 $\Delta_r H_m^\ominus$ 与 $\Delta_r S_m^\ominus$ 受温度的影响很小，则在温度变化不大的情况下，它们可视为常数。因而在两个不同温度下的上述关系式又可表示为：

$$\ln K_{sp2} = -\frac{\Delta_r H_m^\ominus}{RT_2} + \frac{\Delta_r S_m^\ominus}{R} \tag{4}$$

$$\ln K_{sp1} = -\frac{\Delta_r H_m^\ominus}{RT_1} + \frac{\Delta_r S_m^\ominus}{R} \tag{5}$$

若 $T_2 > T_1$，则将式(4)减去式(5)，可得

$$\ln \frac{K_{sp2}}{K_{sp1}} = \frac{\Delta_r H_m^\ominus}{R}\left(\frac{T_2 - T_1}{T_1 T_2}\right)$$

$$\lg \frac{K_{sp2}}{K_{sp1}} = \frac{\Delta_r H_m^\ominus}{2.303R}\left(\frac{T_2 - T_1}{T_1 T_2}\right) \tag{6}$$

应用式(6)中两个不同温度下溶度积常数的测定值，便可计算一个溶解过程的标准溶解热 $\Delta_r H_m^\ominus$。此外，也可以从若干个不同温度得出对应的若干个溶度积常数，以 $\lg K_{sp}$ 对 $1/T$ 作图，由直线斜率 $\left(\dfrac{\Delta_r H_m^\ominus}{2.303R}\right)$ 计算出 $\Delta_r H_m^\ominus$ 值。

PbCl$_2$ 在一定温度下于水中具有以下沉淀 – 溶解平衡：

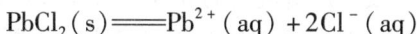

$$PbCl_2(s) \Longrightarrow Pb^{2+}(aq) + 2Cl^-(aq)$$

所以，在该温度下 PbCl$_2$ 的溶解度 S(单位为 mol·L^{-1})与溶度积常数 K_{sp} 以及溶液中氯离子浓度 $c(Cl^-)$ 的关系为：

$$S = \sqrt[3]{\frac{K_{sp}}{4}} \tag{7}$$

$$S = \frac{c(Cl^-)}{2} = c(Pb^{2+}) \qquad (8)$$

所谓莫尔法,就是利用沉淀滴定法测定氯化物中氯离子含量的一种方法。此方法是在中性或弱碱性(pH 范围为 6.5 ~ 10.5)溶液中,以 K_2CrO_4 为指示剂,用 $AgNO_3$ 标准溶液进行滴定。由于 AgCl 沉淀的溶解度比 Ag_2CrO_4 小,因此,溶液中首先析出白色的 AgCl 沉淀,当 AgCl 定量沉淀后,过量一滴 $AgNO_3$ 溶液即与 CrO_4^{2-} 生成砖红色 Ag_2CrO_4 沉淀,指示达到终点。

四、仪器和试剂

仪器: 温度计(0.1℃分度),酸式滴定管(50 mL),容量吸管(10 mL),刻度吸管(1 mL),量筒(50 mL),锥形瓶(100 mL,250 mL),恒温水浴槽,煤气灯(或酒精灯),洗耳球,台秤。

试剂: $PbCl_2$(固体),K_2CrO_4(1.0 mol·L^{-1}),$AgNO_3$ 标准溶液(其浓度应在 0.0500 mol·L^{-1} 附近)。

五、实验内容

在三只 250 mL 锥形瓶中,各加入 $PbCl_2$(固体)5 g 和蒸馏水 200 mL。加热至沸后,放入冷水浴中片刻,待瓶内溶液温度下降到比室温约高 20℃时,将其中一只锥形瓶放在室温下静置,使瓶内溶液的温度与室温达到平衡(15 min 后瓶内液温度应保持不变);随即将第二只锥形瓶放在比室温高 5℃的恒温水浴槽(其操作方法见附录三)中使温度达平衡;并尽快将第三只锥形瓶放在比室温高 10℃的恒温水浴槽中使温度达平衡。在温度平衡过程中需不断地转动锥形瓶,借以搅拌瓶内溶液。静置约 15 min,待瓶内不溶的 $PbCl_2$ 完全聚沉后,用 0.1℃分度的温度计测量瓶内 $PbCl_2$ 饱和溶液的温度。记下溶液温度,立即用洁净干燥的 10 mL 容量吸管小心地伸入瓶内饱和 $PbCl_2$ 溶液中 3~4 cm 深,慢慢吸取上清液 10.00 mL 两份,分别置于两只盛有 25 mL 蒸馏水的 100 mL 锥形瓶中进行稀释。然后用刻度吸管吸取 1.0 mol·L^{-1} K_2CrO_4 溶液 1.00 mL 于其中一只 100 mL 锥形瓶中。接着用浓度约为 0.0500 mol·L^{-1} 的 $AgNO_3$ 标准溶液滴定,当滴定至溶液刚好出现砖红色时即为滴定终点。用同样的方法重复滴定另一份。其他温度的 $PbCl_2$ 饱和溶液中 Cl$^-$ 离子浓度也按上述方法分别滴定两份,所得数据填入下表。

六、数据记录与结果

实验温度 T	K		K		K	
实验次数	1	2	1	2	1	2
试样溶液的体积/mL						
滴定管终读数/mL						
滴定管初读数/mL						
滴定时用去 $AgNO_3$ 标准溶液毫升数/mL						
$AgNO_3$ 标准溶液的浓度/mol·L^{-1}						
试样中 Cl^- 离子的浓度/mol·L^{-1}						
试样中 Pb^{2+} 离子的浓度/mol·L^{-1}						
试样中 Cl^- 离子的平均浓度/mol·L^{-1}						
试样中 Pb^{2+} 离子的平均浓度/mol·L^{-1}						
K_{sp}						
$\lg K_{sp}$						
$\dfrac{1}{T}$/K^{-1}						

根据实验数据，以 $\dfrac{1}{T}$ 为横坐标，$\lg K_{sp}$ 为纵坐标作图，由直线斜率求 $PbCl_2$ 的标准溶解热 $\Delta_r H_m^{\ominus}$。

（编者：王一凡）

实验 11　化学反应焓变的测定
Measurement of Enthalpy Change

一、实验目的

(1)了解化学反应焓变的测定原理,学会焓变的测定方法。

(2)熟练掌握精密温度计的正确使用方法。

二、预习要点

(1)预习化学反应焓变测定的原理和方法及其计算;预习保温杯式量热计的操作要领;预习分析天平、容量瓶及移液管的使用方法。

(2)思考并回答下列问题:

①为什么本实验所用的 $CuSO_4$ 溶液的浓度和体积必须准确,而实验中所用的 Zn 粉则可用台秤称量?

②在计算化学反应焓变时,温度变化 ΔT 的数值,为什么不采用反应前($CuSO_4$ 溶液与 Zn 粉混合前)的平衡温度值与反应后($CuSO_4$ 溶液与 Zn 粉混合)的最高温度值之差,而必须采用 $t - T$ 曲线外推法得到的 ΔT 值?

③本实验中对所用的量热器、温度计有什么要求?是否允许反应器内有残留的洗液或水?为什么?

④影响实验成败的因素有哪些?

三、实验基本原理

化学反应通常是在等压条件下进行的,此时,化学反应的热效应叫做等压热效应 Q_p。在化学热力学中,Q_p 则是用反应体系焓 H 的变化量 ΔH 来表示,简称为焓变。为了有一个比较的统一标准,通常规定 100 kPa 为标准压力,记为 $p^\ominus$。把体系中各固体、液体物质处于 $p^\ominus$ 下的纯物质,气体则在 $p^\ominus$ 下表现出理想气体性质的纯气体状态称为热力学标准态。在标准状态下化学反应的焓变称为化学反应的标准焓变,用 $\Delta_r H^\ominus$ 表示,下标"r"表示一般的化学反应,上标"⊖"表示标准状态。在实际工作中,许多重要的数据都是在298.15 K 下测定的,通常用 298.15 K 下的化学反应的焓变,记为 $\Delta_r H^\ominus$ (298.15 K)。

本实验是测定固体物质锌粉和硫酸铜溶液中的铜离子发生置换反应的化学

反应焓变:

$$Zn(s) + CuSO_4(aq) \longrightarrow ZnSO_4(aq) + Cu(s)$$

$$\Delta_r H_m^\ominus(298.15\ K) = -217\ kJ \cdot mol^{-1}$$

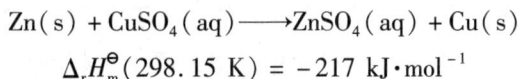

　　这个热化学方程式表示:在标准状态，298.15 K 时，发生了一个摩尔级(或反应进度为 1 mol)的反应，即 1 mol 的 Zn 与 1 mol 的 CuSO₄ 发生置换反应生成 1 mol 的 ZnSO₄ 和 1 mol 的 Cu，此时的化学反应的焓变 $\Delta_r H_m^\ominus$ (298.15 K)称为 298.15 K 时的标准摩尔焓变，其单位为 $kJ \cdot mol^{-1}$。

　　测定化学反应热效应的仪器称为量热计。对于一般溶液反应的摩尔焓变，可用图 3 – 5 所示的"保温杯式"量热计来测定。

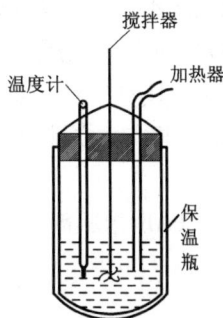

图 3 – 5　保温杯式量热计示意图

　　在实验中，若忽略量热计的热容，则可根据已知溶液的比热容、溶液的密度、浓度、实验中所取溶液的体积和反应过程中(反应前和反应后)溶液温度的变化，求得上述化学反应的摩尔焓变。其计算公式如下:

$$\Delta_r H_m^\ominus \{(273.15 + t)\ K\} = \Delta TCV\rho\ \frac{1}{n} \times \frac{1}{1000}\quad (kJ \cdot mol^{-1})$$

式中: $\Delta_r H_m^\ominus$——在实验温度(273.15 + t)K 时的化学反应摩尔焓变($kJ \cdot mol^{-1}$);

　　　　ΔT——反应前后溶液温度的变化，K;

　　　　C——CuSO₄ 溶液的比热容，$J \cdot g^{-1} \cdot K^{-1}$;

　　　　V——CuSO₄ 溶液的体积，mL;

　　　　ρ——CuSO₄ 溶液的密度，$g \cdot cm^{-3}$;

　　　　n——V 体积溶液中 CuSO₄ 的物质的量，mol。

四、仪器和试剂

　　仪器:台秤，量热器，精密温度计(-5 ~ +50℃，0.1℃刻度)，移液管(50 mL)，洗耳球，移液管架，磁力搅拌器，称量纸。

　　试剂:0.2000 mol·L⁻¹ CuSO₄ 溶液，Zn 粉(A.R)。

　　注:0.2000 mol·L⁻¹ CuSO₄ 溶液的配制与标定如下。

　　(1)取所需量稍多的分析纯 CuSO₄·5H₂O 晶体于一干净的研钵中研细后，倒入称量瓶或蒸发皿中，再放入电热恒温干燥箱中，在低于 60℃ 的温度下烘 1 ~ 2 h，取出，冷至室温，放入干燥器中备用。

（2）在分析天平上准确称取研细、烘干的 $CuSO_4 \cdot 5H_2O$ 晶体 49.936 g，并置于一只 250 mL 的烧杯中，加入约 150 mL 的去离子水，用玻棒搅拌使其完全溶解，再将该溶液倾入 1000 mL 容量瓶中，用去离子水将玻棒及烧杯冲洗 2～3 次，洗涤液全部转入容量瓶中，最后用去离子水稀释到刻度，摇匀。

（3）取该 $CuSO_4$ 溶液 25.00 mL 于 250 mL 锥形瓶中，将 pH 调到 5.0，加入 10 mL $NH_3 \cdot H_2O - NH_4Cl$ 缓冲溶液，加入 8～10 滴 PAR 指示剂，4～5 滴次甲基蓝指示剂，摇匀，立即用 EDTA 标准溶液滴定到溶液恰好由紫红色转为黄绿色时为止。

（4）*PAR 指示剂，化学名称为 4-(2-吡啶偶氮)间苯二酚，结构式为：

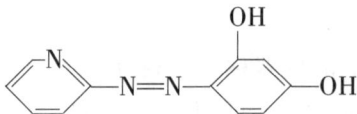

五、实验内容

用 50 mL 移液管准确移取 200.00 mL 的 $0.2000\ mol \cdot L^{-1}$ $CuSO_4$ 溶液注入已经洗净、擦干的量热计中，盖紧盖子，在盖子中央插入一支最小刻度为 0.1℃ 的精密温度计。

双手扶正，握稳量热计的外壳，不断摇动或旋转搅拌子（转速一般为 200～300 $r \cdot min^{-1}$），每隔 0.5 min 记录一次温度数值，直至量热计内 $CuSO_4$ 溶液与量热计温度达到平衡且温度计指示的数值保持不变为止（一般约需 3 min）。

用台秤称取 Zn 粉 3.5 g。开启量热计的盖子，迅速向 $CuSO_4$ 溶液中加入称量好的 Zn 粉 3.5 g，立即盖紧量热计盖子，不断摇动量热计或旋转搅拌子，同时每隔 0.5 min 记录一次温度数值，一直记录到温度上升至最高值，仍继续进行测定直到温度下降或不变后，再测定记录 3 min，测定方可终止。

倾出量热计中反应后的溶液，若使用磁力搅拌器，注意不要将所用的搅拌子丢失。

六、数据记录与结果

（1）反应时间与温度的变化（每 0.5 min 记录一次）

室温 $t =$ _____℃

$CuSO_4$ 溶液的浓度 $c(CuSO_4) =$ _____ $mol \cdot L^{-1}$

$CuSO_4$ 溶液的密度 $\rho(CuSO_4) =$ _____ $g \cdot cm^{-3}$

反应进行的时间 t/min	
温度计指示值 $T/\text{℃}$	
$T/(273.15+t)\,\mathrm{K}$	

设：①$CuSO_4$ 溶液的比热容近似等于水的比热容 $c=4.18\mathrm{J\cdot g^{-1}\cdot K^{-1}}$。

②$\Delta_r H_m^{\ominus}$（理论值）$=-216.8\ \mathrm{kJ\cdot mol^{-1}}$。

（2）作图求 ΔT。

由于量热计并非严格绝热，在实验时间内，量热计不可避免地会与环境发生少量热交换。用作图推算的方法（图3-6），可适当消除这一影响。

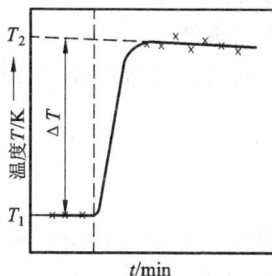

图3-6　反应时间、温度变化的关系

（3）实验误差的计算及误差产生原因分析。

$$相对误差 = \frac{\Delta_r H_m^{\ominus} - \Delta_r H_{m(理)}^{\ominus}}{\Delta_r H_{m(理)}^{\ominus}} \times 100\% = \underline{\qquad\qquad}$$

（编者：王一凡）

实验 12　同离子效应和溶度积原理
Common Ion Effect and Principle of Solubility Product

一、目的要求

(1)验证同离子效应和溶度积原理；
(2)掌握离心机的使用和离心分离操作。

二、预习要点

(1)离心机的使用和离心分离操作。
(2)分步沉淀及沉淀转化的条件。
(3)在产生同离子效应的同时也有盐效应，二者的关系如何？

三、实验基本原理

1. 同离子效应
在弱电解质的离解平衡或难溶电解质的溶解平衡中，加入与弱电解质含有相同离子的强电解质，则弱电解质的离解度降低或难溶电解质的溶解度减小，这种作用称为同离子效应。

2. 溶度积规则
在难溶电解质的饱和溶液中，存在如下的多相离子平衡体系：

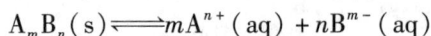

$$A_mB_n(s) \rightleftharpoons mA^{n+}(aq) + nB^{m-}(aq)$$

在一定温度下　　　　　　　$[A^{n+}]^m[B^{m-}]^n = K_{sp}$

此时，K_{sp} 为一常数，称为溶度积常数。

比较难溶电解质的离子积(J)与 K_{sp} 值的大小，可以判断难溶电解质溶液中多相离子平衡的移动方向。即，

$J < K_{sp}$　为不饱和溶液，不生成沉淀；
$J = K_{sp}$　为饱和溶液，沉淀既不增加也不减少；
$J > K_{sp}$　为过饱和溶液，生成沉淀直至达到新的平衡。

四、仪器和试剂

仪器：试管(15×150)，离心管(10 mL)，离心机(4000 r/mim)，搅棒，药匙，量筒(10 mL)。

试剂：HAc(0.2 mol·L^{-1})，NaAc(0.2 mol·L^{-1})，Pb(NO$_3$)$_2$(0.1 mol·L^{-1})，饱

和 PbI_2 溶液，$NaCl(1\ mol\cdot L^{-1})$，$K_2CrO_4((0.1\ mol\cdot L^{-1})$，$KI(0.1\ mol\cdot L^{-1})$，$AgNO_3(0.1\ mol\cdot L^{-1})$，甲基橙指示剂，$Na_2S(0.1\ mol\cdot L^{-1})$，$NH_3\cdot H_2O(6\ mol\cdot L^{-1})$，$HCl(6\ mol\cdot L^{-1})$，$CaCO_3$（固体），蒸馏水。

五、实验内容

1. 同离子效应

(1) 取 1 mL0.2 mol·L^{-1} HAc 溶液加入试管中，滴 1 滴甲基橙溶液，再注入 0.2 mol·L^{-1} NaAc 溶液，观察指示剂颜色的变化，记录并解释现象。

(2) 在试管中注入饱和 PbI_2 溶液 1 mL，然用滴 5 滴 0.1 mol·L^{-1}KI 溶液，振荡试管，记录并解释现象。

2. 溶度积原理

(1) 沉淀的生成与多相离子平衡

在离心试管中滴 10 滴 0.1 mol·L^{-1} $Pb(NO_3)_2$ 溶液，然后滴 5 滴 1 mol·L^{-1} NaCl 溶液，振荡离心试管，待沉淀完全后，离心分离（离心机的使用见本书前文）。在分离后的溶液中，注入少许 0.1 mol·L^{-1} K_2CrO_4 溶液，有什么现象，并解释。

(2) 沉淀的转化

取一支试管，加入浓度为 1 mol·L^{-1} 的 NaCl 溶液 2 滴，再加入 0.1 mol·L^{-1} $AgNO_3$ 溶液 1 滴，摇匀。观察现象，继续滴加 0.1 mol·L^{-1}KI 溶液 3 滴，搅拌，观察沉淀颜色的变化。记录并解释现象，写出离子方程式。

(3) 分步沉淀

在试管中滴入 2 滴 0.1 mol·L^{-1} Na_2S 溶液和 2 滴 0.1 mol·L^{-1} K_2CrO_4 溶液，用水稀释至 5 mL 然后逐滴滴入 0.1 mol·L^{-1} $Pb(NO_3)_2$ 溶液，振匀，加热煮沸数秒钟，放置数分钟，观察首先生成的沉淀的颜色。待沉淀沉降后，继续向清液中滴加 $Pb(NO_3)_2$ 溶液，会出现什么颜色的沉淀？根据有关溶度积数据加以说明。

(4) 沉淀的溶解

①取少许 $CaCO_3$ 固体置于试管中，加水 2 mL 观察其是否溶解。再加入 6 mol·L^{-1} HCl 溶液数滴，观察并解释现象，写出离子方程式。

②取 0.1 mol·L^{-1} $AgNO_3$ 溶液 10 滴，滴入 1 mol·L^{-1} NaCl 溶液 3 ~ 4 滴，观察现象。

③再逐滴滴入 6 mol·L^{-1} $NH_3\cdot H_2O$，有什么现象？写出化学方程式，并说明原因。

六、实验记录与结果

序　号	实验内容	实验现象	解释及结论

（编者：李战辉）

实验 13　醋酸离解常数的测定
Determination of Ionization Constant of Acetic Acid

一、实验目的

(1)掌握弱电解质离解常数的测定方法。

(2)学习使用 pH 计,了解电位法测定溶液 pH 的原理和方法。

(3)学习碱式滴定管的使用。

(4)巩固容量瓶和容量吸管的使用。

二、预习要点

(1)不同浓度的 HAc 溶液离解常数是否相同?

(2)若 HAc 溶液的温度有明显变化,离解常数有什么变化?

(3)同离子效应和盐效应对弱酸弱碱的离解常数有何影响?

(4)同离子效应和盐效应对弱酸弱碱的解离度有何影响?

(5)温度一定时,弱酸弱碱的离解常数与其解离度、浓度有何关系?

(6)影响本实验结果准确性的因素有哪些?

三、实验基本原理

醋酸(HC_3COOH,HAc)是弱电解质,在水溶液中存在以下离解平衡:

$$HAc + H_2O \Longrightarrow H_3O^+ + Ac^-$$

根据化学平衡原理,

$$K_{a(HAc)} = \frac{[H_3O^+][Ac^-]}{[HAc]} \tag{1}$$

$K_{a(HAc)}$ 为 HAc 的离解常数。

将式(1)两边同取负对数,得:

$$pK_{a(HAc)} = pH + \lg\frac{[HAc]}{[Ac^-]}$$

当 $[HAc] = [Ac^-]$ 时,由式(2)得

$$pK_a = pH + \lg 1$$

即,

$$pK_a = pH$$

若用 NaOH 溶液滴定 HAc 溶液,发生如下反应:

$$HAc + OH^- \Longrightarrow Ac^- + H_2O$$

根据反应式可知，当 NaOH 用量等于完全中和 HAc 所需用量的一半时，溶液中

$$[HAc] = [Ac^-]$$

若测得此时溶液的 pH，即可计算出醋酸离解常数的近似值

$$K_a = c(H^+)$$

四、仪器和试剂

酸度计，容量瓶(100 mL)，碱式滴定管(50 mL)，锥形瓶(250 mL)，烧杯(50 mL)，刻度吸管(5 mL,10 mL)，容量吸管(25 mL,50 mL)，洗耳球；

标准 NaOH 溶液(近似于 $0.2\ mol \cdot L^{-1}$)，待标定醋酸溶液(浓度为 $0.2\ mol \cdot L^{-1}$)，酚酞指示剂，标准缓冲液(pH = 4.00)。

五、实验内容

(1)取 250 mL 锥形瓶 1 只，用移液管准确移取 25.00 mL $0.1\ mol \cdot L^{-1}$ HAc 溶液，加入 2 滴酚酞溶液，用碱式滴定管中的 NaOH 标准溶液滴定至溶液刚出现红色即为终点。记录终点时消耗的 NaOH 的体积。重复上述实验 2 次，要求三次滴定消耗 NaOH 体积的偏差不超过 0.10 mL。计算三次消耗 NaOH 溶液的平均体积。

(2)取 100 mL 烧杯 1 个，用移液管准确移取 25.00 mL $0.1\ mol \cdot L^{-1}$ HAc 溶液，用碱式滴定管中的 NaOH 标准溶液滴定至操作 1 所消耗 NaOH 平均体积的一半，搅拌均匀，再用 pH 计测定其 pH。重复上述实验二次，计算三次测定 pH 的平均值和醋酸的离解常数。

六、实验记录与结果

1. HAc 溶液浓度的测定

滴定序号	I	II	III
NaOH 溶液的浓度/mol·L^{-1}			
HAc 溶液的用量/mL			
NaOH 溶液的用量/mL			
三次 NaOH 溶液用量的平均体积/mL			

2. HAc 离解常数的测量

温度_____℃

测定序号	1	2	3
V_{HAc}/mL	25.00	25.00	25.00
$\frac{1}{2}$(三次消耗 NaOH 体积平均值)			
pH			
K_a			
$\overline{K_a}$			
相对平均偏差(d_r)			

（编者：刘绍乾　冯志明）

实验 14　丙酮碘化的反应速率
Determination of Iodination Rate of Acetone

一、实验目的

(1)验证浓度、温度对反应速率影响的理论。

(2)测定反应级数和速率常数 k 值。

(3)根据 Arrhenius 方程式,学会用作图法测定反应活化能 E_a。

(4)练习在水浴中保持恒温操作。

二、预习要点

(1)预习确定具有幂函数形式速率方程的化学反应级数的初始速率法。预习求算化学反应速率常数 k 值的代入法和根据 Arrhenius 方程式用作图法测定反应活化能 E_a 的原理。预习恒温水浴槽的工作原理与水银接点温度计的使用方法(见前文)。预习时应注意实验操作中的有关事项:每次在锥形瓶中最后倒入 I_2 溶液时,应迅速振荡,使反应液混合均匀;应准确记录碘的颜色刚刚消失时的时间。

(2)思考并回答下列问题:

①要保持总体积为 50 mL,还要保持 H^+ 离子和 I_2 在原来的混合物中的浓度,你怎样使反应混合物中丙酮的物质的量浓度增加一倍?

②为什么在各种不同浓度时的反应速率不同,而速率常数则基本不变?

③影响实验成败的因素有哪些?

三、实验基本原理

在本实验中,我们要研究碘和丙酮之间反应的动力学性质:

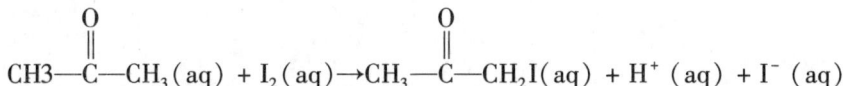

$$CH3-\overset{\overset{\displaystyle O}{\|}}{C}-CH_3(aq) + I_2(aq) \rightarrow CH_3-\overset{\overset{\displaystyle O}{\|}}{C}-CH_2I(aq) + H^+(aq) + I^-(aq)$$

推测,上述反应的速率不仅取决于两个反应物的浓度,还依赖于溶液的氢离子浓度。因此,反应速率方程可表示为:

$$v = -\frac{dC_{I_2}}{dt} = kC_{丙酮}^m C_{I_2}^n C_{H^+}^P$$

丙酮的碘化是一个颇不寻常的反应,很容易通过实验进行研究。首先,碘有颜色，有利于随着碘浓度的变化进行观察。其次，也是反应非常重要的特性——碘浓度的反应分级数为 0。这就意味着由于 $C_{I_2}^0 = 1$，只要 C_{I_2} 本身不为 0，不管它的值多大，反应速率完全与 C_{I_2} 无关。对此，我们将通过实验来验证。

既然反应速率不依赖于 C_{I_2}，我们只需将 I_2 作为限制性试剂，加入到大量过量的丙酮和 H^+ 离子中，然后测定已知初始浓度的 I_2 完全反应所需的时间。若丙酮和 H^+ 的浓度比 I_2 的浓度高得多，它们的浓度在反应过程中就没有明显的变化。根据速率方程，反应速率实际上保持恒定，直到所有的 I_2 完全消失。那时，反应也就停止。在此情况下，如果初始浓度为 $C_{I_{20}}$ 的溶液的颜色完全消失用去 t 秒钟，则反应速率就是：

$$v = -\frac{\mathrm{d}C_{I_2}}{\mathrm{d}t} = \frac{C_{I_{20}}}{t}$$

在我们所设的条件下，虽然反应速率在反应进行过程中为一常数，但我们还是可以用改变丙酮或 H^+ 离子初始浓度的方法来改变反应速率。例如，如果我们将混合物 I 中丙酮浓度增加一倍，并保持 C_{H^+} 和 C_{I_2} 原先的数值，那么，混合物 II 的反应速率就跟混合物 I 不同：

$$v_{II} = k[2A]^m[I_2]^0[H^+]^P$$
$$v_{I} = k[A]^m[I_2]^0[H^+]^P$$

则
$$\frac{v_{II}}{v_{I}} = \frac{[2A]^m}{[A]^m} = \left[\frac{2A}{A}\right]^m = 2^m$$

这样丙酮的反应分级数 m 就可求了。用类似的方法，我们还可以求出 H^+ 离子的反应分级数 P，也能证实 I_2 物质的反应分级数为 0。求出各物质的反应分级数之后，我们就可以估算室温下丙酮碘化反应的速率常数 k 值了。这些即为本实验的主要任务。

测定反应的活化能 E_a 是本实验的选作部分。根据 Arrhenius 方程式，速率常数 k 与温度 T、活化能 E_a 之间有如下关系：

$$\lg k = \frac{-E_a}{2.303RT} + 常数$$

其中 R 为气体常数（$8.314 \mathrm{J \cdot K^{-1} \cdot mol \cdot L^{-1}}$）。测定在不同温度时的 k 值，以 $\lg k$ 对 $\frac{1}{T}$ 作图，可得一直线，其斜率为 $\frac{-E_a}{2.3.3R}$，最终可估算出反应的活化能 E_a。

四、仪器和试剂

仪器：温度计，恒温水浴槽，刻度吸管（10 mL），洗耳球，秒表，锥形瓶（125 mL），烧杯（100 mL），量筒（50 mL），试管。

试剂: 丙酮(4 mol·L^{-1}), HCl(1 mol·L^{-1}), I$_2$(0.005 mol·L^{-1}), 蒸馏水。

五、实验内容

选两个常规试管,倒入蒸馏水后,将试管对准一个白色背景进行俯视时,应出现相同的颜色。

分别取 4 mol·L^{-1}丙酮、1 mol·L^{-1}HCl、0.005 mol·L^{-1} I$_2$ 溶液各 50 mL,倾入洁净干燥的三个 100 mL 烧杯里,用表面皿盖好。用量筒量取 10.0 mL 浓度为 4 mol·L^{-1}的丙酮溶液,倒入一个干净的 125 mL 锥形瓶中;再量出 10.0 mL 1 mol·L^{-1} HCl,加到盛有丙酮的锥形瓶里;再加进 20.0 mL 蒸馏水。倒干量筒,甩出剩余的水,再用它量出 0.005 mol·L^{-1}I$_2$ 溶液 10.0 mL。小心! 不要让碘液溅到手上和衣服上。

用秒表记下时间,准确到 1 s。将 I$_2$ 液倒入锥形瓶,并迅速振荡,使试剂混合均匀。因为有碘,反应混合物会出现黄色,并且这个颜色随着碘和丙酮的反应而慢慢退去。在一个试管中倒入反应混合物,到试管的 3/4 处。在另外一支试管中倒入同样高度的蒸馏水,在一张照亮的白纸上俯视试管,记下碘的颜色刚刚消失时的时间,测出试管中反应混合物的温度。

重复上面的实验,用反应过的溶液代替蒸馏水作对照,两次实验所需的时间相差大约在 20 s 之内。

在新的混合物中,应该保持总体积为 50 mL,而且一定要使 H$^+$ 和 I$_2$ 的浓度和第一个实验中一样。用新的混合物做两次实验,两次反应的时间相差应大约在 15 s 以内,温度变化与第一次反应相比应小于 1℃。

再改变反应混合物的组成。这一次对反应的观测,将为你提供关于 H$^+$ 离子的反应分级数的资料。用这种混合物重复上述实验,来确定反应时间。两次实验的时间相差在 15 s 以内,ΔT 在 1℃ 以内。求出反应分级数 P。

最后,再用一种方法改变反应混合物组成,使你能够证明对 I$_2$ 浓度的反应分级数为 0。同样重复测定一次反应速率。

已知反应速率所依据的每种物质的反应分级数,从速率和你所研究的每个混合物的浓度的数据,计算速率常数 k。若温差仅在 1 ~ 2℃ 之内,则每个反应混合物的 k 应大致相同。

六、数据记录与结果

实验数据及计算列于下面各表。

1. 反应速率的数据

混合物	$4\ mol \cdot L^{-1}$ 丙酮 的体积(mL)	$1\ mol \cdot L^{-1}$ HCl 的体积(mL)	$0.005\ mol \cdot L^{-1}\ I_2$ 的体积(mL)	H_2O 的体积(mL)	反应时间(s)		温度
					反应1	反应2	
I	10	10	10	20			
II							
III							
IV							

2. 测定丙酮、H^+离子和 I_2 的反应分级数

混合物	[丙酮]	[H^+]	$[I_2]^0$	$v = \dfrac{[I_2]^0}{平均时间}$
I	$0.8\ mol \cdot L^{-1}$	$0.2\ mol \cdot L^{-1}$	$0.001\ mol \cdot L^{-1}$	
II				
III				
IV				

$$\frac{v_{II}}{v_{I}} = [\quad]^m,\ 求\ m$$

现写出适于反应混合物直及Ⅲ及Ⅳ的速率方程：

$v_{III} =$

$v_{IV} =$

用混合物Ⅲ、Ⅳ的速率对混合物Ⅰ、Ⅱ的速率的比值，求出 H^+ 离子或 I_2 的反应分级数：

$$\frac{v_{III}}{v_{-}} = [\quad]^p \qquad p =$$

$$\frac{v_{IV}}{v_{-}} = [\quad]^n \qquad n =$$

3. 测定速率常数 k

通过 B 部分的测定，已知 m、P、n，只需将表中的反应级数、起始浓度和测得的速率代入速率方程，便可计算 k。

混合物	I	II	III	IV	平均值
k	——	——	——	——	

4. 选作: 测定活化能 E_a

选择一个你用过的, 而且反应时间合适的反应混合物, 在恒温水浴槽(见附录三)中测定不同温度下的反应时间。

所用的反应混合物＿＿＿＿＿＿＿＿＿＿

10℃ 左右的反应时间＿＿＿＿＿s, 温度＿＿＿＿℃

40℃ 左右的反应时间＿＿＿＿＿s, 温度＿＿＿＿℃

室温时的反应时间＿＿＿＿＿s, 温度＿＿＿＿℃

用 C 部分中的数据, 计算每个温度下的速率常数 k:

	速率 v	k	$\lg k$	$1/T(\mathrm{K}^{-1})$
10℃ 左右	——	——	——	——
40℃ 左右	——	——	——	——
室　温	——	——	——	——

作出 $\lg k$ 对 $\dfrac{1}{T}$ 的图, 求出通过各点的最佳直线的斜率。

斜率＿＿＿＿＿＿＿＿＿＿＿＿＿

根据 Arrhenius 方程

活化能 $E_a = -2.303 \times 8.314 \times$ 斜率 $=$ ＿＿＿＿＿＿＿＿＿＿ J

（编者: 王一凡）

实验 15　食醋中总酸度的测定
Determination of Total Acidity in Vinegar

一、实验目的

(1)测定食醋中的总酸度。
(2)巩固酸碱滴定法的基本操作及其应用。

二、预习要点

(1)为什么酸碱滴定法只能测定食醋中的总酸度? 而不能测定其中各种酸的含量?
(2)若不进行空白试验,则实验结果会怎样?
(3)为什么酸碱滴定法测定食醋中的总酸度时要用酚酞作指示剂? 若用甲基橙作指示剂,会带来什么误差?
(4)用酸碱滴定法测定食醋中总酸度时,为什么要用煮沸过并已冷却的蒸馏水?

三、实验基本原理

因为醋酸的 $K_a = 1.8 \times 10^{-5}$,其 $c \cdot K_a > 10^{-8}$,故可以用 NaOH 标准溶液直接滴定。反应式为

$$NaOH + HAc \rightleftharpoons NaAc + H_2O$$

因为此反应等量点时溶液的 pH = 8.2,故可以用酚酞作指示剂,用 NaOH 标准溶液滴定至溶液呈微红色在 30 秒钟内不褪色为终点。

食醋中含有 3% ~ 5% (W/V)的醋酸和少量的其他有机酸如甲酸、丁酸、乳酸等。在测量食醋的总酸度时,全部用醋酸的含量来表示。

四、仪器和试剂

仪器:容量瓶(100 mL),锥形瓶(250 mL),容量吸管(25 mL),刻度吸管(10 mL),量筒(100 mL),碱式滴定管(50 mL),洗耳球,滴定管架,洗瓶。
试剂:食醋,新煮沸的蒸馏水,NaOH 标准溶液。

五、实验内容

吸取 10.00 mL 食醋注入 100 mL 容量瓶中,加蒸馏水稀释至刻度,混匀。

从容量瓶中吸取 25.00 mL 稀释后的食醋溶液，置于 250 mL 锥形瓶中，加入 30 mL 新煮沸过并已冷却的蒸馏水及 1～2 滴酚酞指示剂，用 NaOH 标准溶液滴至溶液呈微红色并在 30 s 内不褪色即为终点。同时进行空白试验。按上述方法重复测定两次。

根据反应式可知等量点时，$n(\text{NaOH}) = n(\text{HAc})$，可得下列的计算公式：

$$\text{HAc\%}(\text{W/V}) = \dfrac{c(V_1 - V_2) \times \dfrac{M_{\text{HAc}}}{1000}}{10.00 \times \dfrac{25.00}{100.00}} \times 100$$

式中　C——NaOH 标准溶液的准确浓度($\text{mol} \cdot \text{L}^{-1}$)；

　　　V_1——样品溶液消耗 NaOH 标准溶液的体积(mL)；

　　　V_2——空白溶液消耗 NaOH 标准溶液的体积(mL)；

　　　M_{HAc}——HAc 的摩尔质量。

六、实验记录与结果

按照下表处理实验数据：

项　　目	测定序号			
	1	2	3	空　白
V_{HAc}/mL				
$V_{\text{NaOH}}(初)/\text{mL}$				
$V_{\text{NaOH}}(末)/\text{mL}$				
$V_{\text{NaOH}}(消耗)/\text{mL}$				
HAc/%(W/V)				
HAc/%(W/V)(平均)				
偏差(d)				
相对平均偏差(d_r)				

（编者：刘绍乾）

实验 16　碳酸氢钠、碳酸钠混合碱分析
Analysis of the Mixture of Sodium Carbonate and Sodium Bicarbonate

一、实验目的

(1)进一步巩固酸碱滴定法滴定操作。

(2)熟悉测定 $NaHCO_3$、Na_2CO_3 混合碱中各组分含量的双指示剂滴定法。

二、预习要点

(1)锥形瓶中的混合碱溶液在开始滴定前放置太久对分析结果有影响吗? 为什么?

(2)若是测定 Na_2CO_3 与 NaOH 混合样品的各组分含量,也用双指示剂法, 其结果应如何计算?

三、实验基本原理

用酸滴定 Na_2CO_3 时,中和反应实际上分两步进行。

第一步: $CO_3^{2-} + H^+ \longrightarrow HCO_3^-$

第二步: $HCO_3^- + H^+ \longrightarrow H_2O + CO_2 \uparrow$

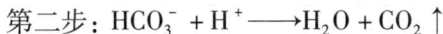

第一个等量点为酸式盐 $NaHCO_3$ 的水溶液,pH 约为 8.3,可用酚酞溶液作指示剂;第二个等量点为 H_2CO_3 的饱和水溶液,可用甲基橙或其他在 pH 3.5 ~ 4 之间发生颜色改变的指示剂。

在混合碱样品溶液中,加入酚酞指示剂,以 HCl 标准溶液滴定,当酚酞的红色褪去时,记下所消耗 HCl 的体积,设为 V_1。此时 Na_2CO_3 被中和成 $NaHCO_3$。再向溶液中加入甲基橙指示剂,继续用 HCl 标准溶液滴定至溶液由黄变为橙色。记下由滴定开始至甲基橙变色时共消耗的 HCl 体积,设为 $V_{总}$, 则样品中所含的 $NaHCO_3$ 和 Na_2CO_3 的量分别为:

$W(NaHCO_3) = (V_{总} - 2V_1) \times c(HCl) \times 0.08400$

$W(Na_2CO_3) = 2V_1 \times c(HCl) \times 0.05300$

在双指示剂法中,由于第一个突跃较小,方法误差较大,只能用作工业分析。

四、仪器和试剂

仪器:酸式滴定管(50 mL),容量吸管(20 mL),锥形瓶(250 mL),滴定

台，洗瓶，洗耳球。

　　试剂：$NaHCO_3 - Na_2CO_3$ 混合样品溶液，HCl 标准溶液，酚酞指示剂，甲基橙指示剂。

五、实验内容

　　用容量吸管吸取样品溶液 25.00 mL 置于 50 mL 锥形瓶中，加酚酞指示剂 2 滴，用 HCl 标准溶液慢慢滴定，不断振摇均匀，滴定至红色刚好消失，记下 HCl 溶液消耗的体积 (V_1)，然后向被滴定的溶液中，加入甲基橙指示剂 1 滴，继续用 HCl 标准溶液滴定，直至溶液颜色由黄色变为橙色，记录所消耗的 HCl 溶液总体积 ($V_总$)。

　　重复测定两次，且使每两次体积相差不超过 0.1 mL。根据下式计算样品溶液中所含的 $NaHCO_3$ 和 Na_2CO_3 的质量浓度 ($g \cdot L^{-1}$)。

$$\rho_{NaHCO_3} = (V_总 - 2V_1) \times c(HCl) \times 84.00 / V_样$$

$$\rho_{Na_2CO_3} = 2V_1 \times c(HCl) \times 53.00 / V_样$$

六、实验记录与结果

项　目　　内　容	1	2	3
$V_{样品}$/mL			
$V_{HCl}(初)$/mL			
$V_N(末)$/mL			
$V_1(消耗)$/mL			
$V_总(末)$/mL			
$V''_总(消耗)$/mL			
ρ_{NaHCO_3}/$g \cdot L^{-1}$			
偏差(d)			
相对平均偏差(d_r)			
$\rho_{Na_2CO_3}$/$g \cdot L^{-1}$			
偏差(d)			
相对平均偏差(d_r)			

（编者：钱　频）

实验 17 氧化还原与电极电势
Oxidation-Reduction and Electrode Potential

一、实验目的

(1)定性比较电极电势高低,掌握电极电势与氧化还原反应的关系。

(2)了解影响氧化还原反应的因素。

(3)熟悉常见的氧化剂和还原剂。

二、预习要点

(1)复习有关氧化还原的基本概念、影响电极电势的因素、能斯特(W. Nernst)方程式及非标准电极电势的计算。掌握利用 pH 计测定原电池电动势的方法。

(2)思考并回答下列问题:

①Fe^{3+} 离子能把 Cu 氧化成 Cu^{2+} 离子,而 Cu^{2+} 离子又能将 Fe 氧化成 Fe^{2+} 离子,这两个反应有无矛盾? 为什么?

②H_2O_2 为什么既可以作氧化剂又可以作还原剂? 写出有关电极反应,说明 H_2O_2 在什么情况下可以作氧化剂,在什么情况下可以作还原剂。

③为什么稀 HCl 不能和 MnO_2 反应,而浓 HCl 则能反应? 这里除 H^+ 离子浓度改变外,Cl^- 离子浓度的改变对反应是否也有影响?

与本实验相关的电极电势数据如下:

标准电极	$\varphi^{\ominus}/V$	标准电极	$\varphi^{\ominus}/V$
$\varphi^{\ominus}(Fe^{3+}/Fe^{2+})$	0.771	$\varphi^{\ominus}(Cu^{2+}/Cu)$	0.324
$\varphi^{\ominus}(I_2/I^-)$	0.536	$\varphi^{\ominus}(Cl_2/Cl^-)$	1.358
$\varphi^{\ominus}(Br_2/Br^-)$	1.066	$\varphi^{\ominus}(H_2O_2/H_2O)$	1.776
$\varphi^{\ominus}(MnO_4^-/Mn^{2+})$	1.507	$\varphi^{\ominus}(O_2/H_2O_2)$	0.695
$\varphi^{\ominus}(MnO_2/Mn^{2+})$	1.224	$\varphi^{\ominus}(Zn^{2+}/Zn)$	-0.762

三、实验基本原理

氧化还原反应是一类重要的化学反应。无论是氧化还原化学反应还是原电池中的电化学反应,都发生了氧化剂和还原剂之间的电子转移。氧化还原反应进行的方向是,电极电势代数值较大的电对中的氧化态物质是较强的氧化剂,它能氧化电极电势代数值较小的电对中的还原态物质。氧化还原电对电极电势的高低可以用 Nernst 方程式表示:

$$\varphi = \varphi^{\ominus} + \frac{RT}{nF}\ln\frac{\left[\text{氧化态}\right]}{\left[\text{还原态}\right]}$$

电极电势的大小,不但取决于电极的本质,而且也和溶液中离子的浓度、气体的压力和温度有关。

四、仪器和试剂

仪器:pH 计(见伏特计使用,mV 测量部分),试管(10 mm×75 mm),酒精灯,烧杯(50 mL),铜片,锌片,铁钉。

试剂:$ZnSO_4$(1 mol·L^{-1}),$CuSO_4$(1 mol·L^{-1}、0.001 mol·L^{-1}),$FeSO_4$(1 mol·L^{-1}),KI(0.1 mol·L^{-1}),$FeCl_3$(0.1 mol·L^{-1}),CCl_4、$K_3[Fe(CN)_6]$(0.1 mol·L^{-1}),KBr(0.1 mol·L^{-1}),$FeSO_4$(0.1 mol·L^{-1}),Br_2 水,$KSCN$(0.1 mol·L^{-1}),I_2 水,HNO_3(12 mol·L^{-1}、2 mol·L^{-1}),Zn 粒,H_2SO_4(1 mol·L^{-1}),KIO_3(0.1 mol·L^{-1}),$NaOH$(6 mol·L^{-1}),$KMnO_4$(0.01 mol·L^{-1}),Na_2SO_3(0.1 mol·L^{-1}),$ZnSO_4$(0.2 mol·L^{-1}),$H_2C_2O_4$(2 mol·L^{-1}),$MnSO_4$(0.2 mol·L^{-1}),NH_4F(3 mol·L^{-1}),H_2O_2(0.1 mol·L^{-1}),$NaBiO_3$(固体),$Na_2S_2O_3$(0.05 mol·L^{-1}),$HgCl_2$(0.1 mol·L^{-1}),$SuCl_2$(0.2 mol·L^{-1})*,Cl_2 水。

(* 使用 6 mol·L^{-1}HCl 溶液配制 $SnCl_2$ 溶液,用前新配)。

五、实验内容

1. 定性比较电极电势的高低

(1)在 50 mL 小烧杯中,按下表顺序及用量加入相应试剂,连接好盐桥及导线(如图 3-7),用伏特计依次测定这 4 组原电池的电势差。

根据测量数据,试比较 $\varphi_{Cu^{2+}(1\,mol\cdot L^{-1})/Cu}$,$\varphi_{Cu^{2+}(0.001\,mol\cdot L^{-1})/Cu}$,$\varphi_{Fe^{2+}(1\,mol\cdot L^{-1})/Fe}$,$\varphi_{Zn^{2+}(1\,mol\cdot L^{-1})/Zn}$ 的相对大小。在第一和第二个原电池中,铁电极所发生的反应是否相同?

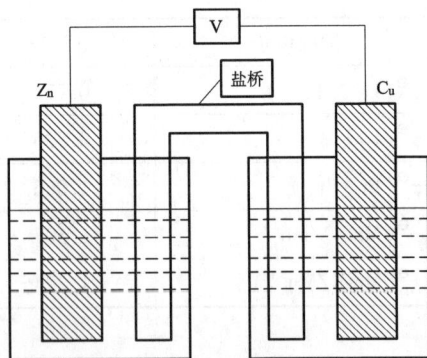

图 3-7 原电池装置图

甲杯(30 mL)	乙杯(30 mL)
1. $Zn \mid ZnSO_4(1 \ mol \cdot L^{-1})$	$FeSO_4(1 \ mol \cdot L^{-1}) \mid Fe$
2. $Fe \mid FeSO_4(1 \ mol \cdot L^{-1})$	$CuSO_4(1 \ mol \cdot L^{-1}) \mid Cu$
3. $Fe \mid FeSO_4(1 \ mol \cdot L^{-1})$	$CuSO_4(0.001 \ mol \cdot L^{-1}) \mid Cu$
4. $Cu \mid CuSO_4(0.001 \ mol \cdot L^{-1})$	$CuSO_4(1 \ mol \cdot L^{-1}) \mid Cu$

(2)在试管中加入 $0.1 \ mol \cdot L^{-1}$ KI 溶液 5 滴和 $0.1 \ mol \cdot L^{-1}$ $FeCl_3$ 溶液 3 滴,混匀后,再加入 5 滴 CCl_4,充分振荡,观察 CCl_4 层的颜色有何变化?试管中发生了什么反应?再往溶液中加入 $0.1 \ mol \cdot L^{-1}$ $K_3[Fe(CN)_6]$ 溶液 1 滴,观察现象,写出反应式。

用 $0.1 \ mol \cdot L^{-1}$ KBr 溶液代替 $0.1 \ mol \cdot L^{-1}$ KI 溶液进行相同的实验,能否发生反应?为什么?

在试管中加入 $0.1 \ mol \cdot L^{-1}$ $FeSO_4$ 溶液 5 滴,再加入 2 滴溴水,振荡后滴加 $0.1 \ mol \cdot L^{-1}$ KSCN 溶液 $1 \sim 2$ 滴,此溶液呈何色?说明试管中发生的反应。

用碘水代替溴水进行相同的实验,能否发生反应?为什么?

根据以上实验,定性比较 I_2/I^-,Br_2/Br^- 和 Fe^{3+}/Fe^{2+} 三个氧化还原电对的电极电势的高低,并指出何者为最强氧化剂?何者为最强还原剂?

2. 浓度对氧化还原反应的影响

(1)在 1(1)实验中,第 4 组原电池是:

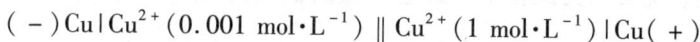

$(-)Cu \mid Cu^{2+}(0.001 \ mol \cdot L^{-1}) \parallel Cu^{2+}(1 \ mol \cdot L^{-1}) \mid Cu(+)$

所测两极间电势差为多少?试用 Nernst 方程式解释产生电势差的原因。

(2)在两支各盛有一粒锌的试管中,分别加入 $12 \ mol \cdot L^{-1}$ HNO_3 及 $2 \ mol \cdot L^{-1}$ HNO_3 溶液各 5 滴,观察所发生的现象,不同浓度的 HNO_3 溶液与 Zn 作用的反应产物及反应速度有何不同?

(3)在试管中加入 $0.1 \ mol \cdot L^{-1}$ KI 溶液 5 滴和 $0.1 \ mol \cdot L^{-1}$ $K_3[Fe(CN)_6]$ 溶液 2 滴,混匀后再加入 5 滴 CCl_4,充分振荡,观察 CCl_4 层的颜色有无变化?然后再加入 $0.2 \ mol \cdot L^{-1}$ $ZnSO_4$ 溶液 $2 \sim 3$ 滴,充分振荡,观察现象,试管中发生了什么变化?根据 $\varphi^\ominus$ 值判断 I^- 离子能否还原 $[Fe(CN)_6]^{3-}$ 离子?加入 Zn^{2+} 离子生成 $Zn_2[Fe(CN)_6]$ 白色沉淀:

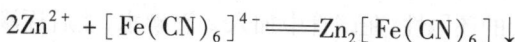

$$2I^- + 2[Fe(CN)_5]^{3-} \rightleftharpoons 2[Fe(CN)_6]^{4-} + I_2$$

$$2Zn^{2+} + [Fe(CN)_6]^{4-} \rightleftharpoons Zn_2[Fe(CN)_6] \downarrow$$

试说明沉淀生成对氧化还原反应的影响。

3．介质酸度对氧化还原反应的影响

(1)在试管中加入 0.1 mol·L^{-1} KI 溶液 5 滴，再加入 2 滴 1 mol·L^{-1} H$_2$SO$_4$ 溶液酸化，然后加入 0.1 mol·L^{-1} KIO$_3$ 溶液 1 滴，振荡并观察现象，试管中发生了什么反应？接着往试管中滴加 6 mol·L^{-1} NaOH 溶液，使溶液呈碱性，振荡后又有何现象？写出各反应式并说明介质对氧化还原反应方向的影响。

(2)取三支试管，各加入 0.01 mol·L^{-1} KMnO$_4$ 溶液 1 滴，然后向第一支试管中加入 1 mol·L^{-1} H$_2$SO$_4$ 溶液 2 滴；第二支试管中加入蒸馏水 2 滴；第三支试管中加入 6 mol·L^{-1} NaOH 溶液 2 滴，分别摇匀后再各加入 0.1 mol·L^{-1} Na$_2$SO$_3$ 溶液 1～2 滴，观察各试管中溶液颜色的变化，写出反应式，并用 Nernst 方程式解释介质对氧化还原反应的影响。

4．催化剂对氧化还原反应速度的影响

H$_2$C$_2$O$_4$ 溶液和 KMnO$_4$ 溶液在酸性介质中能发生如下反应：

$$5H_2C_2O_4 + 2MnO_4^- + 6H^+ \!=\!=\! 2Mn^{2+} + 10CO_2 + 8H_2O$$

此反应的电动势虽然较大，但反应速度较慢。Mn^{2+} 离子对此反应有催化作用。随着反应自身产生的 Mn^{2+} 离子增加，反应速度加快。若加入 F$^-$ 离子将 Mn^{2+} 离子掩蔽起来，则反应速度仍旧较慢。

取三支试管，各加入 2 mol·L^{-1} H$_2$C$_2$O$_4$ 溶液 5 滴及 1 mol·L^{-1} H$_2$SO$_4$ 溶液 1 滴，然后向第一支试管中加入 0.2 mol·L^{-1} MnSO$_4$ 溶液 1 滴，向第三支试管中加入 3 mol·L^{-1} NH$_4$F 溶液 2 滴，最后向三支试管中各加入 0.01 mol·L^{-1} KMnO$_4$ 溶液 1 滴，混匀，观察三支试管中紫红色褪去的快慢情况。必要时，可用小火加热，进行比较。

5．常见的氧化剂和还原剂的反应

(1)H$_2$O$_2$ 的氧化性：在试管中加入 0.1 mol·L^{-1} KI 溶液 5 滴，再加入 1 滴 1 mol·L^{-1} H$_2$SO$_4$ 溶液酸化，然后逐滴加入 0.1 mol·L^{-1} H$_2$O$_2$ 溶液，振荡试管并观察现象，写出反应式。

(2)KMnO$_4$ 的氧化性：在试管中加入 0.01 mol·L^{-1} KMnO$_4$ 溶液 5 滴，再加入 5 滴 1 mol·L^{-1} H$_2$SO$_4$ 溶液酸化，然后逐滴加入 0.1 mol·L^{-1} FeSO$_4$ 溶液，振荡试管并观察现象，写出反应式。

(3)NaBiO$_3$ 的氧化性：在试管中加入 0.2 mol·L^{-1} MnSO$_4$ 溶液 2 滴，再加入 5 滴 2 mol·L^{-1} HNO$_3$ 溶液酸化，然后加入少许固体 NaBiO$_3$，搅拌，静置片刻，观察上清液的颜色变化，写出反应式。

(4)Na$_2$S$_2$O$_3$ 的还原性：在试管中加入 2 滴碘水，然后逐滴加入 0.05 mol·L^{-1} Na$_2$S$_2$O$_3$ 溶液，观察试管中溶液颜色的变化，写出反应式。

（5）$SnCl_2$ 的还原性：在试管中加入 0.1 mol·L^{-1} $HgCl_2$ 溶液 2 滴，然后再加入 0.2 mol·L^{-1} $SnCl_2$ 溶液 1～2 滴，观察生成沉淀的颜色，写出反应式。继续滴加 $SnCl_2$ 溶液 2 滴，振荡后静置片刻，沉淀转变为何色？解释所观察到的现象，写出反应式。

（6）KI 的还原性：在试管中加入 0.1 mol·L^{-1} KI 溶液 2 滴，逐滴加入 Cl_2 水，边加边振荡试管，注意观察溶液颜色变化。继续滴加 Cl_2 水，溶液颜色又有何变化？写出反应式。

六、实验记录与结果

按照下表要求记录并解释实验现象：

实验项目	实验现象	现象解释及相关化学反应方程式
1		
2		
3		
⋮		

（编者：刘绍乾　王曼娟）

实验 18　维生素 C 药片中维生素 C 含量的测定
Quantitative Determination of Vitamin C

一、实验目的

通过维生素 C 的含量测定，掌握直接碘量法的原理及操作。

二、预习要点

(1)为什么维生素 C 含量可以用碘量法测定？
(2)维生素 C 本身就是一个酸，为什么测定时还要加入 HAc？
(3)维生素 C(药片)试样溶解时为何要加入新煮沸的冷蒸馏水？

三、实验基本原理

维生素 C(Vitamin C)又名抗坏血酸(Ascorbic acid)。分子式为 $C_6H_8O_6$，分子量 176.1 易溶于水，略溶于乙醇。因为分子中含有烯二醇基

($-\overset{OH}{\underset{}{C}}=\overset{OH}{\underset{}{C}}-\overset{}{\underset{}{C}}-$)，故具有强还原性，能与碘直接作用，维生素 C 分子中的烯

二醇基($-\overset{OH}{\underset{}{C}}=\overset{OH}{\underset{}{C}}-$)被氧化成二酮基($-\overset{O}{\underset{}{C}}=\overset{O}{\underset{}{C}}-$)。

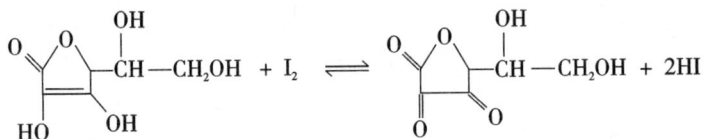

此反应进行很完全。使用淀粉作为指示剂，用直接碘量法可测定药片、注射液、蔬菜、水果中维生素 C 的含量。

四、仪器和试剂

仪器：分析天平，酸式滴定管(50 mL)，锥形瓶(250 mL)，量筒(10 mL，100 mL)，烧杯(150 mL)，容量瓶(100 mL)，容量吸管(25 mL)，滴定台，洗耳球，洗瓶，玻棒。

试剂：标准碘溶液(0.05 mol/L)，维生素 C(药片)，稀醋酸(2 mol/L)；淀

粉指示液[0.5%(W/V)水溶液]。

五、实验内容

(1)准确称取维生素 C 样品 0.6 ~ 0.8 g，置于 150 mL 烧杯中，加入新煮沸的冷却蒸馏水约 50 mL，2 mol·L^{-1} HAc 溶液 30 mL，使其溶解。然后小心转入 100 mL 容量瓶中，用少量蒸馏水洗涤烧杯 2 ~ 3 次，洗液并入容量瓶中，加水稀释至刻度，摇匀，备用。

(2)用容量吸管吸取上述维生素 C 样品溶液 25.00 mL，置于 250 mL 锥形瓶中，加入 0.5%(W/V)淀粉指示液 1 mL，立即用 I$_2$ 标准溶液滴定至溶液呈稳定蓝色，半分钟内不退色，即为终点。

六、数据记录与结果

维生素 C 的百分含量按下式计算：

$$维生素 C\% = \frac{c(I_2)V(I_2) \times M(C_6H_2O_5) \times 10^{-3}}{W(样品)} \times 100\%$$

取 3 份平行样的数据，分别计算百分含量，求出含量百分平均值、偏差及相对偏差。

项目 内容	1	2	3	空白
V_{Vc}/mL				
V_{I_2}(初)/mL				
V_{I_2}(末)/mL				
V_{Iw}(消耗)/mL				
维生素 C/%				
维生素 C/%(平均)				
偏差(d)				
相对平均偏差(d_r)				

（编者：李春云）

实验19　漂白粉中有效氯含量的测定
Determination of Available Chlorine in Bleaching Powder

一、实验目的

(1)掌握间接碘量法的原理和方法。

(2)巩固碱式滴定管、容量瓶和移液管的使用方法。

二、预习要点

(1)漂白粉中有效氯含量测定的基本原理是什么?

(2)吸取漂白粉试液前,为什么必须摇匀?

(3)漂白粉的漂白原理是什么?

(4)为什么用 $Na_2S_2O_3$ 标准溶液滴定待测溶液到淡黄色时才加淀粉指示剂?若在这之前加入淀粉指示剂对实验结果将有什么样的影响?

(5)漂白粉在工业上和日常生活中的用途有哪些?

(6)查阅资料,例举漂白粉含量测定的其他方法。

三、实验基本原理

漂白粉的主要成分是次氯酸钙[$CaCl(OCl)$]。次氯酸钙遇酸产生 Cl_2,Cl_2具有漂白作用。释放出来的氯称有效氯,漂白粉的质量由有效氯决定。

常用间接碘量法测定漂白粉中的有效氯,即在酸性溶液中,加入过量 KI,然后用 $Na_2S_2O_3$ 标准溶液滴定生成的 I_2。反应过程为:

$$CaCl(OCl) + 2H^+ \!=\!=\!= Ca^{2+} + H_2O + Cl_2 \uparrow$$

$$Cl_2 + 2I^- \!=\!=\!= I_2 + 2Cl^-$$

$$2S_2O_3^{2-} + I_2 \!=\!=\!= 2I^- + S_4O_6^{2-}$$

在 H_2SO_4 溶液中,反应可迅速完成,由上述反应可知,Cl_2 与 I_2 反应的物质的量之比为 $1:1$,而 I_2 与 $Na_2S_2O_3$ 反应的物质的量之比为 $1:2$。所以,可以由 $Na_2S_2O_3$ 标准溶液的浓度和所消耗的体积计算漂白粉中有效氯的含量。

四、仪器和试剂

仪器: 移液管(25 mL),碱式滴定管(50 mL),容量瓶(250 mL),锥形瓶(250 mL),量筒(10 mL,50 mL),研钵,洗耳球,洗瓶,滴定管架。

试剂：$Na_2S_2O_3$ 标准溶液，漂白粉（市售），H_2SO_4（$3\ mol\cdot L^{-1}$），淀粉溶液（$0.5\%\ W/V$），KI 固体。

五、实验内容

1. 试样的溶解

准确称取漂白粉 1 g 左右，置于研钵中，加水少许，调成糊状，将上层清液转移至 250 mL 容量瓶中。如此处理数次，并定量地转入容量瓶中，用水稀释至刻度，混匀。

2. 测定

移取 25.00 mL 均匀试液于锥形瓶中，加 80 mL 蒸馏水稀释，加 KI 固体 1 g，$3\ mol\cdot L^{-1}\ H_2SO_4$ 溶液 10 mL，摇匀。用 $Na_2S_2O_3$ 标准溶液滴定溶液呈淡黄色后，再加入 2 mL 淀粉溶液，继续用 $Na_2S_2O_3$ 标准溶液滴定至蓝色刚好消失为终点。按上述方法重复测定两次。记录数据，按下式计算漂白粉中有效氯的含量。

$$Cl_2\%\ 含量 = \frac{\frac{1}{2}c(Na_2S_2O_3) \times V(Na_2S_2O_3) \times \dfrac{M(Cl_2)}{1000}}{W(样品) \times \dfrac{25.00}{250.00}} \times 100\%$$

六、实验记录与结果

按照下表处理实验数据：

项　　目	测定序号			
	1	2	3	空白
$V_{漂白粉}$/mL	25.00	25.00	25.00	25.00（蒸馏水）
$V_{Na_2S_2O_3}$（初）/mL				
$V_{Na_2S_2O_3}$（末）/mL				
$V_{Na_2S_2O_3}$（消耗）/mL				
Cl_2 含量/%（W/W）				
Cl_2 含量/%（W/W）（平均）				
偏差（d）				
相对平均偏差（d_r）				

（编者：刘绍乾）

实验 20 氟离子选择性电极测定自来水中微量氟
Determination of Trace Fluorine in Tap Water
by Fluorine-Ion-Selective Electrode

一、实验目的

（1）了解氟离子选择电极的结构、作用原理及特点。
（2）掌握直接电位法测定离子浓度的原理及方法。

二、预习要求

（1）熟悉氟离子选择电极测定基本原理。
（2）掌握 pH – mV 计（离子计）操作方法。
（3）掌握标准曲线在半对数坐标纸的绘制方法。
（4）思考并回答下列问题：
①用离子选择电极法测定溶液中的离子浓度时，为什么要控制溶液的离子强度？
②总离子强度缓冲液（TISB）的作用是什么？
③测定时为什么用塑料烧杯为宜？
④塑料烧杯未烘干，将给测定结果带来什么误差？
⑤实验成败的影响因素有哪些？

三、实验基本原理

氟的含量是环境监测中一个重要指标。环境中氟化物的污染主要来源于矿山开采、金属冶炼以及工业生产，如金属铝、玻璃、陶瓷、钢铁、磷肥、搪瓷等工业生产的废水废气。有文献报道，人体摄入氟总量的 50% ~70% 来自于饮用水，因此，饮用水中的氟是人体中氟的主要来源。摄入适量的氟能维持机体正常的钙、磷代谢，还可以促进机体的生长发育，有防龋齿的作用。但是，氟对人体有益的剂量非常小。饮用水中氟的适宜含量为 $0.5 \text{ mg} \cdot \text{L}^{-1}$ 左右，低于 $0.5 \text{ mg} \cdot \text{L}^{-1}$ 易得龋齿，而高于 $1.5 \text{ mg} \cdot \text{L}^{-1}$ 时，人体会因摄入过量的氟引起急性或者是慢性氟中毒（早期氟中毒的症状是氟斑牙），破坏钙、磷的正常代谢，大量的氟进入人体后与钙结合成氟化钙，沉积在骨中，使骨中氟化钙增加数十倍，骨密质增高，血钙减少。一方面因缺钙产生四肢抽搐、腰痛，一方面因刺激甲状旁腺，使其功能亢进，导致骨脱钙，产生骨质稀疏。严重时出现氟骨症。除此之外，氟还可抑制机体某些酶的活性，高氟还可阻碍 DNA 的合成，使蛋白质合成受阻。在所有人体元素中，人体对氟的含量最为敏感，从满足需要到由于过多而导致中毒之间的量相差很少，即氟对人体的安全范围比其他微量元素窄得多，所以要更加注重饮水及食物中氟含量对人体健康的影响。因此，监测饮用水中氟离子的含量至关重要。

氟离子选择性电极法已被确定为测定饮用水中氟含量的标准方法。离子选择性电极是一种电化学传感器，它可将溶液中特定离子的活度转换成相应的电位信号。氟离子选择电极的敏感膜由 LaF_3 单晶片制成，为改善导电性能，晶体中还掺杂了少量 0.1% ~ 0.5% 的 EuF_2 和 1% ~ 5% 的 CaF_2。膜导电由离子半径较小、带电荷较少的晶体离子氟离子来担任。Eu^{2+}、Ca^{2+} 代替了晶格点阵中的 La^{3+}，形成了较多空的氟离子点阵，降低了晶体膜的电阻。将氟离子选择电极插入待测溶液中，待测离子可以吸附在膜表面，它与膜上相同离子交换，并通过扩散进入膜相。膜相中存在的晶体缺陷，产生的离子也可以扩散进入溶液相，这样在晶体膜与溶液界面上建立了双电层结构，产生相界电位，氟离子活度的变化符合能斯特方程：

$$E_{氟电极} = K - \frac{2.303RT}{F}\lg a_{F^-} \tag{1}$$

当氟离子选择电极（作指示电极）与饱和甘汞电极（E_{SCE}，参比电极）插入被测溶液中组成工作电池时，工作电池可图解如下：

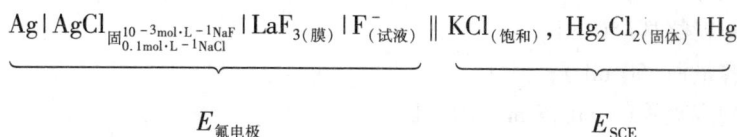

$$\underbrace{Ag|AgCl_{\substack{固10^{-3}mol\cdot L^{-1}NaF \\ 0.1mol\cdot L^{-1}NaCl}}|LaF_{3(膜)}|F^-_{(试液)}}_{E_{氟电极}} \parallel \underbrace{KCl_{(饱和)}, Hg_2Cl_{2(固体)}|Hg}_{E_{SCE}}$$

电池电动势（E）为：

$$E_{cell} = E_{SCE} - E_{氟电极} = E_{SCE} - (K - \frac{2.303RT}{F}\lg a_{F^-}) \tag{2}$$

在测定条件下，E_{SCE} 和 K 均为常数，将两常数合并为常数 K'，则式（2）可表达为：

$$E_{cell} = K' + \frac{2.303RT}{F}\lg a_{F^-} \tag{3}$$

在 25℃时，将常数 R、F 代入式（3）得：

$$E_{cell} = K' + 0.0592\lg a_{F^-} \tag{4}$$

从式（4）可知，电池的电动势与试液中 F^- 离子活度的对数成线性关系。这就是离子选择性电极测定 F^- 离子的理论依据。

用氟电极测定 F^- 离子时，最适宜的 pH 范围为 5.5 ~ 6.5。pH 过低，由于形成 HF，影响 F^- 离子活度；pH 过高，可能由于单晶膜中 La^{3+} 的水解，形成 La $(OH)_3$，而影响电极的响应。故通常用 pH≌6 的柠檬酸钠缓冲溶液来控制溶液的 pH。

Al^{3+}、Fe^{3+}、Si^{4+} 离子对测定严重干扰，其他常见离子无影响。加入大量柠檬酸钠，可消除干扰离子和酸度的影响。

用离子选择电极测量的是溶液中离子的活度。我们通过控制标准溶液和试液有相同的离子强度，则通过标准曲线，可测得溶液中 F^- 离子的浓度。通常在溶液中加入大量硝酸钠后，就可达到控制溶液总离子强度的目的。在此，柠檬酸钠—硝酸钠混合溶液（pH≌6）又称为总离子强度缓冲溶液（TISB）。

本法的最低检出浓度为 0.05 mg/L 氟，测量上限为 1900 mg/L 氟。

本实验采用标准曲线法测量溶液中 F^- 离子浓度。

四、仪器及试剂

仪器：
①数字式 pH – mV 计或离子计；
②氟化镧单晶膜电极；
③饱和甘汞电极；
④电磁搅拌器；
⑤搅拌磁子；
⑥塑料烧杯（50 mL）；
⑦容量瓶（50 mL）；
⑧刻度吸管（1 mL,5 mL,10 mL）；
⑨移液管（25 mL）。

试剂：
①氟标准溶液：准确称取 0.2210 gNaF（于 500~600℃ 干燥 40~50 min，干燥器内冷却），置于烧杯中，用水溶解，转移至 1 L 容量瓶中，以蒸馏水稀至刻度，摇匀。此溶液每毫升含 100 μg 氟。贮于塑料瓶中，供作标准曲线用。

②总离子强度缓冲液：称取 58.8 g 二水合柠檬酸钠和 85 g 硝酸钠，加蒸馏水溶解，以 HCl(1+1) 调节 pH≌6（试纸检验），转入 1 L 容量瓶中，以蒸馏水稀至刻度，摇匀，此溶液浓度为 0.2 $mol \cdot L^{-1}$ 柠檬酸钠 – 1 $mol \cdot L^{-1}$ 硝酸钠。

五、实验内容

（1）电极的清洗：将少量蒸馏水（或去离子水）倒入塑料烧杯中，加入搅拌磁子，插入氟离子选择电极和甘汞电极，将电极分别连在 pH – mV 计上，按 pH

－mV 计说明书(见附录四)进行操作,并打开电磁搅拌器,在搅拌溶液的情况下清洗电极,直洗至溶液电池电动势为 －200 mV 以下为止(即测量时,选择开关为 －mV,饱和甘汞电极接"＋"端,氟电极接"－"端时,读数在 200 mV 以上,或根据氟电极说明书所注操作)。

(2)标准曲线的绘制:用刻度吸液管分别加入 10、25、50、100、250、500 μg 氟于一系列 50 mL 容量瓶中,各加入 10.00 mL 总离子强度缓冲液,用水稀至刻度,摇匀。则相应浓度分别为 0.20、0.50、1.00、2.00、5.00、10.0 mg·L^{-1}氟。转入 50 mL 塑料烧杯中。放入搅拌磁子,连接好电极,搅拌溶液 1 min,停止搅拌后,读取稳定的电动势值。按照从低浓度到高浓度的顺序依次测量各标准溶液对应的电动势,每测量 1 份标准溶液,无需用去离子水清洗电极,只需用滤纸吸干电极上的水珠。记录各次测得的电位数据。在半对数坐标纸上绘制 $E - \lg c(F^-)$标准曲线。

(3)水样的测定:吸取水样 25.00 mL,置于 50 mL 容量瓶中,加入 10.00 mL 总离子强度缓冲液,用水稀至刻度,摇匀。转入 50 mL 塑料烧杯中,放入搅拌磁子,连接好电极,搅拌溶液 1 min,停止搅拌后,测出稳定的电位值。在标准曲线上查得其浓度。

六、数据记录与结果

F 标准(mg·L^{-1})	0.20	0.50	1.00	2.00	5.00	10.0
电动势/mV						
水样/mV						

计算:氟化物$\left[F^- / (mg \cdot L^{-1}) \right] = \dfrac{\text{测得氟量}(\mu g)}{25 \text{ mL}}$

(编者:邓凯佳　刘绍乾)

实验 21   消毒液中过氧化氢含量的测定
Determination of Hydrogen Peroxide in Disinfection Solution

一、实验目的

(1)掌握 $KMnO_4$ 法测定 H_2O_2 含量的原理。

(2)掌握用 $KMnO_4$ 标准溶液直接滴定 H_2O_2 的操作方法。

(3)巩固移液管、容量瓶以及酸式滴定管的使用方法。

二、预习要点

(1)用 $KMnO_4$ 法测定 H_2O_2 时,能否用 HNO_3 或 HCl 控制酸度?

(2)测定 H_2O_2 含量时,为什么第一滴 $KMnO_4$ 的颜色褪得较慢,以后反而逐渐加快?

(3)查阅资料,熟悉 $KMnO_4$ 和 H_2O_2 在日常生活以及医学上的用途。

三、实验基本原理

在临床上,双氧水是常用的消毒剂,药用双氧水中 H_2O_2 的含量约3%。由于 H_2O_2 中氧的氧化数为 -1,故它既可以作为氧化剂,也可以作为还原剂。H_2O_2 在稀硫酸溶液中,在室温条件下,能定量地被 $KMnO_4$ 氧化,其反应式为:

$$5H_2O_2 + 2MnO_4^- + 6H^+ \rightleftharpoons 2Mn^{2+} + 5O_2 \uparrow + 8H_2O$$

开始反应时速度慢,滴入第一滴溶液不易褪色,待 Mn^{2+} 生成后,由于 Mn^{2+} 的自动催化作用加快了反应速度,故能顺利地滴定到终点。也可以在滴定前,往锥形瓶中加 $2 \sim 3$ 滴 $1 \ mol \cdot L^{-1}$($MnSO_4$)作催化剂,以加快反应速度。因此可以用高锰酸钾法测定过氧化氢的含量。

$KMnO_4$ 是强氧化剂,易被其他还原性物质还原,因此,$KMnO_4$ 标准溶液不能直接用市售的 $KMnO_4$ 试剂配制,而是先配制近似浓度的溶液。在中性溶液中,$KMnO_4$ 的还原产物为难溶的 MnO_2,所以,在标定 $KMnO_4$ 溶液之前,需将 $KMnO_4$ 溶液煮沸使其与还原性杂质充分反应后,过滤除去 MnO_2,再用一级标准物质 $Na_2C_2O_4$ 标定 $KMnO_4$ 溶液。其反应式为

$$5C_2O_4^{2-} + 2MnO_4^- + 16H^+ \rightleftharpoons 2Mn^{2+} + 10CO_2 \uparrow + 8H_2O$$

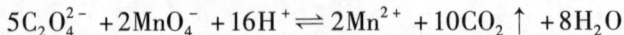

同样,该反应速率较慢,也需要 Mn^{2+} 的催化作用才可快速反应。为了加速反应,也可将反应体系加热到 $80 \sim 90℃$,并始终控制滴定时的温度高于60℃。

四、仪器和试剂

仪器：电子天平、干燥器、称量瓶、25 mL 移液管、容量瓶（1000 mL、250 mL）、酸式滴定管、锥形瓶、量筒、砂芯漏斗、吸滤瓶、抽气泵、滤纸等。

试剂：药用双氧水（3%）、$Na_2C_2O_4$（AR，s）、0.05 mol·L^{-1} $KMnO_4$ 溶液、3 mol·L^{-1} H_2SO_4 溶液。

五、实验内容

1. $KMnO_4$ 溶液的标定

（1）将 3 g $Na_2C_2O_4$（$M_r = 134$ g·moL^{-1}）放入称量瓶，于 110～120℃烘干 1 h，冷却，放置于干燥器中，准确称取 1.6～1.7 g（±0.1 mg）$Na_2C_2O_4$ 置于烧杯中，加适量蒸馏水溶解，用玻璃棒转移至 250 mL 容量瓶中，用少量蒸馏水淋洗烧杯内壁 2～3 次，淋洗液全部转入容量瓶中，加蒸馏水稀释至容量瓶的标线，盖好瓶塞，充分摇匀备用。

（2）用移液管移取 25.00 mL（1）中配好的 $Na_2C_2O_4$ 标准溶液置于 250 mL 锥形瓶中加入 20 mL3 mol·L^{-1} H_2SO_4 溶液。加热溶液至 80～90℃，趁热用 $KMnO_4$ 溶液慢慢滴定（滴定速率过快，$KMnO_4$ 将与 Mn^{2+} 反应生成黑色的 MnO_2）。控制反应温度高于 60℃，边滴边摇动锥形瓶，直至溶液显淡红色且维持 30 s 不退色，即达到滴定终点。记录消耗 $KMnO_4$ 标准溶液的体积（由于 $KMnO_4$ 溶液颜色较深，很难观察到弯月面，所以通常读取液面最高点的数据）。平行测定 3 次。根据滴定所消耗的 $KMnO_4$ 溶液体积和 $Na_2C_2O_4$ 的质量，按照下式计算 $KMnO_4$ 溶液的准确浓度：

$$c(KMnO_4) = \frac{2}{5} \times \frac{m(Na_2C_2O_4) \times 1000 \times 25.00}{250.00 \times V(KMnO_4) \times M(Na_2C_2O_4)} (mol \cdot L^{-1})$$

2. 药用双氧水中 H_2O_2 含量的测定

用移液管吸取 10.00 mL 药用双氧水，置于 250.00 mL 容量瓶中，加蒸馏水稀释至标线，摇匀。用移液管吸取 25.00 mL 稀释后的药用双氧水待测液，置于 250.00 mL 锥形瓶中，加 20 mL 3 mol·L^{-1} H_2SO_4 溶液，摇匀。边摇动锥形瓶边用 $KMnO_4$ 标准溶液滴定，直至溶液显淡红色且维持 30s 不褪色，即达到滴定终点。记录消耗 $KMnO_4$ 标准溶液的体积。平行测定 3 次。根据滴定所消耗的 $KMnO_4$ 溶液体积，按照下式计算 H_2O_2 的含量：

$$\rho(H_2O_2) = \frac{5}{2} \times \frac{c(KMnO_4) \times V(KMnO_4) \times M(H_2O_2)}{10.00 \times \dfrac{25.00}{250.00}} (g \cdot L^{-1})$$

六、实验记录与结果

1. $KMnO_4$ 溶液的标定

项　目	测定序号		
	1	2	3
$m(Na_2C_2O_4)/g$	25.00	25.00	25.00
$V_{KMnO_4}(初)/mL$			
$V_{KMnO_4}(末)/mL$			
$V_{KMnO_4}(消耗)/mL$			
$c(KMnO_4)/mol \cdot L^{-1}$			
$(KMnO_4)/mol \cdot L^{-1}(平均值)$			
偏差(d)			
相对平均偏差(d_r)			

2. 药用双氧水中 H_2O_2 含量的测定

项　目	测定序号		
	1	2	3
$V_{H_2O_2}/mL$	25.00	25.00	25.00
$V_{KMnO_4}(初)/mL$			
$V_{KMnO_4}(末)/mL$			
$V_{KMnO_4}(消耗)/mL$			
$H_2O_2/\%(W/V)$			
$H_2O_2/\%(W/V)(平均值)$			
偏差(d)			
相对平均偏差(d_r)			

（编者：刘绍乾）

实验 22 配合物的生成和性质
The Formation and Properties of Coordination Compounds

一、实验目的

(1)了解配合物的生成及配离子的相对稳定性。

(2)了解沉淀反应、氧化还原反应等对配位平衡的影响。

二、预习要点

(1)预习有关配合物的组成、稳定性等基本概念。

(2)影响配合物的稳定性的主要因素有哪些?

三、实验基本原理

配合物由内界和外界两部分组成。中心离子与配位体组成配合物的内界,其余处于外界。内界和外界在水溶液中完全离解,配离子本身在溶液中只部分离解。例如:

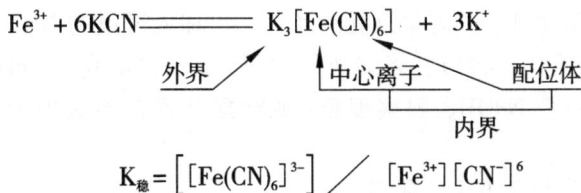

$$Fe^{3+} + 6KCN \Longrightarrow K_3[Fe(CN)_6] + 3K^+$$

外界 / 中心离子 配位体

内界

$$K_稳 = \left[[Fe(CN)_6]^{3-} \right] \Big/ [Fe^{3+}][CN^-]^6$$

$K_稳$ 是配离子的稳定常数,可用于判断配位反应进行的程度。

简单离子形成配位化合物后,其颜色、溶解度、电极电位等都会发生变化,利用这些变化可以检验有关离子。

四、仪器和试剂

试剂: $H_2SO_4(1)$, $NaOH(0.1, 2)$, $NH_3 \cdot H_2O(2)$, $NaCl(0.1)$, $BaCl_2(0.1)$, $HgCl_2(0.1)$, $FeCl_3(0.1)$, $AgNO_3(0.1)$, $CuSO_4(0.1)$, $NiSO_4(0.1, 0.5)$, $Na_2S_2O_3(0.1)$, $KSCN(0.1)$, $NH_4F(1)$, $KBr(0.1)$, $KI(0.1)$, $K_3[Fe(CN)_6]$ (0.1), 碘水, CCl_4, 丁二肟(1%), $K_4P_2O_7(2)$, $(NH_4)_2Fe(SO_4)_2$ 固体。(上述物质括号内数据单位为 $mol \cdot L^{-1}$)

$$Fe^{3+} + nSCN^- \rightleftharpoons [Fe(SCN)n]^{3-n}$$

黄色　　　无色　　　　浅红 → 血红色

$$n = 1 \sim 6$$

$$Ni^{2+} + 2 \begin{array}{c} CH_3-C=N-OH \\ | \\ CH_3-C=N-OH \end{array} \rightleftharpoons \begin{array}{c} CH_3O \underline{\quad} H \underline{\quad} O \quad CH_3 \\ | \qquad\qquad | \\ C=N \qquad N=C \\ \quad\diagdown Ni \diagup \\ C=N \qquad N=C \\ | \qquad\qquad | \\ CH_3O \underline{\quad} H \underline{\quad} O \quad CH_3 \end{array} \downarrow +2H^+$$

浅绿　　　浅黄　　　　　　　　　　　　　　　　　　鲜红

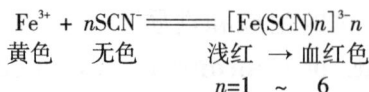

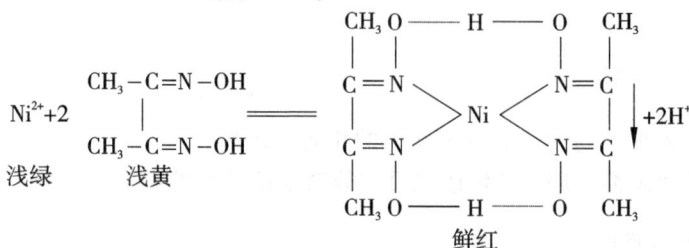

仪器： 试管(10 mm × 75 mm)，烧杯(50 mL, 400 mL)

五、实验内容

1. 配离子的生成和组成

(1)在试管中加入一滴 0.1 mol·L^{-1} HgCl$_2$ 溶液，逐滴加入 0.1 mol·L^{-1} KI 溶液，观察现象(有什么生成?)继续加入过量的 KI 溶液，观察现象(又生成了什么)。

(2)在一支试管中加入 5 滴 0.1 mol·L^{-1} CuSO$_4$ 溶液并逐滴加入 2 mol·L^{-1} NH$_3$·H$_2$O 至沉淀生成，并观察沉淀的颜色。再加入过量的 NH$_3$·H$_2$O 至沉淀完全溶解(生成了什么产物?)。将溶液分成两份，一份加 0.1 mol·L^{-1} BaCl$_2$，另一份加 0.1 mol·L^{-1} NaOH，观察现象(观察配合物在溶液中的存在形式是什么)。

2. 简单离子和配离子的区别

在两支试管中分别加入 1 滴 0.1 mol·L^{-1} FeCl$_3$ 溶液和 2 滴 0.1 mol·L^{-1} K$_3$[Fe(CN)$_6$]溶液，再各加入 0.1 mol·L^{-1} KSCN 溶液，观察现象有何不同，解释原因。

3. 配位平衡与氧化还原反应

(1)在两支试管中分别加入 2 滴 0.1 mol·L^{-1} FeCl$_3$ 溶液和 2 滴 0.1 mol·L^{-1} K$_3$[Fe(CN)$_6$]溶液，再各加入 1 滴 0.1 mol·L^{-1} KI 和 4 滴 CCl$_4$，振荡，观察 CCl$_4$ 层颜色。比较二者有何不同?

(2)在两支试管中分别加入 2 滴碘水，然后分别加入少量(NH$_4$)$_2$Fe(SO$_4$)$_2$ 固体和少量 0.1 mol·L^{-1} K$_4$[Fe(CN)$_6$]溶液，比较二者有何不同，并解释原因。

4. 配位平衡与沉淀反应

在试管中加入 1 滴 0.1 mol·L^{-1} AgNO$_3$ 溶液，加入数滴 0.1 mol·L^{-1} NaCl

溶液，观察现象，再滴加 2 mol·L^{-1} NH$_3$·H$_2$O 至沉淀溶解（生成什么产物？）然后滴加少量 0.1 mol·L^{-1} KBr 溶液，观察现象。再逐滴加入 0.1 mol·L^{-1} NaS$_2$O$_3$ 溶液，观察沉淀的溶解，然后滴加 0.1 mol·L^{-1} KI 溶液，观察有何变化。

通过以上实验，定性比较 AgCl，AgBr，AgI 溶解度的大小和 [Ag(NH$_3$)$_2$]$^+$，[Ag(S$_2$O$_3$)$_2$]$^{3-}$ 配离子稳定性的大小。

5. 配合物之间的转化

在试管中加入 1 滴 0.1 mol·L^{-1} FeCl$_3$ 溶液，然后再加入几滴 0.1 mol·L^{-1} KSCN 溶液观察现象（产生什么物质？），再滴加 1 mol·L^{-1} NH$_4$F 溶液，观察颜色的变化。

6. 配位平衡与介质的酸碱性

在试管中加入 4 滴 0.1 mol·L^{-1} NiSO$_4$ 溶液，逐滴加入 2 mol·L^{-1} NH$_3$·H$_2$O 至生成的沉淀刚好溶解，观察现象（产生什么物质？）。把溶液分成两份，分别试验其与 1 mol·L^{-1} H$_2$SO$_4$，0.1 mol·L^{-1} NaOH 溶液的反应，观察现象。

7. 螯合物的形成

（1）在试管中加入 1 滴 0.1 mol·L^{-1} NiSO$_4$，0.5 mL 蒸馏水和 1 滴 2 mol·L^{-1} NH$_3$·H$_2$O，然后加入 1 滴 1% 丁二肟溶液，观察沉淀的颜色（此为鉴定 Ni^{2+} 的反应）。

（2）往试管中加入 1 滴 0.1 mol·L^{-1} CuSO$_4$ 溶液，然后逐滴加入 2 mol·L^{-1} K$_4$P$_2$O$_7$ 溶液，观察沉淀的颜色。继续加入 K$_4$P$_2$P$_7$ 溶液，生成蓝色透明溶液 [Cu(P$_2$O$_7$)$_2$]$^{6-}$。

（编者：钱　频）

实验 23　磺基水杨酸合铜(Ⅱ)配合物的组成
Determination ofComposition and Stability Constant of Coordination Compound Formed from Sulfosalicylic Acid and Copper(Ⅱ)

一、实验目的

1. 了解配合物稳定常数测定的基本方法。
2. 了解影响配合物稳定常数测定准确性的基本因素。
3. 巩固溶液的配制和标定方法。
4. 掌握酸度计、分光光度计等基本仪器的使用方法。

二、预习要点

1. 学习郎伯－比尔定律及其使用的注意事项。
2. 学习分光光度计、酸度计等基本仪器的使用方法。
3. 如果溶液中同时有几种不同组成的有色配合物存在,能否用本实验方法测定它们的和稳定常数?
4. 如果被测的配合物稳定性太低或者太高,对测定结果是否有影响?
5. 本实验中,为何能用体积比代替物质的量比为横坐标作图?

三、实验基本原理

1. 前言

配合物在化学工业、原子能工业、半导体材料工业、制药工业、湿法冶金、电镀行业、皮革轻工、分析化验方面都有着广泛的应用。目前配位化学已经突破了无机化学的范围,与其他学科相互渗透,形成了许多崭新的而富有生命力的边缘学科,成为当代化学学科中最活跃的领域之一。配合物稳定常数作为衡量配合物稳定性的指标,对于了解配合物的形成、结构以及中心原子与配体间的成健本质等方面有重要意义,也是配合物实际应用的必要条件。

配合物稳定常数的测定是通过实验测定的一系列数据,再通过适当的数学处理,进而求出稳定常数。已报道的测定配离子稳定常数的方法有数十种(见表 3－1)。

表 3 - 1　配合物稳定常数测定常用方法一览表

序	实验方法	数据处理法
1	酸度计法	pH 电位法
2	分光光度法	连续变化法(等摩尔可系列法、浓比递度法)
3	分光光度法	对应溶液法
4	阳离子选择电极法	pM 电位法
5	浓差电池法	pM 电位法
6	极谱法	pM 电位法
7	溶剂萃取法	分配法
8	溶解度法	分配法
9	离子交换法	分配法

　　分光光度法是研究溶液中配合物的组成、稳定性以及反应机理的主要手段之一,方法应用仅次于电位法,测得的常数的准确度也仅次于电位法。用分光光度法测定配合物的组成及稳定常数,常用的方法有连续变化法、等摩尔系列法、平衡移动法等。本实验采用等摩尔系列法测定。

　　以往该实验的实验方法大致为:实验依据 Cu^{2+} 与磺基水杨酸在 pH = 5 左右形成 1:1 配合物,溶液显亮绿色,分别配制 0.05 $mol \cdot L^{-1}$ 硝酸铜溶液和 0.05 $mol \cdot L^{-1}$ 磺基水杨酸溶液,以不同体积混合,用氢氧化钠溶液和硝酸溶液反复调节控制体系 pH = 4.5 ~ 5,在波长为 440 nm 的单色光下测定溶液吸光度,根据吸光度数据以及下面的计算方法,可以计算该配合物的条件稳定常数 $K_{条件}$。在此实验条件下,磺基水杨酸不吸收,Cu^{2+} 的吸收也可以忽略,形成的配合物有一定的吸收。但是,由于实验中要在精密 pH 试纸和酸度计的监测下用强碱和强酸来回调节溶液的 pH,且在铜(Ⅱ)与磺基水杨酸的物质的量之比大于 1 时,氢氧化钠稍过量便会造成 Cu^{2+} 水解,溶液配制过程冗长繁复,而各溶液的 pH 值波动大,而酸度直接影响所测得的条件稳定常数的大小,导致实验的重现性差。因此,本实验在以往实验的基础上进行了改进,使用一种合适的缓冲溶液替代反复调节 pH 的过程,并且将铜离子的浓度降低至 0.010 $mol \cdot L^{-1}$,既解决了 Cu^{2+} 容易沉淀的问题,又大大减少了溶液配制时间,同时 pH 值的波动性很小,可提高结果重现性。

　　2.基本原理

　　(1)配合物的浓度与吸光度的关系

　　当一束具有一定波长的单色光通过一定厚度的有色物质的均匀溶液时,光的一部分被有色溶液吸收，一部分透过溶液，还有一部分被溶液容器的表面反射。所以透过溶液的光(透射光)的强度(I_t)比原来入射光的强度(I_0)有所减弱(图 3-8)。按照 Lambert-beer 定律，溶液中有色物质对光的吸收程度(用 $\lg \dfrac{I_0}{I}$ 即吸光度 A 来表法)与液层的厚度(l)及有色物质的浓度(c)成正比：

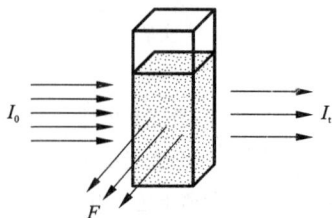

图 3-8　溶液对光的吸收示意图

$$A = \lg \frac{I_0}{I} = \varepsilon l c \tag{1}$$

式中：ε 为比例常数，称为摩尔吸光系数。它是每一种有色物质的特征常数；l 为液层厚度，等于所用比色皿的内空厚度；c 为配合物的浓度，单位为 mol·L^{-1}。从式(1)可知，如果液层的厚度 l 不变，吸光度便只与有色物质的浓度成正比。

如果中心离子 M 与配位 L 在溶液中都是无色的，或者对实验所选定的波长的光不吸收，而他们所形成的配合物 MI$_n$(省去电荷)是有色的，而且在一定条件下只生成这一种配合物，那么溶液的吸光度就与该配合物的浓度成正比。在此前提条件下，便可从测得的吸光度来确定该配合物的组成和稳定常数。

有关符号的说明：

① M——金属离子，T_M——金属离子总浓度，[M]——游离金属离子浓度

② L——配体，T_L——配体总浓度，[L]——游离配体浓度

③ ML$_n$——第 n 级配合物，[ML$_n$]——第 n 级配合物浓度

④ A——溶液吸光度，ε——溶液的摩尔吸光系数，l——比色皿厚度/cm

根据 Lambert-beer 定律有：$A = \varepsilon \times [ML_n] \times l$

(2)配合物组成的确定方法

首先配制等摩尔系列溶液，该溶液具有以下特点：

溶液中金属离子(M)与配体(L)的物质的量之和恒定不变，即 $n_M + n_L$ 恒定，而两者的摩尔分数连续变化。配制等摩尔系列溶液的方法很简单，将原始浓度相等的金属离子与配体溶液，各取不同体积混合，再用蒸馏水定容即可。

然后，在特征波长(440 nm)下测定上述等摩尔系列溶液的吸光度，并绘制吸光度-配合物组成图(见图 3-9)，该图纵坐标为吸光度 A，横座标左端为金属离子浓度最大值，T_M 由左至右递减；横座标右端为配体浓度最大值，T_L 由右至左递减，但横座标上任何一点 $T_M + T_L$ 的值相等。

图中曲线出现一个峰，在峰的两侧分别作两条切线与曲线相切，两切线交点为 B，过 B 点向横轴作垂线，该垂线与曲线交点为 C，与横轴交点为 D。对于配位反应

$$M + nL \rightleftharpoons ML_n \qquad (2)$$

B、C、D 三点分别具有以下性质：

B 点对应的纵坐标 A_1 是假设该配合物完全不离解时溶液所具有的吸光度，称为吸光度理论值。

C 点对应的纵坐标 A_2 是该配合物的实验吸光度，称为吸光度实验值（实验曲线的峰值）。显然，$A_2 < A_1$。因配合物在溶液中不可能不发生

图 3-9 吸光度－配合物组成图

解离，即溶液中配合物的实际浓度肯定比该配合物完全不发生解离时的浓度低，所以，$A_2 < A_1$。

D 点所在位置的横坐标对应的 T_L/T_M 比值，就是配位数 n 值。

例如，如果在系列混合溶液中，溶液吸光度最大处所对应的 B 点其横坐标为 $T_L = 0.5$（则 $T_M = 1 - 0.5 = 0.5$），那么在此溶液中 L 和 M 的物质的量之比（即为 T_L 与 T_M 之比）为 1∶1，因而配合物的组成也就是 1∶1，即形成 ML 配合物。从图 18-2 可以看出，在极大值 C 左边的所有溶液中，对于形成 ML 配合物来说，M 离子是过量的，配合物的浓度由 L 决定。而这些溶液中 T_L 都小于 0.5，所以它们形成的配合物 ML 的浓度也都小于与极大值 C 相对应的溶液中的 ML 浓度，因而其吸光度也小于 A_2。处于极大值 C 右边的所有溶液中，L 是过量的，配合物的浓度由 M 决定。而这些溶液的 T_M 也都小于 0.5，因而形成的 ML 的浓度也都小于与极大值 C 相对应的溶液中的 ML 浓度。所以，只有在 $T_L = T_M = 0.5$ 的溶液中，也就是其组成（M∶L）与配合物组成相一致的溶液中，配合物浓度最大，因而吸光度也最大。

（3）配合物稳定常数的测定方法

根据（2）式，配合物第 n 级累积稳定常数表达式为

$$\beta_n = [ML_n]/[M][L]^n \qquad (3)$$

配合物的组成 n 已确定，只要求出各物种的平衡浓度 $[ML_n]$、$[M]$ 和 $[L]$，代入（3）式，便可求 β_n 值。

① $[ML_n]$ 实验值 $[ML_n]_{实}$ 的确定：

如上所述，$[ML_n]_{理}$ 对应于 A_1（B 点），$[ML_n]_{实}$ 对应于 A_2（C 点）故有

$$\frac{A_2}{A_1} = \frac{\varepsilon [ML_n]_{实} \times l}{\varepsilon [ML_n]_{理} \times l} = \frac{[ML_n]_{实}}{[ML_n]_{理}} \qquad (4)$$

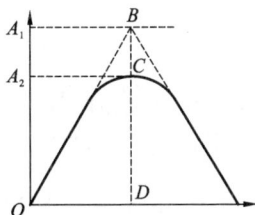

其中$[ML_n]_理$就是D点所对应的T_M值，又已知D点处，$T_L/T_M = n$，故$T_M = T_L/n$，故有

$$\frac{A_2}{A_1} = \frac{[ML_n]_实}{T_M} = [ML_n]_实 \times \frac{n}{T_L}$$

所以
$$[ML_n]_实 = \frac{A_2 \times T_L}{A_1 \times n} \tag{5}$$

② $[M]$实验值$[M]_实$的确定：

由于在实验条件下，只生成一种配合物，所以

$$[M]_实 = T_M - [ML_n]_实 = T_M - \frac{A_2 \times T_L}{A_1 \times n} \tag{6}$$

③ $[L]$实验值$[L]_实$的确定：

$$[L]_实 = T_L - n[ML_n]_实 = T_L - \frac{A_2 \times T_L}{A_1} \tag{7}$$

将式(5)、(6)、(7)式代入式(3)式有

$$\beta_n = [ML_n]/[M][L]^n$$
$$= \frac{A_2 \times T_L}{A_1 \times n} \Big/ \Big\{ (T_M - \frac{A_2 \times T_L}{A_1 \times n}) \times (T_L - \frac{A_2 \times T_L}{A_1})^n \Big\} \tag{8}$$

式中：A_1为D点处配合物的理论吸光度，A_2为D点处配合物的实际吸光度，T_M为D点处金属离子的浓度，T_L为D点处配体的浓度，n为D点处配体浓度与金属离子浓度之比。所以，根据式(8)即可求算配合物的积累稳定常数β_n，此处的β_n实际上就是配合物$[ML]$的条件稳定常数$K_{条件}$。

四、仪器和试剂

仪器：分光光度计、pH 计，容量瓶(50 mL)、烧杯(50 mL)、搅拌器。

试剂：$0.0100\ mol \cdot L^{-1}$硝酸铜溶液：在分析天平上称取 1.208 g 三水硝酸铜用水溶解并转移至 500 mL 容量瓶中，用水定容。

$0.22\ mol \cdot L^{-1}$硝酸钾溶液：22.2 g 硝酸钾溶于 1 L 水中。

$0.0100\ mol \cdot L^{-1}$磺基水杨酸 – 氢氧化钠混合溶液：在分析天平上称取 1.271 g 二水合 5 – 磺基水杨溶于适量水中，定量加入已知准确浓度的氢氧化钠溶液，使得氢氧化钠的浓度为 $0.0200\ mol \cdot L^{-1}$，转移至 500 mL 容量瓶中，用水定容。

$1.0mol \cdot L^{-1}$六亚甲基四胺 – 硝酸缓冲溶液(pH = 5.6)：称取 70.1 g 六亚甲基四胺溶于适量水中，加入 50 mL $2.0\ mol \cdot L^{-1}$硝酸，加水稀释至 500 mL。

所用试剂均为分析纯，实验用水为去离子水。

五、实验内容

（1）按等摩尔系列法，用 $0.010\ mol \cdot L^{-1}$ 硝酸铜溶液和 $0.010\ mol \cdot L^{-1}$ 5 – 磺基水杨酸 – 氢氧化钠混合溶液，以及六亚甲基四胺 – 硝酸缓冲溶液，在 13 个 50 mL 容量瓶（编号依次为 1～13）依表 3 – 2 所列体积比配制混合溶液，并用硝酸钾溶液定容。

（2）在分光光度计上选 440 nm 波长，用 2 cm 比色皿，用 13 号溶液作参比，分别测定每个混合溶液的吸光度 A，并记录数据。

六、实验记录与结果

以吸光度 A 为纵坐标，中心离子的摩尔分数 $c(Cu^{2+})/c$ 为横坐标，作 $A - c$ $(Cu^{2+})/c$ 图，依据该图所得数据以及式（8）计算配合物 $[ML]$ 的条件稳定常数 $K_{条件}$。根据文献[1]给出的 5 – 磺基水杨酸的质子化常数 $lgK_1^H = 11.6$，$lgK_2^H = 2.6$，可计算出在 pH = 5.60 时配体（5 – 磺基水杨酸）的酸效应系数[2] $\alpha_{L(H)} = 1.0 \times 10^6$，而 $lgK(CuL) = lgK_{条件} + lg\alpha_{L(H)}$，由此推算铜（Ⅱ）– 磺基水杨酸配合物的 $lgK(CuL)$，与文献[3]值 $lgK(CuL) = 9.52$ 对比，评估本实验结果的准确度。

表 3 – 2　配合物的组成与稳定常数测定数据记录　温度_____

溶液编号	1	2	3	4	5	6	7	8	9	10	11	12	13
磺基水杨酸 – 氢氧化钠混合溶液体积 V_L /mL	0.0	2.0	4.0	6.0	8.0	10.0	12.0	14.0	16.0	18.0	20.0	22.0	24.0
硝酸铜溶液体积 V_M /mL	24.0	22.0	20.0	18.0	16.0	14.0	12.0	10.0	8.0	6.0	4.0	2.0	0.0
缓冲溶液体积 V_B /mL	10.0	10.0	10.0	10.0	10.0	10.0	10.0	10.0	10.0	10.0	10.0	10.0	10.0
溶液吸光度 A 值													

参考文献

[1] 武汉大学. 分析化学(第 6 版). 北京：高等教育出版社，2016：219.

[2] 杨春，梁萍，张颖，刘晓莉. 无机化学实验. 天津：南开大学出版社，2007：133.

[3] Speight J. G. Lange's handbook of chemistry (Sixteenth Edition); McGraw – Hill: New York, America, 2005: 1.377.

（编者：刘绍乾）

实验 24　水的纯度及总硬度测定
Determination of Purity and Hardness of Water

一、目的要求

(1)掌握电导率仪测定水纯度的方法。

(2)熟悉应用螯合滴定测定水的总硬度的原理和方法。

(3)进一步了解 EDTA 在螯合滴定法中的应用。

二、预习要点

(1)熟悉电导率仪的工作原理及使用方法。

(2)掌握螯合滴定的特点以及作为金属指示剂所必需的条件。

(3)了解滴定中所加条件试剂的作用。

加酸除去 CO_2，可避免钙、镁的碳酸氢盐在加入碱性缓冲溶液时产生碳酸钙或镁的不溶物而使测定结果偏低。

三乙醇胺能掩蔽水样中可能存在的 Al^{3+}、Fe^{3+} 等离子，使滴定终点更易判断。

在 pH = 10.0 的氨性缓冲溶液中进行测定，使在滴定终点时游离出来的指示剂呈蓝色，便于滴定终点的观察。

三、实验基本原理

1. 电导率仪测定水的纯度

天然水中溶有无机盐离子，所以实际上水是一种极稀的电解质溶液。尽管浓度极小，但亦有导电现象，离子浓度越大，导电能力越强。

导体导电能力的大小，一般用电阻 R 或电导 G 表示，二者互为倒数，即

$$G = 1/R \tag{1}$$

电阻的 SI 单位为欧姆，符号为 Ω；电导的 SI 单位为西门子，符号为 S。

温度一定时，两极间溶液的电阻与两极间的距离 l 成正比，与电极面积 A 成反比，即

$$R \propto l/A$$

或

$$R \propto \rho l/A \tag{2}$$

比例常数 ρ 称为电阻率，SI 单位为欧·米，符号为 $\Omega \cdot m$。电阻率的倒数，称为电导率 κ

$$1/\rho = \kappa \tag{3}$$

电导率的 SI 单位为西门子·米$^{-1}$，符号为 S·m^{-1}。

将(2)式、(3)式代入 (1)式得

$$G = \kappa \frac{A}{l}$$

$$\kappa = G \frac{l}{A}$$

由上式可知，当 $l/A = 1$ 时，$\kappa = G$，所以 κ 在数值上等于相距为 1 单位长度和大小为 1 单位面积的两个电极间的溶液的电导。

水中无机盐离子浓度越高，导电性越好，电导率越大，则水的纯度越低；反之，电导率越小，则水的纯度越高。

2. 螯合滴定法测定水的总硬度

水的硬度主要是由于水中存在钙盐和镁盐所引起的。当然，其他金属盐亦会使水的硬度增大，但这些金属盐含量不高。因此水的硬度一般只考虑钙盐和镁盐所造成的硬度。水的硬度的测定可分为水的总硬度的测定和钙镁硬度的测定两种。总硬度的测定是滴定 Ca^{2+}，Mg^{2+} 总量，并以 Ca 进行计算。后一种是分别测定 Ca^{2+} 和 Mg^{2+} 的含量，若水中的钙盐和镁盐是可溶性的酸式碳酸盐，加热就可分解成碳酸盐沉淀，经过过滤便可将钙、镁离子从水中除去。含有这些钙、镁的酸式碳酸盐所形成的水的硬度称为暂时硬度。

$$Ca(HCO_3)_2 \xrightarrow{\triangle} CaCO_3 \downarrow + H_2O + CO_2 \uparrow$$

$$Mg(HCO_3)_2 \xrightarrow{\triangle} MgCO_3 \downarrow + H_2O + CO_2 \uparrow$$

若水中含有钙和镁的硫酸盐或氯化物，虽然经过加热也不能使钙盐和镁盐产生沉淀而从水中除去，水中含有这些由硫酸盐和氯化物所形成的硬度称为永久硬度。

水的暂时硬度与永久硬度的总和称为水的总硬度。

水的总硬度的测定方法很多，最常用的是 EDTA 螯合滴定法。该方法是以铬黑 T 作指示剂，用 EDTA 标准溶液进行滴定。测定水中 Ca^{2+}，Mg^{2+} 离子的总含量，并以 Ca 进行计算，即可算出总硬度。

测定在 pH = 10.0 的氨性缓冲溶液中进行。另外，考虑到 EDTA 能与很多金属离子生成螯合物，为防止干扰，应加入掩蔽剂。

铬黑 T(HIn^{2-}) 可与水中的 Ca^{2+}、Mg^{2+} 生成红色配合物，但该配合物不及 EDTA 与 Ca^{2+}、Mg^{2+} 形成的螯合物稳定，滴定中 EDTA 从指示剂配合物中夺取 Ca^{2+}、Mg^{2+} 形成更稳定的螯合物，使指示剂 HIn^{2-} 游离出来，溶液由红色恰好变为蓝色，即达终点。

上述反应过程可表示为

$$Ca^{2+}(Mg^{2+}) + HIn^{2-} \longrightarrow CaIn^-(MgIn^-) + H^+$$
$$\quad\quad\quad\quad\quad (蓝色) \quad\quad\quad\quad\quad (红色)$$

$$Ca^{2+}(Mg^{2+}) + H_2Y^{2-} \longrightarrow CaY^{2-}(MgY^{2-}) + 2H^+$$
$$\quad\quad\quad\quad\quad (无色) \quad\quad\quad\quad\quad (无色)$$

$$CaIn^-(MgIn^-) + H_2Y^{2-} \longrightarrow CaY^{2-}(MgY^{2-}) + HIn^{2-} + H^+$$
$$\quad (红色) \quad\quad\quad\quad (无色) \quad\quad\quad\quad (无色) \quad\quad\quad (蓝色)$$

水样总硬度可用下列方法表示:

(1)以每升水样中含有 CaO 的毫克数表示:

$$水的总硬度(mg \cdot L^{-1}) = \frac{c_{(EDTA)} \cdot V_{(EDTA)} \times M_{(CaO)}}{V(水样)} \times 1000$$

(2)以每升水样中含有 CaO 的数量表示,国际上规定每升水中含有 CaO 10 mg 称为 1 度。

$$水样总硬度(度数) = \frac{c_{(EDTA)} \cdot V_{(EDTA)} \times M_{(CaO)}}{10 \times V(水样)} \times 1000$$

(3)以每立方米水样中含有 $CaCO_3$ 的毫克数表示(ppm):

$$水的总硬度(ppm) = \frac{c_{(EDTA)} \cdot V_{(EDTA)} \times M_{(CaCO_3)}}{V(水样)} \times 10^6$$

各国对水的硬度表示方法不同。德国硬度是水质硬度表示比较早的一种方法,它以度(°)为计,它表示十万份水中含有一份 CaO,即一升水中含有 10 毫克 CaO 为 1°,这称为一个硬度单位,水质分类是 $0 \sim 4°$ 为很软的水,$4 \sim 8°$ 为软水,$8 \sim 16°$ 为中等硬水,$16 \sim 30°$ 为硬水,$30°$ 以上为很硬的水。

四、仪器和试剂

仪器: 电导率仪,酸式滴定管(50 mL),移液管(50 mL),锥形瓶(250 mL),刻度吸管(1.0 mL,5.0 mL,10.0 mL),滴定管架,电炉,烧杯。

试剂: HCl(6 mol·L⁻¹),三乙醇胺(1.5 mol·L⁻¹),Na_2S(0.25 mol·L⁻¹),pH ≌ 10 的 $NH_3 - NH_4Cl$ 缓冲溶液,铬黑 T 指示剂,EDTA 标准溶液。

五、实验内容

用容量吸管吸取水样(自来水)100.0 mL,加入 $1 \sim 2$ 滴 6 mol·L⁻¹ HCl 溶液使之酸化。然后将锥形瓶放在石棉网上微热至沸,除去水中 CO_2,当水样冷却后,用量杯加入 1.5 mol·L⁻¹ 三乙醇胺溶液 5.0 mL;$NH_3 - NH_4Cl$ 缓冲溶液10.0 mL;0.25 mol·L⁻¹ Na_2S 溶液 1.0 mL;再滴入 $2 \sim 3$ 滴铬黑 T 指示剂,然后用0.005 mol·L⁻¹ EDTA 标准溶液滴定至溶液由紫红色恰好变为蓝色,且经振摇在半分钟内不再消失即为终点。按上述方法重复操作两次。

六、实验记录与结果

测定水样总硬度

数据记录与计算		测定序号	1	2	3
水样量/mL			100.0	100.0	100.0
EDTA	初读数/mL				
	终读数/mL				
	净用量/mL				
Ca^{2+}、Mg^{2+}离子的硬度*					
平均值					
相对平均偏差(d_r)					

＊：以每升水中含 10 mg CaO 为 1°。

（编者：李战辉）

实验 25 邻二氮菲分光光度法测定铁
Determination of Iron(II) with Spectrophotometry
Using Ortho-phenanthroline

一、实验目的

(1)理解分光光度法的原理。

(2)掌握邻二氮菲分光光度法测定铁的原理和方法。

(3)了解 722 型分光光度计的组成,掌握其工作原理、操作方法及使用时应注意的事项。

(4)学会绘制标准工作曲线。

二、预习要点

(1)为什么要控制被测溶液的吸光度最好在 0.2 ~ 0.7 范围内?怎样控制?

(2)实验中各种试剂分别起什么作用?

(3)用邻二氮菲法测定铁时,为什么在测定前需要加入盐酸羟胺?

(4)如果用配制已久的盐酸羟胺溶液,对分析结果会带来什么影响?

(5)实验中哪些试剂需准确配制和准确加入?哪些试剂不需准确配制,但要准确加入?

三、实验基本原理

单一波长的光为单色光,由不同波长组成的光称为复色光,白光就是一种复色光。若两种颜色的光按适当的强度比例混合可组成白光,则这两种光称为互补色光。物质对光的吸收具有选择性,若溶液选择性地吸收了某种颜色的光,则溶液呈吸收光的互补光。将复色光色散成单色光,并分取其中某一波长的光就称为分光,光度即光的强度。分光光度法:将复色光色散成单色光,并分取其中某一波长的光,让其通过待测定溶液,经溶液吸收一部分后,测定透过光的强度,从而确定待测溶液浓度的一种分析方法。

Lambert-Bear 定律:当一适当波长的单色光通过溶液时,若液层厚度一定,则吸光度与溶液浓度成正比,如图 3 - 10 所示。

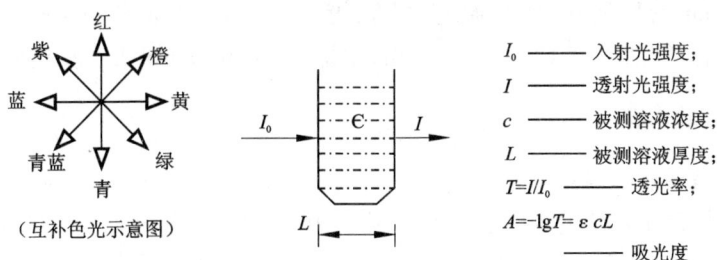

图3-10　吸光光度法示意图

邻二氮菲是目前应用于测定微量铁的较好试剂。此法准确度高，重现性好。在 pH = 2~9 范围内，Fe^{2+} 与邻二氮菲生成极稳定的橘红色配合物，溶液经长时间放置，色泽强度也不发生显著变化，可以很好地服从 Lambert-Bear 定律。方法选择性很高。为了尽量减少其他离子的影响，通常在微酸性(pH = 5)溶液中显色，反应式如下：

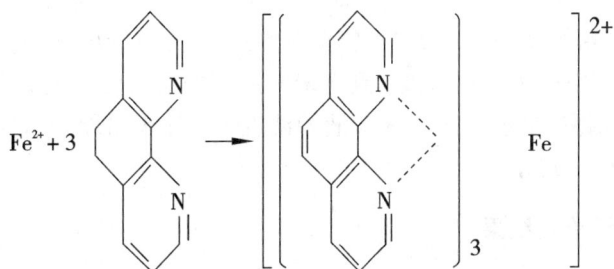

该配合物 $lgK_{稳} = 21.3(20℃)$，最大吸收波长(λ_{max})为 508 nm。Fe^{3+} 离子与邻二氮菲也生成 1:3 的淡蓝色配合物，其 $lgK_{稳} = 14.1$，为保证铁以亚铁状态存在，在显色前要加入还原剂，如盐酸羟胺，将 Fe^{3+} 全部还原为 Fe^{2+}，反应如下：

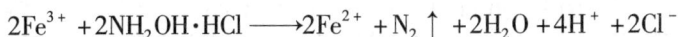

$$2Fe^{3+} + 2NH_2OH \cdot HCl \longrightarrow 2Fe^{2+} + N_2 \uparrow + 2H_2O + 4H^+ + 2Cl^-$$

Cu^{2+}、Co^{2+}、Ni^{2+}、Cd^{2+}、Hg^{2+}、Mn^{2+}、Zn^{2+} 等离子也能与邻二氮菲生成稳定配合物，量少时，不影响 Fe^{2+} 的测定，量大时可用 EDTA 掩蔽或预先分离。

四、仪器和试剂

仪器：722 型分光光度计，容量瓶(25 mL)，刻度吸管(0.5 mL,1 mL,5 mL)，洗耳球。

试剂：标准铁溶液[2.000 mmol·L^{-1}，配制方法：准确称取 0.7842 g(NH_4)$_2$Fe$(SO_4)_2$·6H_2O 置于 250 mL 烧杯中，加入 6 mol·L^{-1} HCl 120 mL 和少量水，溶解后，

转入1000 mL 容量瓶中，加水稀释至刻度，摇匀。]，邻二氮菲溶液（8 mmol·L^{-1}，新鲜配制），盐酸羟胺溶液（1.5 mol·L^{-1}临时配制），NaAc（1 mol·L^{-1}），待测铁溶液。

五、实验内容

1. 标准曲线的制作

在6个25 mL 容量瓶（或比色管）中，用刻度吸管分别加入0.00 mL，0.20 mL，0.40 mL，0.60 mL，0.80 mL，1.00 mL 标准铁溶液（含铁2.000 mmol·L^{-1}），再分别加入1.5 mol·L^{-1}盐酸羟胺溶液0.5 mL，8 mmol·L^{-1}邻二氮菲溶液1.0 mL 和1 mol·L^{-1} NaAc 溶液2.5 mL，加水稀释至刻度，摇匀，在508 nm 波长处，用1 cm 比色皿，以试剂空白（即0.00 mL 铁标准溶液）为参比溶液，依次测定各瓶溶液的吸光度。以吸光度（A）为纵坐标，以 Fe^{2+}离子的浓度（mmol·L^{-1}）为横坐标，绘制标准曲线。

2. 未知液测定

用刻度吸管吸取两份待测铁溶液5.00 mL，分别注入25 mL 容量瓶中，加入1.5 mmol·L^{-1}盐酸羟胺溶液0.5 mL，8 mmol·L^{-1}邻二氮菲溶液1.0 mL，1 mol·L^{-1} NaAc 溶液2.5 mL，加水稀释至刻度，摇匀，在508 nm 波长处，用1 cm 比色皿，以试剂空白为参比溶液，测定未知溶液的吸光度。根据测得的吸光度（A）的大小，在标准曲线上查出未知液的浓度，并计算出原来试样中 Fe^{2+}离子的浓度（mmol·L^{-1}）。

六、实验记录与结果

1. 溶液的配制和吸光度的测量

容量瓶编号	0	1	2	3	4	5	6	7	量　器
铁标准液/mL	0.00	0.20	0.40	0.60	0.80	1.00			1 mL 吸量管
待测铁液/mL							5.00	5.00	5 mL 吸量管
盐酸羟胺/mL				0.5					0.5 mL 吸量管
邻二氮菲/mL				1.0					1 mL 吸量管
NaAc 溶液/mL				2.5					5 mL 吸量管
吸光度 A	参比								

2. 数据的处理

以吸光度(A)为纵坐标,以 Fe^{2+} 离子的浓度($mmol \cdot L^{-1}$)为横坐标,绘制标准曲线,同时测定未知溶液的吸光度,根据测得的吸光度(A)的大小,在标准曲线上查出未知液的浓度,并计算出原来试样中 Fe^{2+} 离子的浓度($mmol \cdot L^{-1}$)。

注意:

①仪器不使用时,应打开试样室盖,以保护光电管。

②在满足分析要求时,灵敏度应尽量选用低挡。

③配制溶液的全部量器专物专用,不能混用、乱用,用毕后立即用蒸馏水淋洗干净。

④要使比色皿中测定溶液与原溶液的浓度保持一致,须用待测溶液淌洗比色皿 2~3 次。将待测溶液灌入比色皿中时不要超过总高度的 4/5,以防溶液溢出,损害仪器。

⑤比色皿盛溶液后,其外壁应用擦镜纸擦净,比色皿的透光面不能用手接触,必须保持十分洁净。不能与其他仪器上的比色皿单个调换。用毕后,比色皿应及时取出、洗净,倒立晾干。

⑥在测定标准系列各溶液吸光度时,最好从稀溶液至浓溶液进行。

⑦作图时,坐标比例应恰当,曲线光滑。

（编者：何跃武）

实验 26　蛋白质含量的分光光度法测定
Determination of Protein by Spectrophotometry

一、实验目的

(1)理解分光光度法的原理。

(2)掌握分光光度法测定蛋白质含量的原理和方法。

(3)进一步熟悉 722 型分光光度计的组成,工作原理、操作方法及使用注意事项。

二、预习要点

(1)实验中各种试剂分别起什么作用?

(2)如何控制测量时溶液的 pH = 2.2 ~ 2.8 范围?

(3)如何选择参比溶液(即空白溶液)?

(4)实验中哪些试剂需准确配制和准确加入?

三、实验基本原理

偶氮胂 M 是一种良好的可见光分光光度法分析用的显色剂,在 pH = 2.2 ~ 2.8 范围内,蛋白质与偶氮胂 M 可生成稳定的蓝色复合物,该复合物的最大吸收波长为 605 nm。通过测定该复合物的吸光度即可求算出蛋白质的含量。在该实验条件下,生物体内的金属离子和阴离子如 K^+、Na^+、Ca^{2+}、Cu^{2+}、Zn^{2+}、Cl^- 以及维生素、肌苷、尿素、葡萄糖等对蛋白质的测定均无影响,因此,可以将样品粉碎、提纯、过滤后直接进行分光光度测定。

四、仪器和试剂

仪器:722 型分光光度计,比色皿(2.0 cm),电子天平,酸度计,高速匀浆器,高速离心机,比色管(210 mL),烧杯(250 mL、500 mL),移液管(2 mL、1 mL),吸量管(1 mL),容量瓶(100 mL),量筒(100 mL),纱布。

试剂:蛋白质标准溶液(约 1 g·L^{-1}),偶氮胂 M 溶液(5.0×10^{-4} mol·L^{-1}),KH_2PO_4 溶液(5.0×10^{-3} mol·L^{-1}),乳化剂 OP 水溶液(0.05%),NaCl 溶液(1.0%),乳酸 – 乳酸钠缓冲溶液(pH = 2.5),含蛋白质的样品(如干花生)。

五、实验内容

1. 含蛋白质样品的预处理

称取 25 g 干花生，用含 KH_2PO_4（5.0×10^{-3} mol·L^{-1}）和 NaCl（1.0%）的溶液（pH = 7.2）在室温下浸泡 4 ~ 8 小时（溶液加至刚好淹没全部花生后再过量 20 mL 左右）。然后用匀浆器匀浆，浆液于 4℃ 下静置过夜。用三层纱布过滤，并用 30 mL pH = 7.2 的缓冲溶液分多次洗涤滤渣，以水稀释滤液至 100 mL。取适量滤液在 12000 r·min^{-1} 转速下离心分离约 20 分钟，在 4℃ 下保存离心后所得的上层清液。

2. 样品中蛋白质含量的测定

取 7 支 10 mL 比色管，将其依次编号后，按照下表中的数据加入各种溶液，最后将 7 支比色管均用去离子水稀释至刻度，摇匀后放置 15 分钟。以 1 号为参比溶液在 605 nm 处测得各溶液的吸光度。以 5、6、7 号溶液的吸光度对标准溶液的浓度作图，得一直线，延长此直线与横坐标相交，交点的绝对值即为样品清液中蛋白质的含量。

六、实验记录与结果

实验编号	1	2	3	4	5	6	7
乳酸－乳酸钠缓冲溶液/mL	2.00	2.00	2.00	2.00	2.00	2.00	2.00
乳化剂 OP/mL	1.00	1.00	1.00	1.00	1.00	1.00	1.00
偶氮胂 M/mL	0.80	0.80	0.80	0.80	0.80	0.80	0.80
样品清液/mL		0.50	0.50	0.50	0.50	0.50	0.50
蛋白质标准溶液/mL					0.20	0.40	0.60
吸光度 A							

（本实验参考《医学基础化学实验》（双语版）冯清主编一书中实验二十"蛋白质的分光光度法测定"）

（编者：刘绍乾）

实验 27　气体密度法测定二氧化碳的分子量
Determination of Molar Mass of Carbon Dioxide
Using Gas Density Method

一、实验目的

(1)了解用相对密度法测定气体分子量的原理和方法。

(2)掌握启普发生器、洗气瓶和干燥塔的装配和使用方法。

(3)进一步熟悉天平的使用方法。

二、预习要点

(1)预习用相对密度法测定 CO_2 分子量的原理和方法。

(2)思考并回答下列问题:

①用启普发生器制取 CO_2 时为什么要让 CO_2 通过水、浓硫酸和无水氯化钙?

②怎样证明锥形瓶中已充满二氧化碳?

③为什么(CO_2 + 瓶 + 塞子)的质量要在分析天平上称量,而(水 + 瓶 + 塞子)的质量可以在托盘天平上称量?

④为什么在计算锥形瓶的容积时不考虑空气的质量,而在计算 CO_2 的质量时要考虑空气的质量?(考虑有效数字的问题)

⑤使用启普发生器,有哪些应注意的地方?

三、实验基本原理

根据阿佛加德罗定律,在同温同压下,同体积的任何气体含有相同数目的分子。因此,在同温、同压下,任何相同体积的两种气体(A 和 B)的质量之比,等于它们的分子量之比,即

$$\frac{W_A}{W_B} = \frac{M_A}{M_B} \tag{1}$$

W_A/W_B 的比值为 A 气体对 B 气体的相对密度(常用 D_B 表示)。

为测定上的方便,设定其中一气体为标准气体。通常用氢气(最轻的气体)或空气(最常见的气体)为标准,根据(1)式来求另一气体的分子量。

本实验以空气(空气的平均分子量为 29.0)为标准气体,来求二氧化碳的分子量,即

$$M_{CO_2} = \frac{W_{CO_2}}{W_{空气}} \cdot M_{空气} = \frac{W_{CO_2}}{W_{空气}} \cdot 29.0 \tag{2}$$

W_{CO_2}、$W_{空气}$分别为同体积的二氧化碳和空气的质量。M_{CO_2}为二氧化碳分子量，若将一个玻璃容器例如锥形瓶（当然其中充满了空气）先进行称量（G_1），然后将其充满二氧化碳并在同一温度和大气压下称量（G_2），两者的质量之差（$G_2 - G_1$）也就是同体积二氧化碳与空气的质量之差，即 $W_{CO_2} - W_{空气} = G_2 - G_1$，所以

$$W_{CO_2} = (G_2 - G_1) + W_{空气} \tag{3}$$

至于空气的质量 $W_{空气}$，可以利用理想气体状态方程式，由瓶的容积 V，实验时的温度 t 和大气压 p 计算出来，即

$$W_{空气} = \frac{29.0pV}{R(273.16 + t)} \tag{4}$$

为了求出瓶的容积 V，可将瓶内充满水并称重（G_3）。充满水的瓶（带塞）的质量和充满空气的瓶（带塞）的质量之差便是水与空气的质量之差，即 $W_{水} - W_{空气}$，实际上也就是水的质量（略去 $W_{空气}$）。$W_{水}$ 除以水的密度（室温下近似为 1.00 g/mL）便得瓶的容积 V。

这样，将 V、p、t 值代入式（4）求出 $W_{空气}$，进一步由式（3）求出 W_{CO_2}，最后将 W_{CO_2} 和 $W_{空气}$ 代入式（2）便计算出二氧化碳的分子量。

为了测定 CO_2 分子量，首先要解决 CO_2 气体的制备与净化的问题，其次要解决 CO_2 气体的收集和称量的问题，其中 CO_2 气体的体积与质量的准确测量是实验的关键，上述方法很巧妙地解决了普通实验条件下测量少量气体体积和质量的难题，在其他实验中值得借鉴。

四、仪器和试剂

仪器：制备 CO_2 的仪器（启普发生器），洗水瓶（2 个），干燥塔（2 个），玻璃纤维，橡皮管，分析天平，托盘天平，锥形瓶（250 mL），橡皮塞，温度计（0 ~ 100℃），气压计。

试剂：H_2SO_4（浓），HCl（6 mol·L^{-1}），$CaCO_3$（大理石或石灰石），无水氯化钙。

五、实验内容

1. 二氧化碳气体的制备

制备 CO_2 的实验装置如图 3 – 11 所示（由实验员负责装配好）。在启普发生器的球形容器中盛有 $CaCO_3$（石灰石或大理石），安全漏斗中注入适量的

6 mol·L⁻¹ HCl。打开活塞，则盐酸从底部上升与石灰石接触，发生反应，产生 CO_2 气体，气体经过装水的洗气瓶除去氯化氢和其他可溶性杂质，经过装有浓 H_2SO_4 的洗气瓶，装有无水氯化钙和玻璃纤维的干燥塔，进一步除去气体中的水分和其他杂质，最后经导气管导出。先排气 4～5 min，使整套装置中的空气被 CO_2 驱出后，关闭活塞待用。

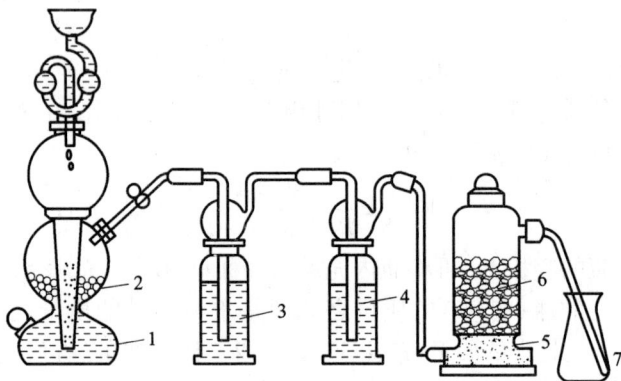

图 3－11　制备和收集二氧化碳的装置图

1—盐酸；2—石灰石；3—水；4—浓硫酸；5—玻璃纤维；6—无水氯化钙；7—CO₂ 接收瓶

2. 二氧化碳分子量的测定

取一只洁净而干燥的 250 mL 锥形瓶，配上合适的橡皮塞，塞紧后在塞子上作一记号，以固定塞子塞入瓶口的位置。然后在天平上称量(空气＋瓶＋塞子)的质量(G_1)，记下数据。

打开橡皮塞，将导气管插入锥形瓶的底部(为什么?)再打开启普发生器的活塞，通入 CO_2，约 4～5 min(CO_2 的流速以每秒在浓 H_2SO_4 中鼓泡 4～5 个为宜)，慢慢地抽出导气管，迅速用塞子塞住瓶口(至记号处)。在天平上称量 CO_2＋瓶＋塞子)的质量(G_2)。

重复通入 CO_2 气体和称量的操作，直到两次质量相差不超过 1～2 mg(表明瓶内空气已完全被 CO_2 所代替)，记下质量。

往瓶中装水至橡皮塞刚好塞到原来的位置上，在托盘天平上称量(水＋瓶＋塞子)的质量(G_3)，称准至 0.1 g。(G_3－G_1)即得水的质量(空气的质量在此可以忽略不计)。

记下实验时的温度 t 和大气压力 p。

六、实验记录与结果

实验时的室温 $t/℃$ _____

实验时的大气压力 p/kPa _____

（空气 + 瓶 + 塞子）的质量 G_1/g _____

（CO_2 + 瓶 + 塞子）的质量 G_2/g _____

（水 + 瓶 + 塞子）的质量 G_3/g _____

瓶的容积 $V/mL = G_3 - G_1/1.00$ _____

空气的质量 $W_{空气}$ _____

CO_2 的质量 $W_{CO_2} = (G_2 - G_1) + W_{空气}$ _____

CO_2 分子量 $W_{CO_2} = \dfrac{W_{CO_2}}{W 空气} \cdot 29.9$ _____

百分误差 _____

分析产生误差的主要原因：

（本实验参考了关鲁雄主编的《化学基本操作与物质制备实验》一书。）

（编者：刘绍乾）

实验 28　硝酸钾溶解度与温度的关系
Relation between Solubility of Potassium Nitrate and Temperature

一、实验目的

(1)掌握测定盐类在水中溶解度的方法,了解溶解度与温度的关系。

(2)练习绘制溶解度 – 温度曲线。

二、预习要点

(1)预习本书中关于吸量管的使用、作图等内容并复习盐类溶解度的基础知识。

(2)思考并回答下列问题:

①在实验过程中,搅拌与不搅拌对实验结果有何影响?

②如果实验过程中试管内的水显著蒸发,对实验结果有何影响?

③为什么硝酸钾的称量要准确至 1 mg,水的量取要准确至 0.01 mL?

三、仪器和试剂

仪器:分析天平,温度计(0 ~ 100℃),吸量管(1 mL),洗耳球,小试管(15 mm × 80 mm 四支),玻璃棒(100 mm 四支),带有橡皮塞的玻璃棒,橡皮圈。

试剂:KNO_3(固体,分析纯)。

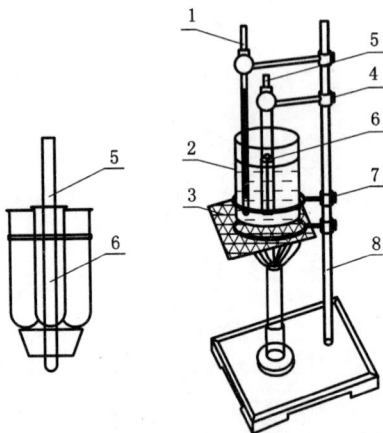

图 3 – 12　实验装置图

1—温度计;2—烧杯(250 mL);3—石棉网;4—铁夹;5—玻璃棒;6—小试管;7—铁圈;8—铁架

四、实验内容

1. 固体硝酸钾的称量

用分析天平称取四份固体硝酸钾,其质量分别为:1.7 ~ 1.8 g, 1.4 ~ 1.5 g, 1.1 ~ 1.2 g, 0.8 ~ 0.9 g(准确至 1 mg)。将硝酸钾分别小心地倒入四支干燥洁净的小试管中,试管编号顺次为 1、2、3、4 号。

2. 仪器的安装

如图 3 – 12 所示，将四支小试管用橡皮圈固定在套有橡皮塞的玻璃棒上（为了增大摩擦力，玻璃棒可套上一段橡皮管）。通过铁夹、铁架将支架垂直悬挂在 250 mL 烧杯中，通过另一个铁夹使温度计的下端与试管底部处于同一水平位置，并紧贴着试管。

3. 蒸馏水的量取

用 1 mL 吸量管分别往每支小试管中注入 1.00 mL 蒸馏水。每支小试管内插入一支小玻璃棒，小心搅拌管内的水，使沾在试管壁上的硝酸钾晶体全部落入水中。

4. 升温溶解

往 250 mL 烧杯中注入热水，注意热水不得溅入小试管内，热水液面应当高于试管内的液面而低于固定试管的橡皮圈。加热水浴，不停地小心搅拌试管内的固体，直至固体全部溶解为止（水浴温度不应高于 90℃，以免溶液过分蒸发）。

5. 冷却并记录温度

停止加热，让水浴自然冷却。首先不断搅拌一号试管内的溶液，注意观察溶液的变化，当刚有晶体出现并不再消失时即记下当时的温度值。然后用相同的方法记下 2、3、4 号试管中晶体开始析出的温度。

如果测定不准确，可以将水浴重新加热升温，使晶体重新溶解，再重复上述操作。

五、实验记录与结果

试管编号	1	2	3	4
KNO_3 晶体的质量/g				
水的质量/g	1.00	1.00	1.00	1.00
溶液中开始析出晶体时的温度/℃				
KNO_3 在各温度时的溶解度(g/100 g 水)				

根据上表数据，以温度为横坐标，以溶解度（100 g 水中溶解 KNO_3 的克数）为纵坐标，绘出硝酸钾的溶解度曲线。

在同一个坐标系中，根据对应温度范围内硝酸钾溶解度文献值绘制另一条溶解度曲线，与上述曲线比较。

（本实验参考了关鲁雄主编的《化学基本操作与物质制备实验》。）

（编者：刘绍乾）

实验 29　荧光分析法测定血清中的镁
Determination of Magnesium in Serum by Fluorescence Analysis

一、实验目的

(1)了解荧光分析法的基本原理。
(2)掌握荧光分析法的基本实验技术。
(3)学习荧光分析法测定血清中镁的原理和实验方法。
(4)学会绘制标准工作曲线。

二、预习要点

(1)学习荧光分析法的基本原理。
(2)8 – 羟基喹啉与镁形成的配合物的激发光谱和发射光谱有何区别和联系?
(3)从分子结构角度分析 8 – 羟基喹啉及其金属配合物的荧光性质。

三、实验基本原理

镁与 8 – 羟基喹啉在 pH = 6.5 的醋酸盐缓冲溶液中可反应生成具有强荧光性质的配合物,而相同条件下,8 – 羟基喹啉本身的荧光强度很弱,实验中可将其作为空白而扣除。在一般测定条件下,血液中的其他物质对此测定无干扰。

镁和 8 – 羟基喹啉的反应如下:

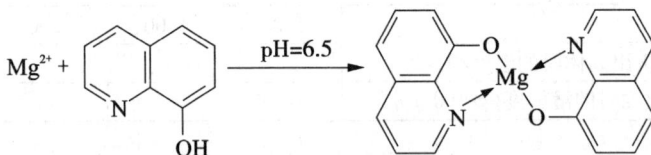

镁与 8 – 羟基喹啉形成的配合物的荧光光谱见图 3 – 13。

四、仪器和试剂

仪器:LS – 55 型荧光光度计,离心机,离心管,比色管,吸量管(1 mL、5 mL)。
试剂:8 – 羟基喹啉乙醇溶液,20 μg·mL^{-1}镁标准溶液,未知血清样品。

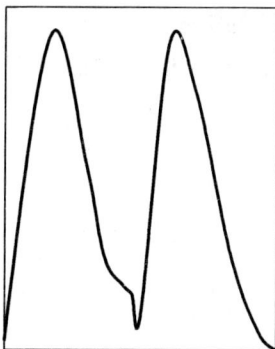

图 3 - 13　8 - 羟基喹啉与镁的配合物的激发光谱和发射光谱图

五、实验内容

1. 本实验采用 LS - 55 型荧光光度计，仪器原理及使用方法见本教材 2.7 "常用测量仪器的使用方法"中荧光光度计部分。

2. 血清样品中镁含量的测定

（1）于 3 个干燥的试管中按照下述方法配制溶液：

①空白溶液：分别加入 0.1 mL 去离子水和 3.9 mL 8 - 羟基喹啉乙醇溶液，摇匀。

②标准溶液：分别加入 0.1 mL Mg^{2+} 标准溶液和 3.9 mL 8 - 羟基喹啉乙醇溶液，摇匀。

③未知溶液：分别加入 0.1 mL 未知血清样和 3.9 mL 8 - 羟基喹啉乙醇溶液，充分摇匀 2 min，在离心机上以 8000 rpm 的转速离心 10 min，取上层清液作测定用。

（2）用上述②的溶液绘制激发光谱和发射光谱。先固定发射波长为 510 nm，在 350 ~ 450 nm 范围内扫描激发光谱，确定最大激发波长 λ_{ex}；再固定激发波长于 λ_{ex} 处，在 450 ~ 600 nm 范围内扫描发射光谱，确定最大发射波长 λ_{em}。

（3）在已确定的 λ_{ex} 和 λ_{em} 处分别测定（1）中 a、b、c 溶液的荧光强度 F_a、F_b、F_c。

六、实验结果与记录

计算未知血清样品中 Mg^{2+} 的含量。

$$c_x = \frac{F_c - F_a}{F_b - F_a} \times c_s$$

式中：c_x——未知血清样品中 Mg^{2+} 的含量；

　　　c_s——标准溶液中 Mg^{2+} 的浓度，

（本实验参考刘毅敏主编《医学化学实验》一书中实验十七"荧光分析法测定血清中的镁"）

（编者：刘绍乾）

实验 30　葡萄糖酸钙含量的测定
Determination of Calcium Gluconate

一、实验目的

(1)了解配位滴定中辅助指示剂的作用原理。

(2)掌握螯合滴定的原理和方法。

(3)进一步了解 EDTA 在螯合滴定中的应用。

二、预习要点

(1)了解滴定中所加条件试剂的作用。

(2)掌握螯合滴定的特点以及金属指示剂的使用条件。

(3)该测定方法为什么要在 pH = 10 的 $NH_3 - NH_4Cl$ 缓冲溶液中进行?

(4)在配制辅助指示剂 MgY^{2-} 时,要求滴定至溶液恰好显蓝色,若滴过量或量不足,对实验结果有影响吗?

三、实验基本原理

以 EDTA 为标准溶液的配位滴定中,常用铬黑 T 作指示剂。但是 Ca^{2+} 与铬黑 T 在 pH = 10 时形成的 $CaIn^-$ 不够稳定,会使终点提前出现而产生误差。由于 CaY^{2-} 比 MgY^{2-} 更稳定,因此,可在实验中加入少量 MgY^{2-} 作为辅助指示剂。当在 Ca^{2+} 试液中加入铬黑 T 和 MgY^{2-} 的混合液后,发生下列反应:

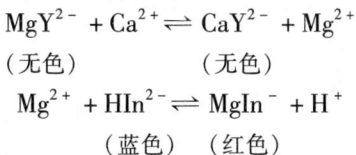

$$MgY^{2-} + Ca^{2+} \rightleftharpoons CaY^{2-} + Mg^{2+}$$
$$\text{(无色)} \qquad\qquad \text{(无色)}$$
$$Mg^{2+} + HIn^{2-} \rightleftharpoons MgIn^- + H^+$$
$$\qquad\qquad \text{(蓝色)}\quad \text{(红色)}$$

滴定过程中,EDTA 先与游离的 Ca^{2+} 发生配位反应,所以滴定终点前,溶液的颜色为 $MgIn^-$ 的红色。终点时,EDTA 与 $MgIn^-$ 中的 Mg^{2+} 反应(因 MgY^{2-} 的稳定性高于 $MgIn^-$ 的稳定性),将铬黑 T 指示剂 HIn^{2-} 置换出来,因此,终点时溶液的颜色从红色变为蓝色,即溶液终点时呈现铬黑 T 指示剂的颜色。反应如下:

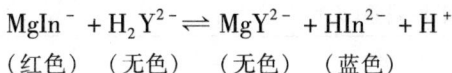

$$MgIn^- + H_2Y^{2-} \rightleftharpoons MgY^{2-} + HIn^{2-} + H^+$$
$$\text{(红色)}\quad \text{(无色)}\qquad \text{(无色)}\qquad \text{(蓝色)}$$

　　在整个滴定过程中 MgY^{2-} 并未消耗 EDTA，而是起了辅助铬黑 T 作指示剂指示滴定终点的作用。

四、仪器和试剂

　　仪器：电导率仪，酸式滴定管(50 mL)，移液管(50 mL)，锥形瓶(250 mL)，刻度吸管(1.0 mL, 5.0 mL, 10.0 mL)，滴定管架，电炉，烧杯。

　　试剂：$NH_3 - NH_4Cl$ 缓冲溶液(pH = 10)，铬黑 T 指示剂，EDTA 标准溶液(0.05 mol·L^{-1})，葡萄糖酸钙($M_r = 448.4$)，$MgSO_4$ 稀溶液。

五、实验内容

　　1. 辅助指示剂的配制

　　在 250 mL 锥形瓶中，加入蒸馏水 20 mL，$NH_3 - NH_4Cl$ 缓冲溶液(pH = 10) 20 mL，稀 $MgSO_4$ 溶液 2 滴，铬黑 T 指示剂 6 滴，用 EDTA 标准溶液滴到溶液恰好显纯蓝色。

　　2. 葡萄糖酸钙含量的测定

　　准确称取样品约 0.5 g 置于锥形瓶中，加蒸馏水 10 mL，微热使其溶解，冷却至室温后，加入辅助指示剂 20 mL，用 EDTA 标准溶液滴定至溶液由红色转变为纯蓝色即为终点。重复滴定 2 次，取三次平行实验的结果，按照下式计算葡萄糖酸钙的含量：

$$葡萄糖酸钙(\%) = \frac{c(EDTA) \times V(EDTA) \times 448.4 \times 10^{-3}}{样品质量} \times 100\%$$

六、实验记录与结果

数据记录与计算	测定序号	1	2	3
Na_2CO_3 标准溶液净用量/mL				
EDTA	初始读数/mL			
	终读数/mL			
	净用量/mL			
葡萄糖酸钙样品质量/g				
葡萄糖酸钙含量/%				
葡萄糖酸钙含量的平均值/%				
相对平均偏差(d_r)				

　　（本实验参考刘毅敏主编《医学化学实验》一书中实验八"葡萄糖酸钙的含量测定"）

（编者：王曼娟）

3.3　无机化合物的提纯、制备实验

实验 31　氯化钠的提纯
Purification of Sodium Chloride

一、实验目的

(1)学习提纯氯化钠的原理和方法及有关离子的鉴定。

(2)掌握溶解、沉淀、常压和减压过滤、蒸发、浓缩、结晶、干燥等基本操作。

(3)熟练掌握使用托盘天平、量筒和 pH 试纸使用方法,以及常用玻璃容器(烧杯、试管等)的洗涤。

二、预习要点

(1)溶解盐的水量过多或过少有何影响?

(2)检验 Ca^{2+} 是否存在时,为何加 HAc 溶液?

(3)在除去 Ca^{2+}、Mg^{2+}、SO_4^{2-} 等离子时,为什么要先加 $BaCl_2$ 溶液,然后再加 Na_2CO_3 溶液?

(4)为什么往粗盐溶液加 $BaCl_2$ 和 Na_2CO_3 后,均要加热至沸?

(5)提纯后的食盐溶液浓缩时为什么不能蒸干?

(6)什么情况下会造成产品收率过高?

(7)实验成败的影响因素有哪些?

三、实验基本原理

粗盐中泥沙等不溶性杂质可用过滤方法除去,Mg^{2+}、Ca^{2+} 和 SO_4^{2-} 等离子的可溶性杂质可用适当的试剂使其生成难溶化合物沉淀而除去。

一般是先在食盐溶液中加入 $BaCl_2$ 溶液,以除去 SO_4^{2-} 离子:

$$Ba^{2+} + SO_4^{2-} = BaSO_4 \downarrow$$

然后在溶液中加入 Na_2CO_3 溶液,除去 Ca^{2+}、Mg^{2+} 和过量的 Ba^{2+} 离子:

$$Ca^{2+} + CO_3^{2-} = CaCO_3 \downarrow$$

$$2Mg^{2+} + 2OH^- + CO_3^{2-} = Mg_2(OH)_2CO_3 \downarrow$$

$$Ba^{2+} + CO_3^{2-} = BaCO_3 \downarrow$$

过量的 Na_2CO_3 可用盐酸中和

$$Na_2CO_3 + 2HCl = 2NaCl + H_2O + CO_2 \uparrow$$

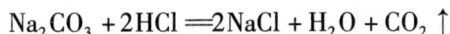

除去杂质后的溶液,再经蒸发,结晶、过滤及干燥可得到精制的氯化钠。

四、仪器和试剂

仪器：托盘天平,烧杯(50 mL),量筒(10 mL,50 mL),漏斗,蒸发皿,洗瓶,试管,玻棒,酒精灯,试管架,铁架台,石棉网,药匙,滴管,布氏漏斗,吸滤瓶,抽气泵。

试剂：研细的粗食盐,$BaCl_2$($1\ mol\cdot L^{-1}$),Na_2CO_3(饱和),HCl($6\ mol\cdot L^{-1}$),HAc($6\ mol\cdot L^{-1}$),$NaOH$($2\ mol\cdot L^{-1}$),$(NH_4)_2C_2O_4$(饱和),镁试剂。

五、实验内容

1. 粗盐的精制

(1)称量和溶解：用托盘天平称取粗盐 5 g,放入 50 mL 小烧杯中,加入蒸馏水 20 mL,加热搅拌使其溶解,此时不溶性杂质沉入底部。

(2)SO_4^{2-} 离子的除去：将盛有溶液的烧杯置于石棉网上加热至近沸,边搅拌边逐滴加入 $1\ mol\cdot L^{-1}$ 的 $BaCl_2$ 溶液 2 mL,继续加热 5 min(使 $BaSO_4$ 沉淀颗粒长大较易于沉淀和过滤)。将烧杯取下,待沉淀沉降后,于上清液中沿烧杯壁加入几滴 $6\ mol\cdot L^{-1}$ HCl 和 2 滴 $1\ mol\cdot L^{-1}$ $BaCl_2$ 溶液,检查 SO_4^{2-} 离子是否除尽。若无新沉淀生成,过滤,弃去沉淀。

(3)Ca^{2+},Mg^{2+} 和 Ba^{2+} 等离子的除去：将上述滤液加热至近沸,边搅拌边滴加饱和 Na_2CO_3 溶液,直至无沉淀生成为止(约 2 mL),静置,过滤,弃去沉淀。

(4)剩余 CO_3^{2-} 离子的除去：将上述滤液加热,在搅拌下滴加 $6\ mol\cdot L^{-1}$ HCl 溶液中和滤液至微酸性(pH 试纸检查 pH 约为 6)。

(5)蒸发、干燥处理：将上述滤液转入洗净的蒸发皿中,蒸发浓缩至有大量结晶析出时,趁热用布氏漏斗抽滤,用滤纸吸干晶体。

2. NaCl 纯度检验

取粗盐与精盐(自制)各 1 g,分别溶解于 5 mL 蒸馏水中。以此两份上清液分别进行纯度的检验。

(1)SO_4^{2-} 离子的检验：取上述清液各 1 mL,分别置于两支试管中,各加入 $1\ mol\cdot L^{-1}$ $BaCl_2$ 溶液 2 滴,若有白色沉淀生成,再加入几滴 $6\ mol\cdot L^{-1}$ HCl 至溶液呈酸性;沉淀若不溶解,表示有 SO_4^{2-} 离子存在。比较两支试管的结果。

（2）Mg^{2+}离子的检验：取上述清液各 1 mL，分别置于两支试管中，并各加入 2 $mol \cdot L^{-1}$ NaOH 溶液 5 滴和镁试剂 2 滴。若有天蓝色沉淀生成，表示有 Mg^{2+}离子存在。比较两支试管的结果。

（3）Ca^{2+}离子的检验：取上述清液各 1 mL，分别置于两支试管中，各加入 6 $mol \cdot L^{-1}$ HAc 溶液，使溶液呈酸性。再分别加入饱和$(NH_4)_2C_2O_4$试液数滴，若有白色沉淀生成表示有 Ca^{2+}离子存在。比较两支试管的结果。

六、实验结果

产品外观：
产量/克：
产率/%：

七、产品纯度检验

检验项目	产品溶液	粗盐溶液
SO_4^{2-} :（ $+BaCl_2$）		
Ca^{2+} :（ $+(NH_4)_2C_2O_4$）		
Mg^{2+} :（ $+NaOH+$镁试剂）		

（编者：肖旭贤）

实验 32　溶胶的制备、净化和性质
Preparation, Purification and Properties of Colloid Solution

一、实验目的

(1)了解溶胶的制备和净化方法。

(2)熟悉溶胶的光学、电学性质。

(3)掌握影响溶胶稳定性的因素。

二、预习要点

(1)复习有关溶胶和大分子溶液的稳定性因素，及其聚沉的方法，并比较两者的联系和区别。

(2)思考并回答下列问题：

①写出制备 $Fe(OH)_3$ 溶胶、Sb_2S_3 溶胶的反应式和胶粒结构。

②如果条件改变：①把 $FeCl_3$ 溶液加到冷水中；②把 $1\ mol \cdot L^{-1}\ Na_2S$ 溶液加到浓酒石酸锑钾溶液中。能否得到 $Fe(OH)_3$ 和 Sb_2S_3 溶胶？为什么？

③溶胶是热力学不稳定体系，但在一定条件下可以长期稳定存在，为什么？如何破坏胶体？举出两个日常生活中或生产上应用和破坏胶体的例子。

三、实验基本原理

溶胶是一种多相分散体系。它是介于真溶液和悬浊液之间的分散系，溶胶具有以下基本特征：①是多相体系，相界面大，胶粒大小在 $1 \sim 100\ nm$ 之间；②是动力学稳定体系；③是热力学不稳定体系。制备溶胶方法有分散法和凝聚法。制得的溶胶中，常会带有一些低分子量溶质及电解质等杂质，可用透析法使溶胶净化。

溶胶的光学性质是因为溶胶胶粒的直径略小于入射光波长，胶粒对光产生散射作用而造成的。当一束光线通过溶胶，在光束的垂直方向可以观察到一个光柱，这种溶胶对光的散射作用称为 Tyndall 效应。

由于胶粒表面带有电荷，因此具有双电层结构，所以溶胶分散相粒子在外电场作用下，可以向带电荷相反的电极方向泳动，这种现象称为电泳。

溶胶是热力学不稳定体系，溶胶有自动聚结变大的趋势而沉降。向溶胶中加入一定量的电解质后就能使它聚沉。聚沉能力的大小通常用沉降值表示，沉降值是使溶胶发生聚沉所需电解质的最小浓度值，其单位用 $mmol \cdot L^{-1}$ 表示。

两种带有相反电荷的溶胶相混合也可以发生聚沉。加热也可以降低溶胶的稳定性。

　　向溶胶中加入适量的高分子溶液（如动物胶），可以增加溶胶的稳定性。即高分子溶液对溶胶有保护作用。但是如果加入的量很少，则不但不能起保护作用，反而降低其稳定性，促使其聚沉，这种现象称为敏化作用。

四、仪器和试剂

　　仪器：电泳管及电泳仪，观察 Tyndall 效应的装置（或激光笔），烧杯（100 mL），量筒（10 mL，50 mL），试管，玻棒，酒精灯，透析袋，层析缸，滴管。

　　试剂：$FeCl_3$（0.1 mol·L^{-1}），酒石酸锑钾溶液（0.4%，W/V），$AgNO_3$（0.1 mol·L^{-1}），KSCN（0.1 mol·L^{-1}），NaCl（0.005 mol·L^{-1}），$CaCl_2$（0.005 mol·L^{-1}），$AlCl_3$（0.005 mol·L^{-1}），动物胶溶液（1%，W/V），硫化氢（饱和溶液），NaCl（5%，W/V），$AgNO_3$（0.1 mol·L^{-1}），蒸馏水。

五、实验内容

　　1. 溶胶的制备

　　(1)用水解反应制备 $Fe(OH)_3$ 溶胶：在 250 mL 烧杯中，放入 120 mL 蒸馏水，加热至沸，慢慢地滴入 0.1 mol·L^{-1} $FeCl_3$ 溶液 12 mL，并不断搅拌，加完后继续煮沸 1~2 min，则有红棕色 $Fe(OH)_3$ 溶胶生成。写出反应式及胶粒的结构，制得的 $Fe(OH)_3$ 溶胶留下备用。

　　(2)用复分解反应制备 Sb_2S_3 溶胶：在 100 mL 烧杯中盛放 0.4 mLW/V 酒石酸锑钾溶液 50 mL，然后滴加饱和硫化氢溶液，并适当搅拌，直到溶液变成橙红色溶胶为止，写出反应式，制得的 Sb_2S_3 溶胶留下备用。

　　2. 溶胶的净化——透析

　　将制得的 $Fe(OH)_3$ 溶胶注入透析袋中，用线拴住袋口，置于盛有蒸馏水的烧杯中，每隔 20 min 换一次水，同时分别用 $AgNO_3$ 和 KSCN 试剂检测水中的 Cl^- 和 Fe^{3+} 离子。透析至不能检出 Cl^- 和 Fe^{3+} 离子为止。

　　3. 溶胶的光学性质和电学性质

　　(1)溶胶的光学性质——Tyndall 效应：将制得的 $Fe(OH)_3$ 溶胶和 Sb_2S_3 溶胶，分别放入试管中，用激光笔的光源作为入射光照射溶胶，观察 Tyndall 现象。

　　(2)溶胶的电学性质——电泳：简单的电泳管是一 U 形管（图 3-14）。首先将 U 形管用洗液及蒸馏水洗净烘干，注入制备的 $Fe(OH)_3$ 溶胶；然后分别在两侧管内的溶胶面上小心地加上一层纯水，使溶胶与纯水间保持清晰的界面，在 U 形管的两端各插一根铂电极，接通直流电，缓慢调节电压，过一段时间后，即可看到溶胶界面发生移动，根据移动的方向，判断溶胶胶粒的电性。

用同样的方法进行 Sb_2S_3 溶胶的电泳实验，观察结果。

4. 溶胶的聚沉

(1)取三支干燥试管，每支试管中各加入 Sb_2S_3 溶胶 1 mL；边振荡边向 1

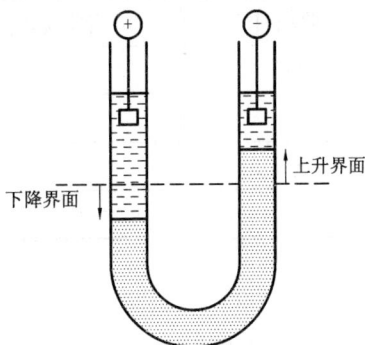

图 3 – 14　电泳管示意图

号试管中滴加 $0.005\ mol \cdot L^{-1}\ NaCl$ 溶液，向 2 号试管中滴加 $0.005\ mol \cdot L^{-1}\ CaCl_2$ 溶液，向 3 号试管中滴加 $0.005\ mol \cdot L^{-1}\ AlCl_3$ 溶液，均加至溶胶刚呈现浑浊为止。记下加入每种电解质溶液引起溶胶发生聚沉所需的最小量，估算每种电解质的聚沉值。简要说明所需电解质溶液的数量与它们的阳离子电荷间的关系。

(2)取一支试管，分别加入 $Fe(OH)_3$ 溶胶和 Sb_2S_3 溶胶各 2 mL，混合并振荡试管，观察出现的现象，并解释之。

(3)取一支试管，加入 Sb_2S_3 溶胶 2 mL，加热至沸，观察有何变化？并加以解释。

5. 动物胶的保护作用和敏化作用

(1)保护作用：取两支试管，在一支试管中加入 1%(w/v)动物胶 2 mL，另一支试管中加入蒸馏水 2 mL，然后在两支试管中各加入 Sb_2S_3 溶胶 2 mL，振荡试管 3 分钟，使之混匀，再向两支试管中各滴加 1 滴 5%(w/v)NaCl 溶液，摇匀，观察两支试管中聚沉现象的差异，并解释两支试管中聚沉现象不同的原因。

(2)敏化作用：取两支试管，各加入 Sb_2S_3 溶胶 3 mL，将其中一支试管稍稍倾斜，滴加 1 滴 1%(w/v)动物胶于试管上部的内壁上，使动物胶缓慢流入溶液；在另一支试管中以同样的方式滴加 1 滴 5%(w/v)的 NaCl 溶液，观察两支试管中的聚沉现象，解释聚现沉现象不同的原因。

六、实验记录与结果

按照下表要求记录并解释实验现象：

实验项目	实验现象	现象解释及相关化学反应方程式
1		
2		
3		
⋮		

（编者：刘绍乾）

实验33　CuSO₄的提纯(设计性实验)
Purification of Copper Sulfate

一、实验目的

(1)了解提纯固体物质的一般原理和方法。

(2)了解重结晶法提纯物质的基本原理,熟悉其操作技术。

(3)掌握研磨、溶解、加热、蒸发浓缩、结晶、常压过滤、减压过滤等基本操作技术。

二、预习要点

(1)查阅资料,收集提纯方法,拟定实验方案。

(2)将拟定的实验方案,按实验目的、基本原理、仪器和试剂、实验内容(包括试剂取量)、注意事项、影响产品纯度和产率的因素、结果处理、参考文献等项目书写成文,交教师审阅。

(3)粗硫酸铜溶解时,加热和搅拌起什么作用?

(4)用重结晶法提纯硫酸铜,在蒸发滤液时,为什么加热不可过猛?为什么不可将滤液蒸干?

(5)滤液为什么必须经过酸化后才能进行加热浓缩?在浓缩过程中应注意哪些问题?

(6)在提纯硫酸铜过程中,为什么要加H_2O_2溶液,并保持溶液的pH约为4?

(7)为了提高精制硫酸铜的产率,实验过程中应注意哪些问题?

三、实验基本原理

重结晶就是利用溶剂对被提纯化合物及杂质在不同温度下溶解度的不同,使杂质在热过滤时被除去或冷却后被留在母液中,从而达到分离提纯的方法。

硫酸铜为可溶性晶体物质。根据物质的溶解度的不同,可溶性晶体物质中的杂质包括难溶于水的杂质和易溶于水的杂质。一般可先用溶解、过滤的方法,除去可溶性晶体物质中所含的难溶于水的杂质;然后再用重结晶法使可溶性晶体物质中的易溶于水的杂质分离。

粗硫酸铜晶体中的杂质主要是硫酸亚铁($FeSO_4$)和硫酸铁$[Fe_2(SO_4)_3]$。当蒸发浓缩硫酸铜溶液时,亚铁盐容易氧化成铁盐,而铁盐易水解,有可能生

成 $Fe(OH)_3$ 沉淀，混杂于析出的硫酸铜晶体中，所以在蒸发浓缩的过程中，溶液应保持酸性。

若亚铁盐或铁盐含量较多，可先用过氧化氢(H_2O_2)将 Fe^{2+} 氧化为 Fe^{3+}，再调节溶液的 pH 约至 4，使 Fe^{3+} 水解为 $Fe(OH)_3$ 沉淀，由过滤除去。

$$2Fe^{2+} + H_2O_2 + 2H^+ = 2Fe^{3+} + 2H_2O$$

$$Fe^{3+} + 3H_2O \xrightarrow{pH \approx 4} Fe(OH)_3 + 3H^+$$

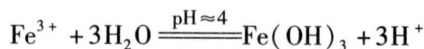

四、仪器和试剂

仪器：台秤(公用)，烧杯(100 mL)，量筒，石棉网，玻棒，酒精灯，漏斗，滤纸，漏斗架，表面皿，蒸发皿，铁三脚，洗瓶，布氏漏斗，油滤装置，硫酸铜回收瓶。

试剂：$CuSO_4 \cdot 5H_2O$(粗)，H_2SO_4(1 mol·L^{-1})，H_2O_2(3%)，pH 试纸，NaOH(0.5 mol·L^{-1})。

五、实验内容

1. 称量和溶解

用台秤称取粗硫酸铜 4 g(大块的硫酸铜晶体应预先在研钵中研细。每次研磨的量不宜过多。研磨时，不得用研棒敲击，应慢慢转动研棒，轻压晶体成细粉末。)，放入洁净的 100 mL 烧杯中，加水 20 mL。然后将烧杯置于石棉网上加热，并用玻棒搅拌。当硫酸铜完全溶解时，立即停止加热。

2. 沉淀

往溶液中加入 3% H_2O_2 溶液 10 滴，加热，逐滴加入 0.5 mol·L^{-1} NaOH 溶液直到 pH = 4(用玻璃棒蘸取待检验溶液点在 pH 试纸上，切忌将试纸投入溶液中检验。)，再加热片刻，放置，使红棕色 $Fe(OH)_3$ 沉降。

3. 常压过滤

将漏斗放在漏斗架上，趁热过滤硫酸铜溶液，滤液承接在清洁的蒸发皿中。从洗瓶中挤出少量水洗涤烧杯及玻璃棒，洗涤用水也应全部滤入蒸发皿中。过滤后的滤纸及残渣投入废液缸中。

4. 蒸发和结晶

在滤液中滴入 2 滴 1 mol·L^{-1} H_2SO_4 溶液，使溶液酸化，然后放在石棉网上加热，蒸发浓缩(切勿加热过猛以免液体溅失)。当溶液表面刚出现一层极薄的晶膜时，停止加热。静置冷却至室温，使 $CuSO_4 \cdot 5H_2O$ 充分结晶析出。

5．减压过滤

将蒸发皿中 $CuSO_4 \cdot 5H_2O$ 晶体用玻棒全部转移到布氏漏斗中，抽气减压过滤，尽量抽干，并用干净的玻棒轻轻挤压布氏漏斗上的晶体，尽可能除去晶体间夹杂的母液。停止抽气过滤，将晶体转到已备好的干净滤纸上，再用滤纸尽量吸干母液，然后将晶体用台秤称量，计算产率。晶体倒入硫酸铜回收瓶中。

六、实验记录与结果

粗硫酸铜的重量 $W_1 = $ _____ g　　精制硫酸铜的重量 $W_2 = $ _____ g。

产率 = _____。

（编者：何跃武）

实验 34　硫酸铝的制备
Preparation of Aluminum Sulfate

一、实验目的

(1)了解碱法制备硫酸铝的方法。
(2)加深对氢氧化铝两性性质的认识。

二、预习要点

(1)复习铝、氢氧化铝和硫酸铝的性质。
(2)思考并回答下列问题：
①将铝酸钠转化为氢氧化铝时，所加的碳酸氢铵起什么作用？
②氢氧化铝的生成过程中，为什么要加热煮沸并搅拌？
③浓缩硫酸铝溶液进行结晶时，为什么不要过分浓缩？
④某学生的实验，产率大于 100%，这是为什么？

三、实验基本原理

本实验从金属铝出发制备硫酸铝晶体$[Al_2(SO_4)_3 \cdot 18H_2O]$。
利用金属铝可以溶于氢氧化钠溶液的特性先制备成铝酸钠溶液[1]。
$$2Al + 2NaOH + 6H_2O = 2Na[Al(OH)_4] + 3H_2 \uparrow$$
再用碳酸氢铵调节溶液的 pH 至 8~9，将其转化为氢氧化铝沉淀。
$$2Na[Al(OH)_4] + NH_4HCO_3 = 2Al(OH)_3 \downarrow + Na_2CO_3 + NH_3 \uparrow + 2H_2O$$
氢氧化铝溶于硫酸并生成硫酸铝溶液。
$$2Al(OH)_3 + 3H_2SO_4 + 12H_2O = Al_2(SO_4)_3 \cdot 18H_2O$$
加热浓缩并冷却结晶，即得硫酸铝晶体。
硫酸铝为白色六角形磷片或针状结晶，易溶于水，难溶于酒精。在空气中易潮解。加热至赤热即分解成 SO_3 和 Al_2O_3。

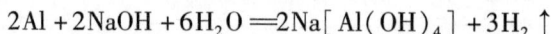

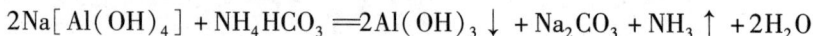

四、仪器和试剂

仪器：托盘天平，抽滤装置，生物显微镜(60 倍)。
试剂：H_2SO_4(6 mol·L^{-1})，NH_4HCO_3(饱和)，无水酒精，NaOH(固体，化学纯)铝片。
其他：pH 试纸，滤纸。

五、实验内容

1. 铝酸钠溶液的制备

用托盘天平快速称取 NaOH 固体 2 g，置于 100 mL 烧杯中，注入蒸馏水 15 mL，微热，搅拌溶液。加入金属铝片 0.5 g，搅拌使其全部溶解（该过程要防止浓碱溅出伤人）。反应完毕后加水约 20 mL，常压过滤，用 5 mL 水荡洗烧杯，洗涤液也转入漏斗中过滤，滤液及洗涤液盛接于 250 mL 烧杯中。

2. 氢氧化铝的生成和洗涤

将上述铝酸钠溶液转入 250 mL 烧杯中，补加水 75 mL，加热至沸，并保持沸腾状态。在不断搅拌下，以细流状缓慢加入饱和 NH_4HCO_3 溶液约 40 mL，pH 约为 9。将沉淀物煮沸 5 min 并不断搅拌（防止暴溅），取上层清液检验是否沉淀完全（方法自拟）。沉淀完全后趁热减压过滤。再用 30 mL 沸水淋洗沉淀，并抽至无水滴出。

3. 硫酸铝的制备

将制得的氢氧化铝沉淀转入到 100 mL 蒸发皿中，边加热边搅拌、边滴加 6 $mol·L^{-1}$ H_2SO_4 溶液至沉淀全部溶解，H_2SO_4 不必过量。继续加热浓缩至溶液体积为原来的二分之一左右（工业上控制溶液密度为 1.38 g/mL），在空气中缓慢冷却结晶。[2]

4. 观察硫酸铝晶体

取少量硫酸铝晶体置玻片上，加 1 滴无水酒精将晶体散开，在 60 倍显微镜下观察晶体形状，或取两滴硫酸铝过饱和溶液，在显微镜下观察晶体的形成和长大过程。

注意：

①也可以将金属铝与 H_2SO_4 反应得硫酸铝溶液。

②如果出现固结块而没有母液，可加少量蒸馏水重新溶解，再冷却结晶。如果冷却到室温仍无结晶出现，可滴加少量无水酒精，即有大量结晶出现。

（摘自《化学基本操作与物质制备实验》，关鲁雄主编）

实验 35　水的净化
Purification of Water

一、实验目的

(1)了解离子交换法制取去离子水的原理和方法。

(2)了解自来水中的主要杂质离子,学习鉴定水中无机杂质离子的方法。

(3)学习使用 DDS – 12A 数字式电导仪。

二、预习要点

(1)离子交换法净化水的原理及流程。

(2)DDS – 12A 数字式电导仪的使用说明(见 2.7.4 DDS – 12A 数字式电导仪使用说明)。

(3)思考并回答下列问题:

①蒸馏水与去离子水有何区别? 制取它们的原理各如何?

②自来水中主要无机杂质离子有哪些? 如何鉴别?

③阳离子交换树脂、阴离子交换树脂和阴阳离子混合柱各起什么作用? 能否将前两者颠倒顺序安装?

④为什么可由水样的电导率估计它的纯度? 电导率数值越大,水样的纯度是否越高? 电阻数值越大说明水的纯度是越高还是越低?

三、实验基本原理

天然水或自来水中常含有无机和有机杂质。可溶性杂质主要是钠、镁、钙、碳酸盐、硫酸盐、氯化物及某些气体,所以水中含有 Na^+、Mg^{2+}、Ca^{2+}、Cl^-、SO_4^{2-}、CO_3^{2-} 等离子。由于生产、科研和化学实验对水质有一定的要求,常采用蒸馏法和离子交换法予以净化。

蒸馏水是液态物质分离和提纯最常用的方法之一。在一定温度下,各种液体的蒸气压不相等,即沸点有所差异,因而可用加热方法使混合液气体先后气化、冷凝而之分离或提纯。水样在蒸馏过程中,常是沸点较低的水(溶剂)先气化而与沸点较高的杂质(溶质)分离。经蒸馏而净化的水叫做蒸馏水。

离子交换法是基于阴、阳离子交换树脂能与其他物质的离子进行选择性的离子交换反应。水样中所含的阴、阳离子经离子交换后得到净化,这种净化的水叫做去离子水。

阳离子交换树脂如聚乙烯磺酸钠型离子交换树脂 $RSO_3^- Na^+$ 经 HCl 转型后，可与水样中的 Na^+、Mg^{2+}、Ca^{2+} 等阳离子进行交换。例如：

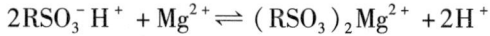

$$2RSO_3^- H^+ + Mg^{2+} \rightleftharpoons (RSO_3)_2 Mg^{2+} + 2H^+$$

阴离子交换树脂如季铵盐型碱性阴离子交换树脂 $RN^+ X^-$ 经 NaOH 转型后，可与水样中的 Cl^-、SO_4^{2-}、CO_3^{2-} 等阴离子进行交换。例如：

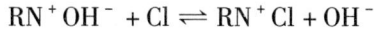

$$RN^+ OH^- + Cl^- \rightleftharpoons RN^+ Cl^- + OH^-$$

经阴、阳离子交换后产生的 H^+ 与 OH^- 结合又生成水。

$$H^+ + OH^- \rightleftharpoons H_2O$$

制取去离子水的流程是：高位槽中的自来水→阳离子交换柱→阴离子交换柱→阴阳离子混合交换柱。

高位槽中的自来水必须先进入阳离子交换柱，再进入阴离子交换柱，次序不能颠倒，否则大量 Mg^{2+}、Ca^{2+} 流过阴离子交换树脂会产生 $Mg(OH)_2$、$Ca(OH)_2$ 沉淀，大大降低阴离子树脂的交换能力。实际生产中，常把阳离子树脂与阴离子树脂串联起来使用。为了进一步提高水质，可在阴离子交换柱后接一个阴阳离子树脂混合柱。为了保证一定水压，各柱应从上部进水，下部出水。

失效的阴、阳离子交换树脂，可分别用稀 NaOH，稀 HCl 溶液再生。

水质检测（水质检测分电导法和化学法）：

（1）电导法

用电导仪测定水样电导。因为水是弱电解质，水中若含有可溶性杂质，将使其导电能力增大。用电导仪测定水样的电导，根据电导的大小，可估计出水样的纯度。

（2）化学法

①用铬黑 T 检验 Mg^{2+}。在 pH = 8 ~ 11 的溶液中，铬黑 T 本身显蓝色，若水样中含有 Mg^{2+}，则与铬黑 T 形成紫红色。

②用钙指示剂检验 Ca^{2+}。游离的钙指示剂呈蓝色，在 pH > 12 的碱性溶液中，它能与 Ca^{2+} 结合显红色。在此 pH 下，Mg^{2+} 不干扰 Ca^{2+} 的检验，因为 pH > 12 时，Mg^{2+} 已生成 $Mg(OH)_2$ 沉淀。

③用 $AgNO_3$ 溶液检验 Cl^-。

④用 $BaCl_2$ 溶液检验 SO_4^{2-}。

四、仪器和试剂

仪器： 离子交换装置，铁架台，蝴蝶夹，乳胶管，T 形玻璃管，弯玻璃管，玻璃纤维，DDS-12A 电导仪（带电极），烧杯（50 mL，5 只），锥形瓶（250 mL，配有孔软木塞,浸泡电极用)。

试剂: HNO_3(0.5 mol·L^{-1}, HCl(5%)), NaOH(4%), NH_3·H_2O(2 mol·L^{-1}), 络黑 T,钙指示剂,强酸型阳离子交换树脂,强碱型阴离子交换树脂。

其他:滤纸碎片。

六、实验内容

1. 树脂的预处理、转型和再生(由实验员准备)

(1)预处理 用清水分别清洗钠型阳离子树脂和氯型阴离子树脂,直至清洗液中无污浊为止。用清水浸泡 2 ~ 4 h,使树脂充分膨胀,然后将水倒去。

(2)转型 用 4% NaOH 溶液浸泡阴离子树脂 8 h,用 5% HCl 溶液浸泡阳离子树脂 8 h。分别将酸、碱倒去,用水反复冲洗,直至清洗阴、阳离子树脂的水的 pH 为 7。用蒸馏水浸泡树脂备用。

(3)再生 树脂使用一段时间后,当从阴离子树脂柱流出来的水的电阻率小于 100 kΩ·cm 时就应再生。阴、阳离子的再生方法与转型相同。混合柱中树脂的再生是将混合树脂浸泡于饱和 NaCl 溶液中,搅拌,静置分层。由于阴离子树脂的密度小,在上层。将它们分离后,分别进行转型处理。

2. 装柱(流程图如图 3 - 15 所示)

在三支交换柱底部塞入少量玻璃纤维,关闭活塞,先各加入数毫升去离子水(或蒸馏水)。用小烧杯盛上约四分之一烧杯已转型的带水阳离子交换树脂,用玻璃充分拌混匀成"糊状",边搅拌边注入第 1 柱中(若在装柱过程中水不够时,再加些蒸馏水),至交换树脂高度为 15 cm(注意:装柱过程中,柱内水面始终要高于树脂,才可能使树脂填充紧密,不留气泡。若在装柱过程中发现树脂层中有少量气泡,应及时用玻棒搅动树脂,赶走气泡,赶不净时,应重新装柱)。用同样方法在第 2 柱中注入已转型的阴离子树脂;第 3 柱中注入体积比为 1:1 的已转型的混合阴、阳离子树脂。

3. 离子交换

将自来水首先注入第 1 柱并拧开活塞,调节活塞使流出液以每分钟 25 ~ 30 滴的流速通过交换柱。用 50 mL 小烧杯盛各交换柱流出来的水。开始流出的约 30 mL 水应弃去,重新控制流速为每分钟 15 ~ 20 滴,收集水样 30 mL,待检验。

4. 水质的检验

对三个交换柱流出来的水,连同自来水、蒸馏水,分别进行下列检验,并记录结果。

(1)电导法 每次测定前,都应以待测水样冲洗电导电极,并用滤纸片仔细吸干。测量时将电极下端的铂片全部浸入水样中,注意勿使电极引线潮湿。

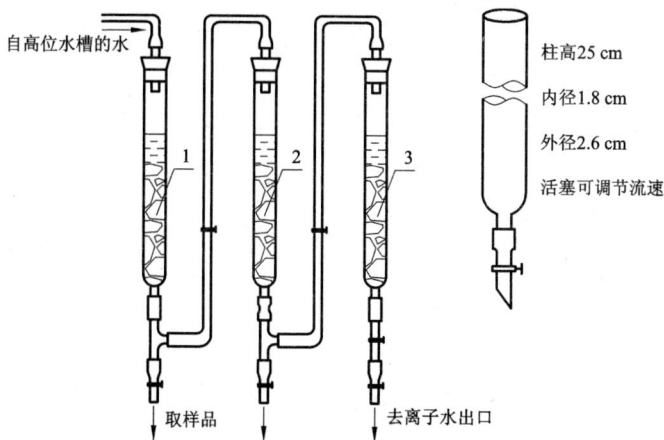

图 3 - 15 离子交换装置示意图
1—阳离子交换柱;2—阴离子交换柱;3—阴阳离子交换柱

（2）化学法

①Mg²⁺ 离子的检验 取水样 1 mL,加入 1 滴 2 mol·L⁻¹NH₃·H₂O,再加入 3~4 颗芝麻大的络黑 T,观察溶液颜色,判断有无 Mg^{2+} 离子。

②Ca²⁺ 离子的检验 取水样 1 mL,加入 8 滴 2 mol·L⁻¹NH₃·H₂O,再加入 3~4 颗芝麻大的钙指示剂,观察溶液颜色,判断有无 C_a^{2+} 离子。

③Cl⁻ 离子的检验 取水样 1 mL,加入 2 滴 0.5 mol·L⁻¹NHO₃,使之酸化,然后加入 1 滴 0.1 mol·L⁻¹AgNO₃,观察是否出现白色混浊。

④SO₄²⁻ 离子的检验 取水样 1 mL,加入 5 滴 0.5 mol·L⁻¹BaCl₂,观察是否出现白色混浊。

将检验结果列于下表：

样 品 名 称	检 测 项 目				
	电导	Mg^{2+}	Ca^{2+}	Cl^-	SO_4^{2-}
蒸馏水					
混合柱流出液					
阴离子交换柱流出液					
阳离子交换柱流出液					
自来水					

（参考关鲁雄主编《化学基本操作与物质制备实验》）

（编者：刘绍乾）

下　篇

英文部分

Part One   Introduction of Basic Chemistry Experiments

Chapter 1   Basic Rules and Safety Requirements

I . Laboratory Objectives

Chemistry is exciting! Each experiment holds many secrets. Look hard and you may see them. Work hard and you can solve them. The word "science" comes from the Latin word "scire", which means "to know". The goal of all science is knowledge. Scientists are men and women who devote their lives to the pursuit of knowledge.

In this class, you are given the opportunity to do what scientists do. The goals of laboratory work in a basic chemistry course are somewhat different from the goals of laboratory work in research. Primarily, we design Laboratory "experiments" (which have predictable results) to emphasize and to reinforce important textbook concepts. Almost as important, we select laboratory "questions" to teach essential experimental skills which will be useful in your future scientific work.

This laboratory manual is designed to present you with experiments that are challenging, coordinated investigations related to your study of chemical principles. These experiments utilize modern techniques and apparatus and emphasize operations performed by practicing scientists. Master the scientists' skills of observation and experiment. These skills are tools to solve the secrets of the unknown. Wherever possible, they are quantitative and related to contemporary problems in pure and applied chemistry. These experiments have also been designed to be efficient, safe, economical and nonpolluting.

II . General Laboratory Rules

Please read and follow these rules before beginning to laboratory work.

(1) During the first week of classes, locker and stockroom arrangements will be

explained. Since the administration of a laboratory involves coordination of the activities of many students, be sure to note and to follow requirements that may require careful attention (locker assignments, payment of fees, lock combination, laboratory hours, etc).

(2) You should wear a laboratory coat. Tie back long hair.

(3) Whenever you are in the chemistry laboratory, follow the directions of your instructor and the laboratory manual. Students are not permitted in laboratories unless an instructor is present.

(4) Your instructor will point out the location and proper use of the safety shower, eyewash, and fire extinguisher. Remember their locations and the locations of emergency exits from the building. Read the chapter on LABORATORY SAFETY now, and again during the middle of the semester.

(5) Read all the directions for each experiment before starting work. Note all warnings about possible dangers that may be involved.

(6) Proceed with your work thoughtfully and cautiously. It is wiser to prevent an emergency than to deal with it after it arises. Do not attempt to do experiments not specifically authorized by your instructor.

(7) Laboratory housekeeping is the responsibility of everyone in the laboratory. Carry out the following responsibilities:

①Never return chemicals to bottles of their origins. If you have taken an excess of a chemical, give it to another student, or if necessary, throw the excess away. It's better to waste a small amount of the chemical than to rise contaminating the entire contents of the bottle.

②Clean up any spills (especially in or near a balance) at once. Report any unusual spill or breakages to your laboratory instructor.

③Dispense malodorous reagents (such as concentrated HCl and NH_3 solutions) in a fume hood, if practicable, set up and conduct all experiments that may release any toxic or flammable gas directly in a hood, so that all gaseous products enter the hood intake.

④Never insert an unclean spatula or pipet into a reagent bottle. Don't stick to objects such as pencils or eyedroppers into reagent bottles and don't lay reagent bottle stoppers down in any way that the part which goes into the bottle which comes into contact with any surface. If you need a few drops of liquid, pour a little into a beaker and then take what you need from the beaker. If a solid has packed hard in a bottle, slap the side of the bottle to loosen it. If this doesn't loosen some solid, ask the instructor to help you. Most reagent bottles for solids have hollow caps. If you need a small amount of the solid, with the cap still in the bottle, shake a little into the cap and take what you need from the cap. These techniques prevent introduction of contamination.

⑤All chemicals in the laboratory must be clearly labeled. This applies not only to the bottles on the shelves but also to chemicals on or in your desk.

⑥many reagents can be recycled, many solid reagents cannot be safely disposed of in trash receptacles. When recycling or waste containers are provided, use them. Do not pour insoluble solids in sinks, solutions that may be discarded in sinks should be flushed thoroughly with water.

⑦Dispose of wastes only in the manner prescribed by your instructor. Ask him or her if you do not know. Serious cuts, fires, and explosions have resulted from improper waste disposal.

⑧Switching reagent bottle stoppers will invariably contaminate the reagent. To avoid this, never have more than one bottle unstoppered at a time. If the stopper is the pennyhead type, hold it between the fingers of the hand you are pouring with, while pouring. If you do this, you can be certain that you are not mixing up stoppers or contaminating the reagent.

(8) Organize your laboratory time so that you have cleaned up your apparatus and be out of the laboratory at the end of the laboratory period

(9) Report all accidents, no matter how minor, to the instructor.

Do not take food into the laboratory. Don't taste or eat anything in the laboratory.

III. Laboratory Notebook

A few general comments are in order about the laboratory notebook that is the primary record of your experimental work. First, although many campus bookstores sell notebooks that are specifically designed as lab notebooks, it is often sufficient to use any notebooks with tightly bound pages. Spiral and three-ring binders are inappropriate for lab notebooks because pages can be easily removed or torn out. All entries about your work must be made directly in your laboratory notebook in ink. Recording data on scraps of paper is an unacceptable practice because the pacer may be lost; the practice is strictly forbidden in your laboratory.

The notebook should begin with a table of contents, set aside the first two or three pages for this purpose. The rest of the pages should be numbered sequentially, and no page should ever be torn out of a laboratory notebook. Your notebook must be written with accuracy and completeness. It must be organized and legible, but it does not need to be a work of art.

Some flexibility in format and style may be allowed, but proper records of your experimental results must answer certain questions.

(1) When did you do the work?

(2) What are you trying to accomplish in the experiment?

(3) How did you do the experiment?

(4) What did you observe?

(5) How do you explain your observations?

A lab record needs to be written in three steps: prelab, during lab, and postlab. It should contain the following sections for each experiment you do.

1. To Be Done Before You Come to the Laboratory

The basic notebook setup discussed here is designed to help you prepare for an experiment in an effective and safe fashion. It includes the date and title of the experiment or project, the balanced chemical reaction you are studying, a statement of purpose, a table of reagents and solvents, the way you will calculate the percent yield, an outline of the procedure to be used, and the answers to any prelab questions. Your instructor will undoubtedly provide specific guidelines for lab notebook procedures at your institution.

(1) Title: Use a title that clearly identifies what you are doing in this experiment or project.

(2) Date(s): Use the date on which an experiment is actually carried out. In some research labs, where patent issues are important, a witnessed signature of the date is required.

(3) Balanced chemical reaction: Write balanced chemical equations that show the overall process. Any details of reaction mechanisms go into the summary.

(4) Purpose statement: Write a brief statement of purpose for the synthesis or analysis, or state the question you are addressing, with a few words on major analytical or conceptual approaches.

(5) Table: Include all reagents and solvents. The table normally lists molecular weights, the number of moles, and grams of reagents, as well as the densities of

liquids you will be using, boiling points of compounds that are liquids at room temperature and melting points of all organic solids, and pertinent hazard warnings.

(6) Method of yield calculation: Outline the computations to be used in a synthesis experiment, including calculation of the theoretical yield.

(7) Procedure and prelab questions: Write a procedural outline in sufficient detail so that the experiment could be done without reference to your lab textbook. This outline is especially important in experiments where you have designed the procedure. Answer any assigned prelab questions.

2. To Be Done During the Laboratory Session(s)

Recording observations during the experiment is a crucial part of your laboratory record. If your results of your experiments observations are incomplete, you cannot interpret the results of your experiments once you have left the laboratory. It is difficult, if not impossible, to reconstruct them at a later time.

Observations must be recorded in your notebook in ink while you are doing an experiment. You mustrecord the actual quantities of all reagents as well as the amounts of crude and purified products you obtain. Mention which measurements (temperature, time, melting point, and so on) you took and which spectra you recorded.

Because basic chemistry is primarily an experimental science, your observations are crucial to your success. Things that seem insignificant may be important in understanding and explaining your results later. Typical laboratory observations might be as follows.

(1) A white precipitate appeared, which dissolved when sulfuric acid was added.

(2) The solution turned cloudy when it was cooled to 10℃.

(3) An additional 10 mL of solvent were required to completely dissolve the yellow solid.

(4) The reaction was heated at 50℃ for 25 min in a water bath.

(5) A small puff of white smoke appeared when sodium hydroxide was added to the reaction mixture.

It is a good idea to cross-index your observations to specific steps in the procedure that you wrote out as part of your prelab preparation. Your instructor will probably provide specific advice on how you should record your observations during the laboratory.

3. To Be Done After the Experimental Work Has Been Completed

In this section of the notebook you evaluate and interpret your experimental results. Entries include a section on interpretation of physical and spectral data, a summary of your conclusions, calculation of the percent yield, and answers to any assigned postlab questions.

(1) Conclusions and summary: In an inquiry-based experiment or project, return to the question being addressed and discuss the conclusions you can draw from analysis of your data. For both inquiry-based experiments and those where you learned about laboratory techniques and the design of basic synthesises, discuss how your experimental results will support your conclusions. Include a thorough interpretation of analytical results, such as UV/Vis analysis. Cite any reference sources that you used and give answers to any assigned postlab questions.

(2) Percent yield: One of most important measure of success in a chemical synthesis is the quantity of desired product synthesized. To be sure, the purity of the product is also crucial, but if a synthetic method produces only small amounts of the desired product, it is not much good. Reactions on the pages of textbooks are often far more difficult to carry out in good yield than the books suggest.

IV. Laboratory Safety

Chemistry is a laboratory science. As part of your laboratory experiences you will handle many chemical substances and manipulate specialized laboratory equipments. Many of these substances pose a health risk if handled improperly, while some of the laboratory equipments can cause severe injury if used improperly. This section is a guide to the safe laboratory practices you will use throughout this course.

1. Prevention of Fires and Explosions

(1) Keep flammable materials away from flames.

(2) Extinguish all flames not in use. Never leave a flame unattended.

(3) Use an electric heater or a water bath for heating flammable liquids. Never heat them over a direct flame.

(4) Avoid, wherever possible, confining mixtures of air and flammable gasesor volatile liquids. Where such cannot be avoided, wrap the container with a cloth or place it behind a shield.

(5) When gases or vapors are generated by heat or chemical reaction, provide for pressure release to prevent explosion. Never cap a vessel being heated.

(6) Mixing of chemicals should always be done in small quantities as explained in your instructions. Chemicals that react vigorously can cause violent explosions if large quantities are mixed.

(7) Keep strong oxidizing agents from coming into contact with strong reducing agents.

(8) In case of fire:

①If your clothing catches fire, immediately drop to the floor and roll to smother the flames and call for help;

②If a compound or solvent catches fire, if you can, quickly cover the flames with a piece of glassware;

③If it is feasible, use a fire extinguisher to put the fire out;

④Do not put water on an organic chemical fire because it will only spread the fire;

⑤If the fire is large, do not take chances, evacuate the lab and the building immediately and tell your instructor or the coordinator what has happened;

⑥If no one in authority is available, pull the fire alarm in the hallway and call 119 from a safe phone;

⑦If the fire alarm sounds for any reason, leave the room immediately and exit the building.

2. Prevention of Poisoning

(1) Regard all chemicals in the laboratory as poisonous and never eat, drink, or taste anything while in the chemistry laboratory. It is always possible that the bottle is mislabeled.

(2) Handle chemicals in such a way that they do not come into contact with your skin. Utilize the tools of the laboratory: spatula, scoopula, piper, funnel, and so on.

(3) When it is necessary to note the odor of a substance, waft the fumes gently with your hand toward your nose. Never smell concentrated fumes.

(4) Note before performing the experiment whether any of the chemicals are especially poisonous. A few of the particularly poisonous chemicals commonly used in the laboratory are as follows:

Acids (especiallywhen concentrated) Alkalies (such as NAOH and KOH) Arsenic (As) and its compounds Hydrogen sulfide (H_2S) and sulfides Carbon tetrachloride (CCl_4) and other chlorinated hydrocarbons Mercury (Hg) and its compounds.

(5) Use mechanical devices for applying suction in pipetting. Never use your mouth for this purpose.

(6) Carry out all experiments involving poisonous, irritating, or objectionable gases or vapors in a fume hood.

(7) If you inhale vapors, leave the area immediately—at least into the hallway. Tell your instructor or the coordinator, they will take you outside into the fresh air, and if necessary provide first aid or take you to get medical attention.

3. Prevention of Cuts

(1) All broken glassware should be discarded immediately by placing it in containers provided for that purpose.

(2) Fire-polish all glass tubing or rods to prevent cuts from the sharp edges.

(3) When inserting glass tubing through a rubber stopper, hold the tubing very close to the rubber stopper with a towel to protect your hand and use glycerin as a lubricant.

(4) Do not attempt to force frozen glass joints, stoppers, or stopcocks. Take them to your instructor for removal.

(5) If you cut yourself, wash the wound immediately with large amount of cool water. If your neighbor has been hurt, be prepared to help him if he is unable to help himself. Apply direct pressure to stop the bleeding as necessary. If the bleeding is profuse, elevate the affected limb. Watch for evidence of shock and contact your instructor or the lab coordinator as necessary.

4. Prevention of Chemical Burns

(1) Concentrated acids and alkalies are particularly dangerous and can produce painful burns on skin and eyes, which may not be healed. These reagents must be dispensed with great care, never add them directly to other chemicals unless you are certain that it is safe to do so. Avoid all contact of the skin with these materials.

(2) Use the corks to stopper all test tubes or flasks that must be shaken, do not use your hand or thumb as a stopper.

(3) White (yellow) phosphorus, bromine, and hydrofluoric acid produce very painful, slow-healing burns. Handle them only with your hands adequately protected.

(4) When diluting concentrated sulfuric acid, add the acid slowly to the water. Never add water to concentrated sulfuric acid. The considerable heat evolved can

cause the acid to spatter and result in serious burns.

(5) If you burn yourself, thermal burns are treated by covering the affected area with cool water or ice. After a while, you can apply a pain-relieving cream. If the bum looks like it is more than just reddening of the skin, seek medical attention.

Chapter 2　　Common Laboratory Techniques and Practices

I . Glassware and Equipments List

A few of glassware and equipments commonly used in the laboratory are as shown in Figure 1 – 1.

Long-stemmed funnel　　　　Wsh bottle　　　　Erlenmeyer flask　　　　Glass-stopped bottle

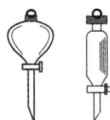 

Sepparatory funnel　　　　Buchner funnel　　　　Filter flask　　　　Volumetric flask

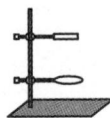 

Ring stand　　　　Tripod　　　　Mhr measuring pipet　　　　Transfer pipet

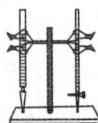

Buret stand　　　　Geiser buret　　　　Mohr buret　　　　Centrifuge tube

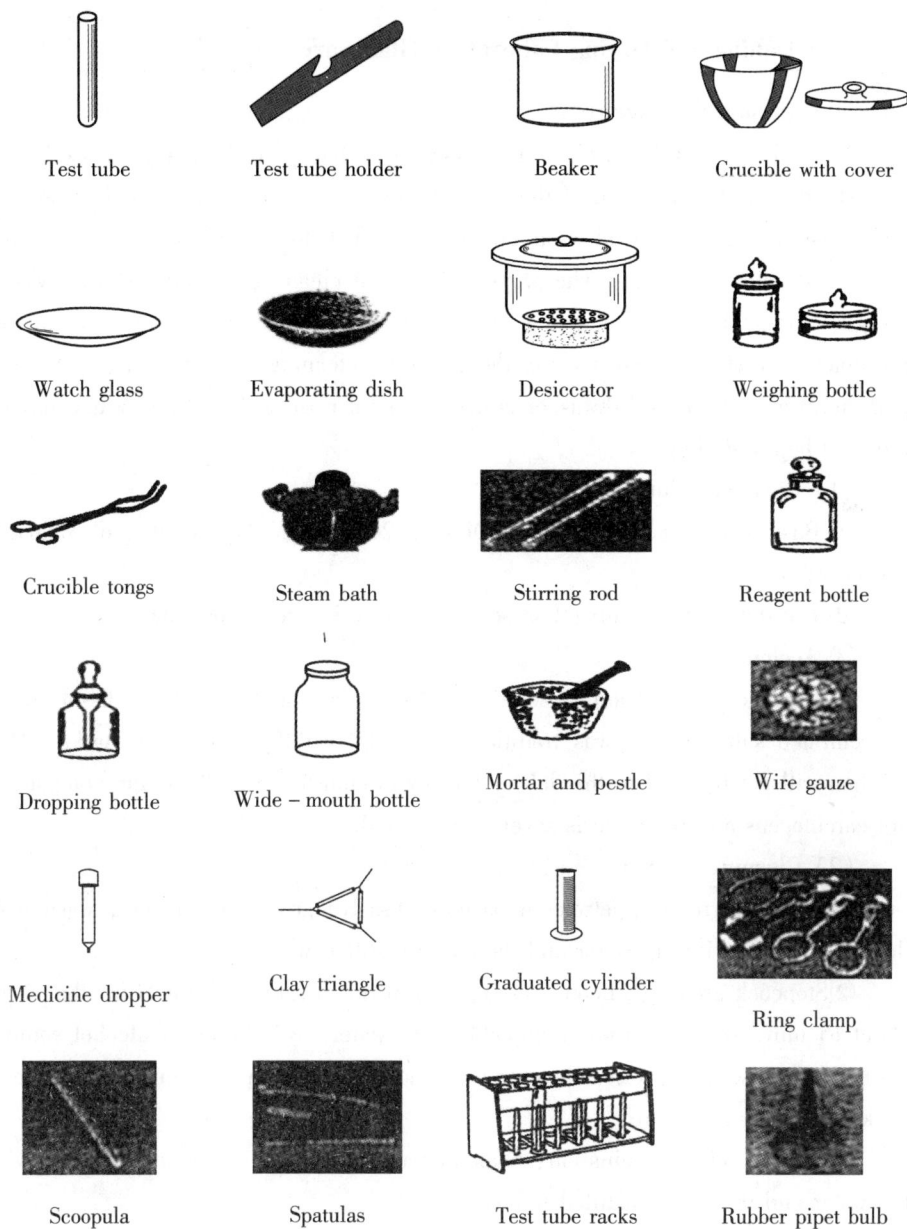

Test tube   Test tube holder   Beaker   Crucible with cover

Watch glass   Evaporating dish   Desiccator   Weighing bottle

Crucible tongs   Steam bath   Stirring rod   Reagent bottle

Dropping bottle   Wide – mouth bottle   Mortar and pestle   Wire gauze

Medicine dropper   Clay triangle   Graduated cylinder   Ring clamp

Scoopula   Spatulas   Test tube racks   Rubber pipet bulb

Figure1 – 1   Glasswave and Equipments List

II. Cleaning and Drying Laboratory Glassware

1. Cleaning Glassware

You will be required to clean all glassware so that it will be ready for use by the next student. First, the bulk of the contents must be disposed of properly (product vial, waste bottle, or special waste container; do not dispose of anything in the sink unless instructed to do so). The proper method for cleaning depends on what was in the glassware. The list of cleaning solutions and methods for specific contaminants presented here should cover most of the glassware cleaning a technician may have to perform. If an automatic dishwasher is used, a final rinse with distilled water may be required before drying.

(1) Cleaning solutions:

① Basic permanganate solution: Dissolve 20 g of $KMnO_4$ and 50 g of NaOH in 1 L of water.

② Hot detergent mixture: Use sparingly to avoid excess foaming.

③ Acetone.

④ Chromic acid cleaning solution: A solution of $K_2Cr_2O_7$, dissolved in concentrated sulfuric acid was traditionally used for difficult cleaning tasks. This solution will not be used in this lab under any circumstances. Chromium compounds are carcinogens and their use is severely restricted.

(2) Cleaning methods.

①Stopcock grease (petroleum base): Dissolve grease in acetone, wash with detergent, rinse with tap water and then with distilled water.

②Stopcock grease (silicone base): Soak it in hot saturated NaOH in ethanol for about 15 min, and then rinse thoroughly with water. A solution of alcohol sodium hydroxide is also our favorite cleanser for removing grease and organic residues from flasks and other glassware.

③Fat and oil contamination: Soak in basic permanganate solution, rinse with tap water and then with distilled water.

④Albuminous "crusts": Soak in chromic acid solution, rinse with tap water and then with distilled water.

The rinsing operation must be carried out thoroughly. If clean, the walls will retain an unbroken film of water, not droplets. The cleaning operation may be simplified by placing stained or contaminated glassware into detergent solution after use. Spectrophotometer cuvettes and other delicate glassware must be handled with extreme care and not subjected to the harsher cleaning agents.

2. Drying Glassware

(1) Oven drying of glassware. Electrically heated ovens (Figure 1 – 2) are commonly used in the laboratory to remove water or other solvents from chemical samples and to dry laboratory glassware. Wet glassware can be dried by heating it in an oven at 120 ℃ for 20 min. Remove the dried glassware from the oven with tongs and allow it to cool to room temperature before using it for a reaction.

(2) Drying wet glassware with acetone. Glassware that is wet after washing can be dried quickly by rinsing it

Figure 1 – 2　Drying oven

in a fume hood with a few milliliters of acetone. Acetone and water are completely miscible, so the water is removed from the glassware. The acetone is collected as flammable waste, any residual acetone on the glassware is allowed to evaporate into atmosphere. Thus there is an environmental cost, as well as the initial purchase and later waste disposal costs, in using acetone for drying glassware.

III. Water Quality Standards

Purified water is used in all industries and science-based organizations, Therefore, international and national standards authorities have established water quality standards for general types of application:

(1) The International Organization for Standardization (ISO);

(2) The American Society for Testing and Materials (ASTM).

Other representative organizations have specified criteria relevant to their particular domains. Prominent among these are:

(1) The National Committee for Clinical Laboratory Standards (NCCLS);

(2) The Pharmacopoeia.

International Organization for Standardization specification to water for laboratory use is ISO 3696:1987 (Table 1 – 1). This standard covers three grades of water as follows.

Table1 －1　International Organization for Standardization
Specification to Water for Laboratory Use

parameter	Grade 1	Grade 2	Grade 3
pH value inclusive range(at 25℃)	N/A	N/A	5.0 to 7.5
Electrical conductivity, 25℃, max/(S · cm － 1)	0.1	1.0	5.0
Oxidizable matter: Oxygen (O2) content, max/(mg · L － 1)	N/A	0.08	0.4
Absorbance at 254 nm and 1 cm optical path length, absorbance units, max	0.001	0.01	not specified
Residue after evaporation on heating at 110℃, max/(mg · kg － 1)	N/A	1	2
Silica(SiO2) content, max/(mg · L － 1)	0.01	0.02	not specified

1. Grade 1

Essentially free from dissolved or colloidal ionic and organic contaminants. It is suitable for the most stringent analytical requirements including those of high performance liquid chromatography (HPLC). It should be produced by further treatment of Grade 2 water for example by reverse osmosis or ion exchange followed by filtration through a membrane filter of pore size 0.2 μm to remove particle matter or re-distillation from a fused silica apparatus.

2. Grade 2

Very low inorganic, organic or colloidal contaminants and suitable for sensitive analytical purposes including atomic absorption spectrometry (AAS) and the determination of constituents in trace quantities. It can be produced by multiple distillation, ion exchange or reverse osmosis, followed by distillation.

3. Grade 3

Suitable for most laboratory wet chemistry work and preparation of reagent solutions. It can be produced by single distillation, by ion exchange, or by reverse osmosis. Unless otherwise specified, it should be used for ordinary analytical work.

IV. Drying of Solids

There are various methods of drying solid materials. When deciding which method to use it is important to know something of the physical properties of the material.

Although the method of air drying takes longer than the others, it is one of the safest for nondeliquescent solids. The damp solid, drained as dry as possible on the filter, is transferred to a watch glass and spread out evenly. This solid can be left to dry overnight in some dust-free place. If possible a second, larger watch glass should be arranged over the product as a precaution against dust, but this cover should allow free evaporation.

If the sample will not decompose by moderate heating and is not volatile, the nonessential water can be removed by placing the sample in a drying oven heated to 105℃ to110℃. The sample is placed in a weighing bottle, which is then placed in a beaker. The weighing bottle lid should also be placed in the beaker and heated along with the sample in order to remove excess moisture from its surface. Because many drying ovens are dirty and rusty, there is always the risk that dirt may fall into your sample. Therefore, use a cover glass supported on glass hooks to cover the beaker. If a procedure does not specify the drying time, dry the sample and then weigh it, repeat the drying and weighing until the weight is constant.

Whenever a sample is dried before weighing, there is always a question whether it is really dry or not. Therefore, samples are weighed to constant weight. They are weighed after an initial drying, then put back in the oven and later weighed again. If the weighings are the same, the sample is dry. If, however, the sample loses weighs during the second drying, then it is dried a third time and reweighed. This process is repeated until two consecutive weighings give the same result.

Vacuum ovens (which may be aspirated with a vacuum pump) speed up the drying process and are especially effective for drying heat-sensitive materials since they can be used at considerably lower temperatures.

It is important to understand that even though a sample contains nonessential water, the water is a constituent of the material " as received", Unless otherwise requested, the analyst should report the percent water removed by drying and the analysis obtained on the dry sample. If an analysis is requested on an "as is" basis, the analyst does not need to dry the material at all.

Unfortunately, letting the residue reach room temperature before weighing introduces a second problem. As the object cools, it can absorb moisture from the air. This problem is solved by placing the residue in a desiccator as it cools. Many types of desiccators are in use, a typical one is shown in Figure 1 – 3. All have two things in common, an airtight seal which keeps the sample isolated from the outside air and a material called a desiccant which absorbs moisture.

The desiccator must be regularly recharged with fresh desiccant, and the ground-glass seal must be greased with the minimum of silicone grease, so it appears transparent. It is removed by sliding it away from you with one hand while holding the bottom portion stationary with the other. At the bottom of the desiccator is a drying agent(often the very inexpensive $CaCl_2$) that takes up any moisture that enters when the desiccator is opened. Samples to be dried should be spread out on a watch glass and labeled with their name and date.

Any object taken from a flame or furnace should be cooled for at least 2 min before it is placed in a desiccator. A general rule is that if heat can no longer be felt on the back of your hand when it is held 5 cm from the object, the object can be put in a desiccator. If hot objects are placed inside, a vacuum forms upon cooling, making it difficult to remove the lid. The desiccator serves as a place for dried samples to cool and be stored prior to weighing them. Transfer to and from the desiccator should normally not be by fingers but by crucible tongs or by strips of lintless paper. Opening a desiccator is as shown in Figure 1 – 3 (a).

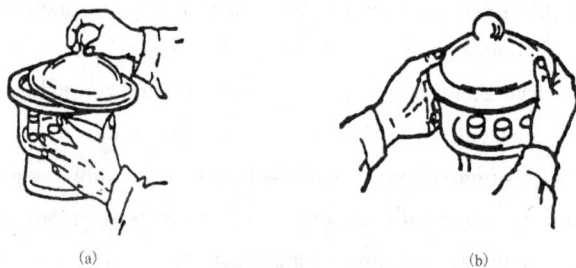

(a)   (b)

Fingure 1 – 3   A laboratory desiccator

(a) Opening a desiccator; (b) Carry a desiccator

Occasionally students are tempted to carry a desiccator by the top. This usually works until the desiccator is about 1 m above the floor, then it falls with disastrous results. Always carry the desiccator as shown in Figure 1 – 3 (b).

It is important to remember that after opening a desiccator it takes at least two hours to re-establish a dry atmosphere.

A vacuum desiccator is used to speed the drying of a sample. The sample mustbe covered with a second watch glass and the desiccator evacuated and filled slowly to avoid blowing the sample about, A vacuum desiccator must be covered with strong adhesive tape, or be enclosed in a special cage, when being evacuated and deevacuated to guard against an implosion.

V. Thermometer and Stopwatch

1. Thermometer

Most laboratory thermometers are mercury-in-glass, that is to say, mercury (a liquid metal with a low freezing point, 235 K or – 38℃ , a high boiling point, 630 K or 357℃ , and a relatively constant rate of expansion with increasing temperature) expands from a bulb into a glass capillary tube. Thermometers should be handled with care because they are expensive, because the glass is sharp when broken, and because mercury is poisonous. Never use a thermometer as a stirring rod; if it must be inserted through a stopper, use glycerine as a lubricant and hold the thermometer with a towel. If a thermometer

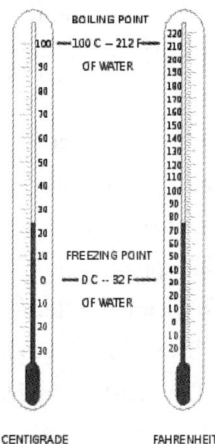

Figure 1 – 4   Thermometer

breaks, give the lower part to your instructor for proper disposal of the mercury.

2. Stopwatch

Strict quality assurance from The Lab Depot for Timers/Controllers and more than 750 000 other laboratory products means that you will always have confidence in your research and development or quality control results. After all, your laboratory testing procedures should always include using only the very best—whether timers, controllers, chemicals, glassware or very sophisticated instrumentation.

Figure 1 – 5   Stopwatch

VI. Chemicals

1. Purity and Grading

We offer wide range of products with most common grading available on each item. It includes Synthesis grade, GR grade, ACS grade, Grading for special application such as HPLC grade, and Biochemistry grade, etc.

The purity grade helps us to classify the large variety of reagents that exist in the market. Purity grade is expressed in the product name by means of a quality denomination that follows to the product nomenclature (i. e. "Guarantee Reagent").

Chemicals have large diversity in grading and types of use summarized as follows.

(1) Technical grade: These reagents are suitable for non-critical tasks in the laboratory such as rinsing, dissolving or are used as raw materials in production tasks.

(2) Synthesis grade: These reagents are suitable for organic synthesis and preparative applications.

(3) Chemically Pure (CP) grade: This grade is almost as pure as Synthesis grade, but application determines whether purity is adequate for the purpose.

(4) Guarantee Reagent (GR): It is the ideal quality for laboratory purposes. Batch to batch reproducibility is specially controlled to guarantee consistent analytical results. The grade is equivalent to Analytical grade (AR).

(5) ACS reagent: Reagents meet the specification of American Chemical Society CACS) Analytical reagents found in most laboratories and are used in a wide variety of analytical techniques for quality control, research and development.

(6) HPLC reagent: Product range is specially made for high performance liquid chromatography.

(7) Spectroscopy grade: Solvents display a high U V permeability and are subjected to strict IR Spectroscopy tests.

(8) Biotech/Biochemistry grade: Highly pure reagents are suitable for biochemical research and analysis.

2. Labeling and Storage of Chemicals

Be certain that all chemicals are correctly and clearly labeled. Post warning signs if chemicals are flammable, highly toxic, and carcinogenic or other special problems exist. Many chemical suppliers also indicate hazards by printing the universally understandable pictograms approved at the UN-sponsored Rio Earth Summit in 1992 on the labels of their reagents (Figure 1 – 6). If the label on the container does not give safety information, obtain the information from some reference sources; your supervisor; a handbook such as the CRC Handbook of Laboratory Safety, Merck Index; or MSDS.

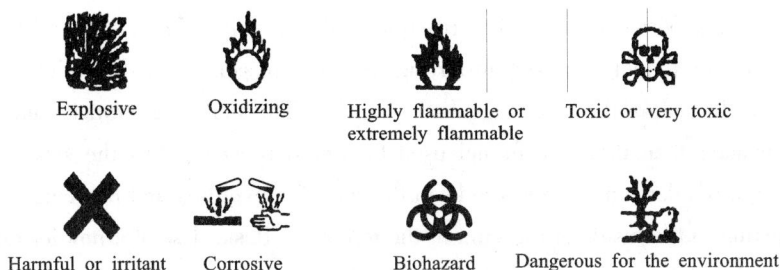

| Explosive | Oxidizing | Highly flammable or extremely flammable | Toxic or very toxic |

| Harmful or irritant | Corrosive | Biohazard | Dangerous for the environment |

Figure 1 – 6   Globally Harmonized System (GHS)
pictograms indicating chemical hazards

Centralized storage of bulk quantities of flammable liquids provides the best method of controlling fire hazards. The flammable liquids must be stored in fire resistant cupboard.

Figure 1 –4 Globally Harmonized System (GHS) pictograms indicating chemical hazards

Chemicals which are light-sensitive should be stored in dark bottles.

Some chemicals are incompatible with others. The incompatible chemicals should bekept segregated.

Chemicals that have high chronic toxicity, including those classified as potential carcinogens, should be stored in ventilated storage areas in unbreakable, chemically resistant containers. Storage vessels containing such substances should carry the label "CAUTION: HIGH CHRONIC TOXICITY or CANCER SUSPECT AGENT".

3. Handling Chemicals

(1) Handling solids.

Once the mass of a reagent has been determined, the reagent must be transferred to the reaction vessel without mishap. There are many local variants in each of these procedures. For example, some prefer to transfer solid with a weighing spoon, some with a finger hold bottle, and some with a paper-strap hold bottle. Students should follow the local preference, but should be aware of other acceptable options: Whatever the technique option chosen, the procedure must be done reproducibly, if analysis quality is to be optimized.

①Hold a container with its label facing your hand. Tilt the bottle towards the vessel to which you are transferring solid, and roll the bottle back and forth until the desired amount of solid is transferred into the new vessel.

②Using a powder funnel. For reactions being run in miniscale round-bottomed flask, a powder funnel aids in transferring the solid reagent from the weighing paper into the small neck of the flask (Figure 1 −7(a)). The stem of a powder funnel has a larger diameter than that of a funnel used for liquid transfer, thus the solid will not clog it. The powder funnel serves to keep the solid from spilling and prevents any solid from sticking to the inside of the joint at the top of the flask. Use of a powder funnel is essential with Williamson microscale glassware because of the very small opening at the top of the round-bottomed flasks and reaction tubes (Figure 1 −7(b)).

③ Transferring solids to a standard taper microscale vial. To transfer solids to a standard taper microscale vial, roll the weighing paper containing the solid into a cone by overlapping two corners on the same side of the paper. Place the rolled end well down into the neck of the microscale vial and gently slide one corner of the paper away from the other to create an opening just large enough for the solid to fall into the reaction vessel, as shown in Figure 1 −8. A microspatula can be used to gently push the solid through the opening in the weighing paper.

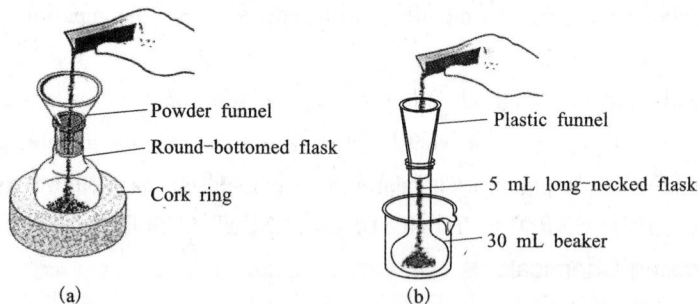

Figure 1 −7　Transferring solids with a powder funnel
(a)Miniscale apparatus; (b)Williamson microscale apparatus

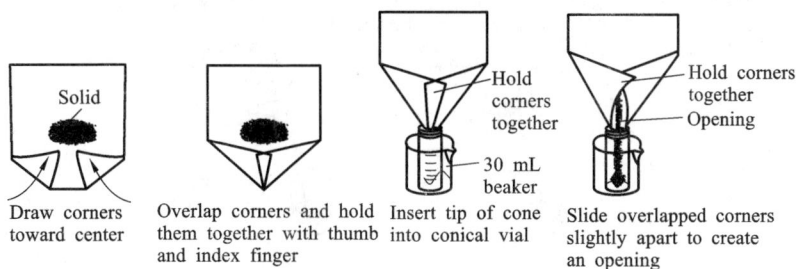

Figure 1 – 8   **Transferring solids with a weighing paper**

(2) Handling liquids.

① To dispense a liquid or solution from a reagent bottle, use the back of your fingers to remove the stopper from a reagent bottle (Figure 1 –9). Hold the stopper between your fingers until the transfer of liquid is complete. Do not place the stopper on your workbench.

② When you are pouring a liquid from a reagent bottle into a beaker, the reagent should

Figure 1 –9   **Removing a stopper from a reagent bottle**

be poured slowly down a glass stirring rod, as shown in Figure 1 – 10(a). When you are transferring a liquid from one beaker to another, you can hold the stirring rod and beaker in one hand as shown in Figure 1 – 10(b). Then no drops will dribble down the outside wall of the beaker.

③ Medicine droppers are great for transferring liquids from one container to another. They come as two parts, a bulb and a glass pipet. They must be used in a vertical position. This means, up and down. The bulb is on top, the tip of the pipet points down (Figure 1 – 11). Never, I repeat NEVER, hold a filled pipet upside down. If you hold it this way, the liquid runs into the bulb. The bulb is dirty and should not be used again.

Droppers from dropper bottles should not come into contact with any surface outside of the dropper bottle itself.

A very rough, but often satisfactory method for estimating volume is by counting drops delivered from a medicine dropper 15 to 20 drops per milliliter, depending on the size of the tip and the surface tension of the liquid.

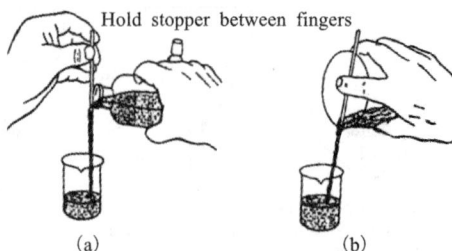

Figure 1 – 11   Pouring a liquid

(a) Pour down a glass stirring rod if possible to avoid dripping;

(b) When pouring from a beaker, hold the stirring rod in this manner

④When you are transferring a liquid to a test tube or graduated cylinder, the container should be held at eye level. Pour the liquid slowly, until the correct volume has been transferred (Figure 1 – 12(a)). When reading a volume of water in a container such as a graduated cylinder, first place the cylinder on the lab bench, you will notice that the water is higher at the edges than in the middle and forms a phenomenon called a meniscus. The meniscus is the apparent downward curvature in

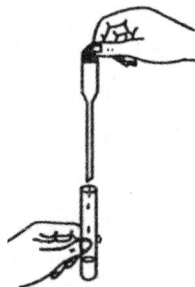

Figure 1 – 9   Using a medicine dropper in this manner

the surface of a liquid contained in any narrow measuring tube, caused in part by surface tension. In graduated cylinders, and also in pipets or burets that are filled from the top, it is necessary to read the bottom of the meniscus with the eye horizontal to this surface (Figure 1 – 12(b)). If the meniscus is not read at eye level, so that the front and rear parts of the graduation mark nearest the meniscus appear to coincide, parallax error in the reading will result. Determine this factor for yourself by comparing three readings of the same meniscus: one with the eye level horizontal, one with the eye directed from somewhat above, and one from somewhat below. Proper lighting is necessary to see the meniscus clearly. Each line on the graduate in Figure 1 – 12 (b) represents 1 mL. By estimating the height of the meniscus between these lines it is possible to measure the volume of the liquid to the nearest 0.1 mL. Thus, the volume of liquid in the Figure 1 – 12(b) is 39.1 mL.

The scale on your graduated cylinder may be different from that shown, but it should be divided into sufficiently small units so that the nearest 0. 1 mL can be estimated accurately.

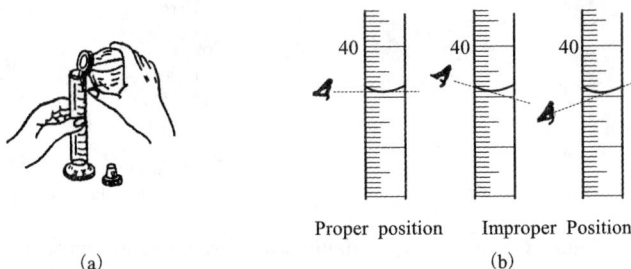

Proper position   Improper Position

(a)   (b)

Figure 1 – 12   A garduated cylinder

(a) Transferring a liquid to a graduated cylinder; (b) The reading of graduated cylinder

VII. Heating and Cooling

1. Heating

(1) Heating Devices.

Most labs use at least one type of heating device, such as ovens, hot plates, heating mantles and tapes, hot-tube furnaces, hot-air guns, muffle furnaces and microwave ovens.

①Hot plates: Hot plates work well for heating flat-bottomed containers such as beakers, Erlenmeyer flasks, and crystallizing dishes used as water baths or sand baths. Laboratory hot plates are normally used for heating solutions to 100 ℃ or above when inherently safer steam baths cannot be used. Any newly purchased hot plates should be designed in a way that avoids electrical sparks. Do not store volatile flammable materials near a hot plate.

②Heating mantles: Heating mantles are commonly used for heating round-bottomed flasks, reaction kettles and related reaction vessels. These mantles enclose a heating element in a series of layers of fiberglass cloth. Fiberglass heating mantles come in a variety of sizes to fit specific sizes of round-bottomed flasks, one size for a 100 mL flask will not work well with a flask of another size.

Both types of heating mantles have no controls and must be plugged into a variable transformer or other variable controller to adjust the rate of heating

(Figure 1 – 13). The variable transformer is then plugged into a wall outlet.

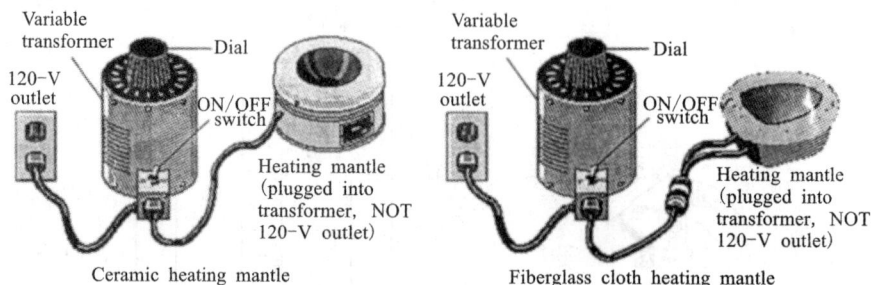

Figure 1 – 13　Heating mantle and variable transformer

A heating mantle with variable electronic control is used for flammable solvents that boil between room temperature and 200℃.

③ Heat guns: A heat gun allows hot air to be directed over a fairly narrow area (Figure 1 – 14 (a)). A heat gun is particularly useful as a heat source for distillations because of the high temperature near the nozzle. Heat guns usually have two heat settings as well as a cool air setting. After use, the gun should be suspended in a ring clamp with the heat setting on cool for a few minutes to allow the nozzle to cool before the gun is set on the bench (Figure 1 – 14(b)).

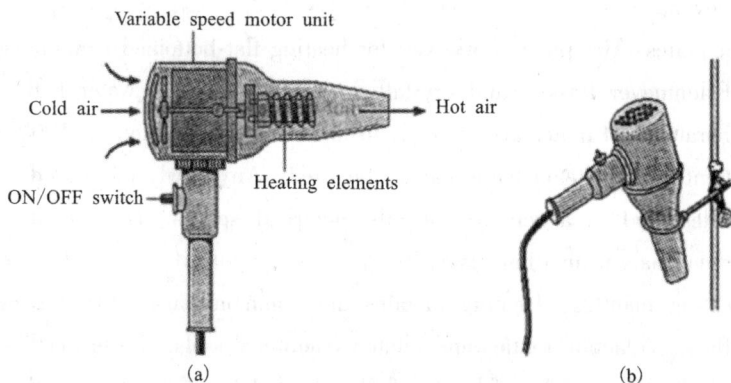

Figure 1 – 14　The use of heat gun

(a) Heat gun; (b) Heat gun cooling in a ring clamp after use

Other uses of heat gunsinclude the rapid removal of moisture from glassware where dry but not strictly anhydrous conditions are needed, and the heating of thin-layer chromatographic plates after they have been dipped in a visualizing reagent that requires heat to develop the color.

④ Laboratory jacks: Laboratory jacks are adjustable platforms that are useful for holding heating mantles, magnetic stirrers and cooling baths under reaction flasks (Figure 1 – 15). The reaction apparatus is assembled with enough clearance between the bottom of the reaction or distillation flask and the bench top to position the heating or cooling device under the flask by raising the platform of the laboratory jack. At the end of the operation, the heating or cooling device can be removed easily by lowering the platform of the laboratory jack.

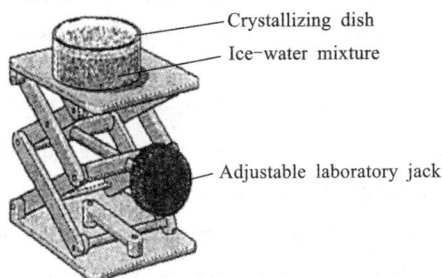

Figure 1 – 15   Laboratory jack with water batl

(2) Heating methods.

① Direct heating.

Alcohol burners are usually used to heat nonflammable solvents, such as water, that boil between 100 ℃ and 200 ℃.

When heating materials in a test tube with a burner, the test tube should be held with a test-tube holder, near the upper end of the tube (Figure 1 – 15(a)), the mouth of the test tube should be directed away from yourself and neighbor, and the heating should be done gently. Fill a test tube one-third full with the liquid to be heated. The test tube must be heated gently and uniformly. Move the test tube in and out of the flame to allow time for heat transfer. Heat the area near the liquid surface slightly more than the bottom of the test tube. Heat the upper part of the test tube to prevent vapor from condensing there.

CAUTION: Never point the open end of the test tube you are heating either toward yourself or anyone working nearby, never heat the bottom of the test tube.

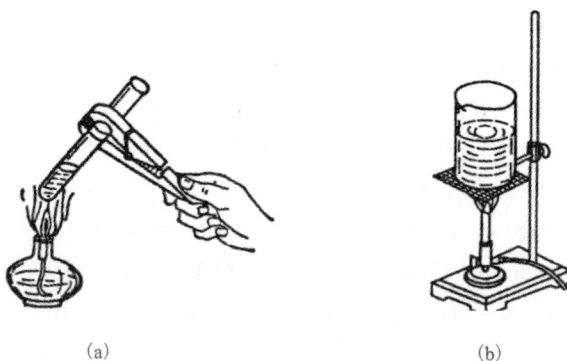

(a)   (b)

Figure 1 – 16   The use of alcohol burner

(a)Heating materials in a test tube；(b)Heating liquids in a beaker

When heating liquids in a beaker or a flask with a burner. The container should be set on wire gauze which evenly spreads the heat from the burner (Figure 1 – 16(b)). This will minimize the possibility of cracking the container or splattering the liquid. Fill a beaker one-half full with the liquid to be heated. When it is important that there is no loss whatever from splashing, it is customary to place a watch glass over the container. When acids or other solutions which emit noxious fumes are evaporated in this way it is imperative to carry out the operation under a fume hood with strong ventilation.

CAUTION: Never heat plastic beakers or graduated glassware in a burner flame, never let a boiling water bath boil dry, add water to it as necessary.

② Water baths.

When a temperature of less than 100℃ is needed, a water bath allows for closer temperature control than can be achieved with the heating methods discussed previously. The water bath can be contained in a beaker or crystallizing dish. Once the desired temperature of the water bath is reached, the water temperature can be maintained by using a low heat setting on a hot plate.

The thermometer used to monitor the temperature of a water bath should always be clamped, as shown in Figure 1 – 17, so that it is not touching the wall or bottom of the vessel holding the water. The reaction vessel should be submerged in the water bath farther than the depth of the reaction mixture the reaction vessel contains.

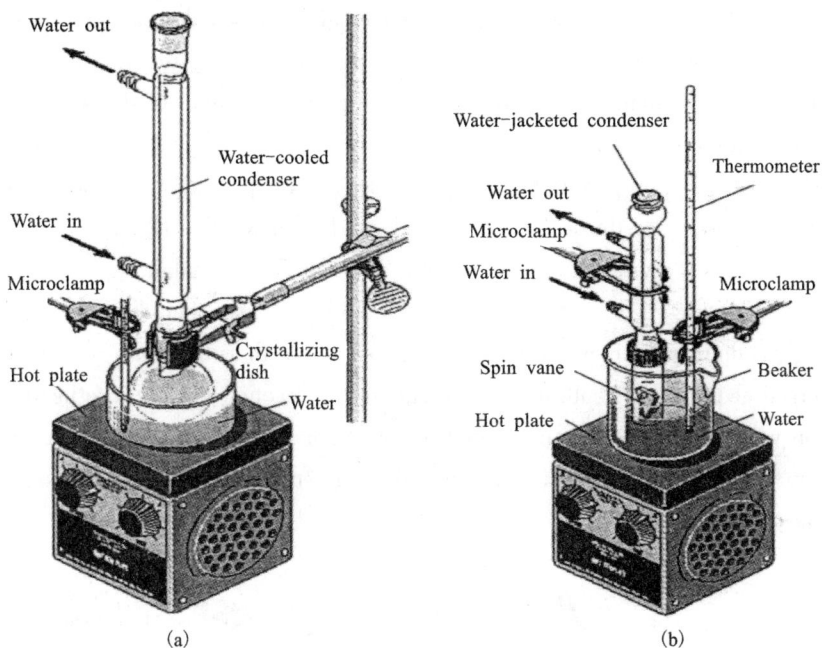

Figure 1 – 17   Water baths

(a) Heating miniscale reflux apparatus in a crystallizing dish;

(b) Heating microscale reflux apparatus in a beaker

③Oil and salt baths.

Electrically heated oil baths are often used to heat small or irregularly shaped vessels or when a stable heat source that can be maintained at a constant temperature is desired. Molten salt baths, like hot oil baths, offer the advantages of good heat transfer, commonly have a higher operating range (e. g. 200℃ to 425℃) and may have a high thermal stability (e. g. 540℃). There are several precautions to take when working with these types of heating devices:

A. Take care with hot oil baths not to generate smoke or have the oil burst into flames from overheating;

B. Always monitor oil baths by using a thermometer or other thermal sensing devices to ensure that its temperature does not exceed the flash point of the oil being used;

C. Wear heat-resistant gloves when handling a hot bath.

④ Sand baths.

A sand bath provides another method for heating microscale reactions. Sand is a poor conductor of heat, so a temperature gradient exists along the various depths of the sand, with the highest temperature occurring at the bottom of the sand and the lowest temperature near the top surface.

One method of preparing the sand bath uses a ceramic heating mantle, such as a thermowell, about two-thirds full of washed sand (Figure 1 – 18(a)). A second method employs a crystallizing dish, heated on a hot plate, containing 1 – 1.5 cm of washed sand (Figure 1 – 18(b)); the sand in the dish should be level, not a mound. A thermometer is inserted in the sand so that the bulb is completely submerged at the same depth as the contents of the reaction vessel. The heating of a reaction vessel can be closely controlled by raising or lowering the vessel to a different depth in the sand, as well as by changing the heat supplied by the heating mantle or hot plate.

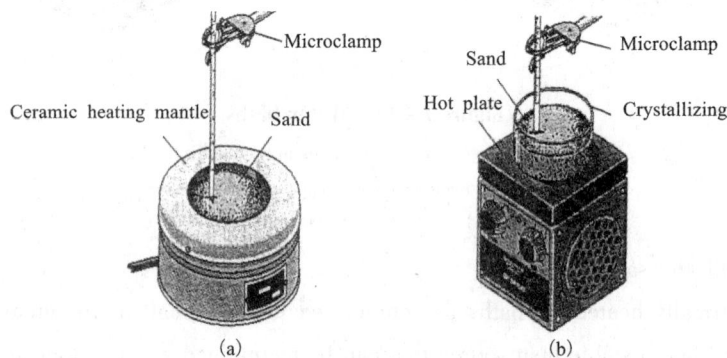

Figure 1 – 18   Sand baths

2. Cooling

Cooling baths are frequently needed in the organic laboratory to control exothermic reactions, to cool reaction mixtures before the next step in a procedure, and to promote recovery of the maximum amount of crystalline solid from a recrystallization. Most commonly, cold tap water or an ice-water mixture serves as the coolant. Effective cooling with ice requires the addition of just enough water to provide complete contact between the vessel being cooled and the ice. Even crushed ice does not pack well enough against the vessel for efficient cooling because the air in the spaces between the ice particles is a poor conductor of heat.

Temperatures from 0℃ to −10℃ can be achieved by the mixing solid NaCl into an ice-water mixture. The amount of water mixed with the ice should be only enough to make good contact with the vessel being cooled.

A cooling bath of 2-propanol and chunks of solid carbon dioxide (dry ice) can be used for temperatures from −10℃ to −78℃. (CAUTION: Foaming occurs as chunks of solid carbon dioxide are added to 2-propanol.) The mixture of 2-propanol and dry ice is contained in a Dewar flask, a double – walled vacuum chamber that insulates the contents from ambient circumstance (Figure 1 – 19).

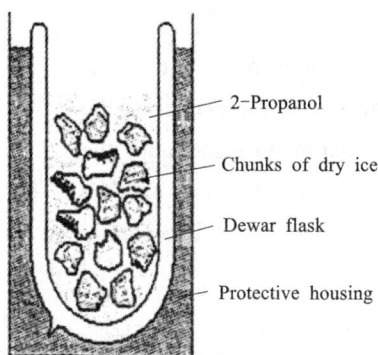

- 2-Propanol
- Chunks of dry ice
- Dewar flask
- Protective housing

Figure 1 – 19   Dewar flasd with a mixture of 2 – propanol and dry ice

VIII. Crystallization and Recrystallization

Crystallization is a technique which chemists use to purify solid compounds. It is one of the fundamental procedures each chemist must master to become proficient in the laboratory. Crystallization is based on the principles of solubility: Compounds (solutes) tend to be more soluble in hot liquids (solvents) than they are in cold liquids. If a saturated hot solution is allowed to cool, the solute is no longer soluble in the solvent and forms crystals of pure compound. Impurities are excluded from the growing crystals and the pure solid crystals can be separated from the dissolved impurities by filtration.

The process of crystallization is as follows. Heat some solvent to boiling, Place the solid to be crystallized in an Erlenmeyer flask. Pour a small amount of the hot solvent into the flask containing the solid, Swirl the flask to dissolve the solid. Place the flask on the steam bath to keep the solution warm. If the solid is still not dissolved, add a tiny amount of solvent and swirl again. When the solid is all in solution, set it on the bench top. Do not disturb it! After a while, crystals should appear in the flask. You can now place the flask in an ice bath to finish the crystallization process. You are now ready to filter the solution to isolate the crystals. Please see the section on vacuum filtration. After your crystals are filtered from the solution, carefully scrape the crystals onto the watch glass. Let the crystals finish drying on the watch glass.

IX. Separating and Purifying

1. Decantation

Sometimes adequate separation of a solid and a liquid can be achieved by decantation especially if the solid is dense. Allow the mixture to stand until the solid has settled; then carefully decant (pour off) the liquid, leaving the solid in the original container (Figure 1 –20).

2. Filtration

Figure 1 –20　Decantation

Filtration is a technique used either to remove impurities from an organic solution or to isolate a solid. The two types of filtration commonly used in chemistry laboratories are gravity filtration and vacuum or suction filtration.

(1) Gravity filtration.

Select the size of filter paper that, when folded, will be a few millimeters below the rim of your glass funnel. Fold a filter paper circle in half and then quarters. Open the folded paper to form a cone thickness of paper on one side and three thicknesses on the other, as shown in Figure 1 –21 (a). Insert the conical filter paper into the funnel and wet it with a few milliliters of the solvent to be used in the following procedure. When properly set up, the liquid will fill the entire stem of the funnel, and the weight of the long liquid column will apply "suction" to speed the filtration. Place a beaker beneath the funnel to collect the filtrate. The tip of the funnel should touch the inside surface of the beaker and extend about 2 cm below the rim. Decant the liquid from the solid by pouring it down a glass stirring rod into the funnel (Figure 1 –21(b)), Be careful to keep the liquid below the top edge of the cone of filter paper at all times; the liquid must not overflow. (If the precipitate has settled, decant the clear solution to save filtering time. Save the decantate if desired.) Then grasp the beaker and stirring rod with one hand as shown, swirl to suspend the precipitate, and pour the suspension into the filter the filtration is slow, carefully set the beaker down and, if necessary, leave the stirring rod in the suspension (never on the bench). When all the primary suspension has been poured out of the beaker, hold the beaker and stirring rod with one hand over the filter and direct a stream of wash water into the beaker to rinse it.

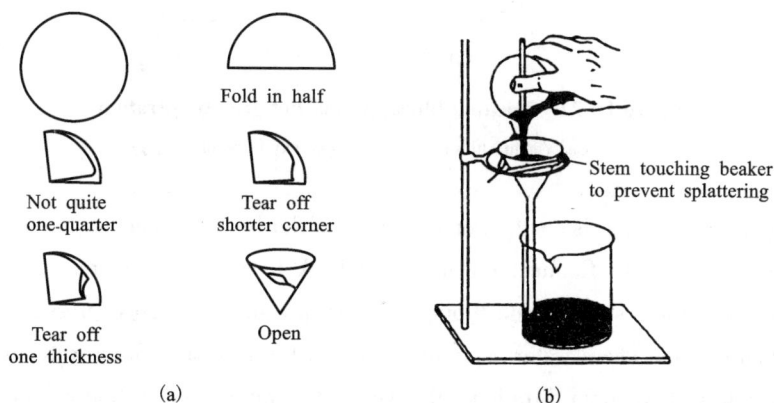

Figure 1 – 21   Gravity filtration

(a) Folding filter paper for gravity filtration; (b) Gravity filtration setup

(2) Vacuum filtration.

Vacuum filtration is used primarily to collect a desired solid, for instance, the collection of crystals in a recrystallization procedure. Vacuum filtration is faster than gravity filtration, because the solvent or solution and air are forced through the filter paper by the application of reduced pressure. The reduced pressure requires that they be carried out in special equipment (Figure 1 –21(a)): Buchner or Hirsch funnel; heavy-walled, side arm filtering flask; rubber adaptor or stopper to seal the funnel to the flask when under vacuum; vacuum source.

Procedure for vacuum filtration is as follows:

① Assemble the apparatus. Check the side arm flask carefully for cracks, since cracks could cause the flask to break when vacuum is applied. Then, clamp the flask securely to a ring stand. Add an adaptor and a Buchner funnel. Place a piece of filter

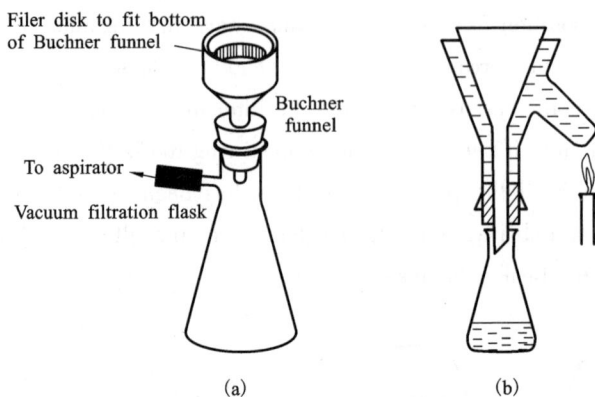

Figure 1 – 22   Vacuum filtration and hot gravity filtration

(a) Vacuum filtration；(b) Hot gravity filtration

paper in the funnel that is small enough to remain flat but large enough to cover all of the holes in the filter. Connect the side arm flask to a vacuum source always with thick-walled tubing, since Tygon tubing will collapse under reduced pressure.

Whenever you use a water aspirator, you run the risk of sucking water from the aspirator into your vacuum filtration unit. The water-trap placed in-line will catch any such water. If you do not need the filtrate, but only need the solid matter collected on the Buchner funnel, this presents no problem. Most of the time this is the case, and usually students do not need a water-trap.

② Wet the filter paper with a small amount of the solvent to be used in the filtration. Turn on the vacuum source.

③ Filter the solution. Pour the mixture to be filtered onto the filter paper. The vacuum should rapidly pull the liquid through the funnel. Watch that particulates do not creep under the edges of the filter paper. If this happens, start over and carefully pour portions of the solution onto the very center of the filter paper.

④Rinse the solids. Rinse the cake with a small amount of fresh, cold solvent to help remove impurities that were dissolved in the filtrate. Disconnect the rubber tubing before turning off the water aspirator. This prevents water from being sucked into the vacuum flask. Carefully remove the filter paper and solid from the Buchner funnel. Usually you will set it on a watch glass and let it air dry for a while. Note：Do not use vacuum filtration to filter a solid from a liquid if it is the liquid that you want and if the liquid is low boiling point. Any solvent which boils at about 125℃ or lower will boil off under the reduced pressure in the vacuum flask.

(3) Hot gravity filtration.

A filtration procedure called"hot gravity filtration"is used to separate insoluble impurities from a hot solution. Hot filtrations (Figure 1 – 22(b)) require fluted filter paper and careful attention to the procedure to keep the apparatus warm but covered so that solvent does not evaporate.

3. Centrifugation

A centrifuge substitutes centrifugal force for gravity in the separation of solids from liquids (Figure 1 – 23). Whenever the centrifuge is used, it must be balanced, or it may become damaged. Therefore, before centrifuging a mixture contained in a test tube or centrifuge tube, prepare another tube to balance it in the centrifuge by filling an identical tube

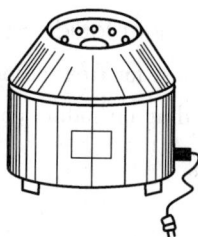

Figure 1 – 23   Electric centrifuge for separating a precipitate from a liquid

with water until the liquid levels in both tubes are the same. Insert the tubes in opposite positions (at 180°) in the centrifuge, close the cover, and set the machine in motion. The time required for centrifugation depends on the particle size of the solid being separated; for example, crystalline solids require less time than colloidal precipitates. Allow the centrifuge to come to rest before removing the tubes. Keep the cover closed and your hands away from the top of the centrifuge while it is rotating.

The solution is either quickly decanted from the tube without disturbing the precipitate or withdrawn by means of a medicine dropper.

X. Test Papers

Wide range test papers (pH, lead acetate, potassium iodide-starch, etc.) enable quick and costeffective semi-quantitative testing in the laboratory.

1. pH Test Paper

Indispensable for lab work! Simply dip paper wick into liquid or wet surface to measure acidity or alkalinity level. This measurement can be determined by the use of paper with color changes, solutions with color changes or pH meters that read pH in digital or analog numbers. Wide range pH test strips provide a distinct color for each pH unit from 1 to 14. Short range pH test strops show a distinct color change for each half pH unit. For instance, strong Acids like sulfuric acid and nitric acid test around 1 (red), while weak acids like acetic acid and carbonic acid test Around 6 (yellow). pH also indicates relative hydrogen ion concentration (increases with acidity).

2. Lead Acetate Test Paper

Used to test for hydrogen sulfide (H_2S) vapors. Paper must be moistened with water prior to use.

3. Potassium Iodide-Starch Test Paper

Used to test for free iodine in solution. Color varies from white through blue to black, depending on iodine concentration.

Chapter 3   Weighing with Balances

The analytical balanceis the most accurate and precise instrument in a laboratory. Objects of up to 100 g may be weighed to 6 significant figures. Early analytical balances were entirely, mechanics with two weighing pans, one for the chemical, and one for the counterweight. Now, most analytical balances are hybrid mechanical and electronic with a single pan for the substance to be weighed. These balances use the substitution method of weighing. That is to say, the counterweight is fixed (hidden within the balance) , and removable weights are mechanically added or subtracted from the sample side of the lever and fulcrum.

A very schematic sketch of the two pan balances is shown in Figure 1 – 24. The product of the mass times the distance from the balance point to the fulcrum determinesthe moment about the fulcrum. When these are equal for both sample and standard (i. e. weight), the pans will be level and the balance beam will be horizontal.

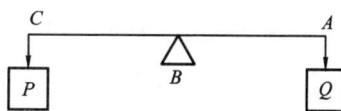

Figure 1 – 24   Schematic sketch of the two pans balance

$$m_Q \times l_{AB} = m_P \times l_{BC}$$

Most laboratories have two different types of balances available for measuring the mass of an object. The first type is the general-purpose balance, such as a DA/DTA electronic top-loading balance or the tray balance, which is usually capable of measuring to two decimal places(i. e. to 0.01 g). The second type is the analytical balance, which is usually electronic in more modern laboratories, and mechanical in older facilities.

I . Tray Balance

The tray balance (Figure 1 – 25) is commonly used for comparative weighing to determine the difference in mass between two objects. It is a versatile and inexpensive way to measure a variety of materials. These scales are perfect for general weighing.

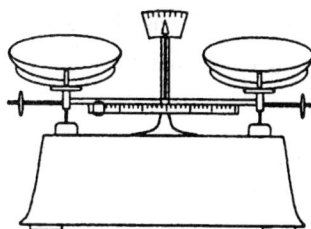

Figure 1 – 25   Tray balance

Tray balance operating instructions are as follows.

(1) The balance should be placed on the stable work table.

(2) The balance ride moves to the left side of "0" point of the staff gauge before using it, when the tray balance at the same time averages graduated arm, pointer aims median, if unbalance, may adjust the lever balance nut, make the balance be balanced state.

(3) When to weigh goods, should put the goods and weights on the location of scale pan respectively. The weight of weighed goods is the total weight of weights and sliding poise.

(4) The balance and the weights should use brushsoftly to clean at any time, and maintain dry.

(5) Discovering balance damaged or off normal, it should go out of service. Deliver it to the factory or the repair department promptly for repairs, after the examination chartered then this one can work.

(6) In order to maintain the balance accurate, you should inspect it once in a year against losing accuracy.

II. DA/DTA Electronic Balance

Use a top loading balance to weigh solid material when a precision of 0.1 g is adequate. For more accurate mass measurements or small amounts, use a DA/DTA electronic balance (Figure 1 - 26).

(1) Check if the balance is turned on. If not, press the on/off button and wait until the display reads 0.0 g.

(2) Place a container or large, creased weighing paper on the balance pan. Push tare button to zero the balance.

(3) Use a spatula to add small portions of the reagent to the container or paper until the desired mass is shown on the digital display. Record the actual amount you use in your notebook.

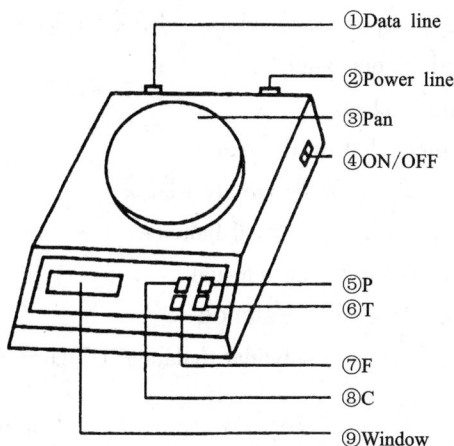

①Data line

②Power line

③Pan

④ON/OFF

⑤P
⑥T

⑦F

⑧C

⑨Window

Fingure 1 – 26   DA/DTA electronic balance

(4) Clean-up: Use the brush provided to clean any spills. Discard any disposable tare containers, weighing paper or Kimwipes in the nearest wastebasket (wastebin).

Ⅲ. Analytical Balance

In certain instances, the less precise (and quicker) top loading or triple beam balances should be used. However, since every analysis involves at least one, accurate weighing step, it is essential that you are able to use the analytical balances accurately and reproducibly. Analytical balances can measure to four or five decimal places (i. e. to 0.0001 or 0.00001 g).

1. General Rules for Use of the Analytical Balance

(1) The balance should be located on a hard, stable, level surface which is free of vibrations and excessive air drafts.

(2) Keep the balance clean. Remove dust, etc. from the pans with a camel hair brush.

(3)Learn the capacity of your balance, and never exceed this capacity.

(4) Objects to be weighed should be at room temperature.

(5) Strategiesmust be developed to ensure that moisture is not transferred to the object being weighed during handling,

(6) Chemicals are never placed directly on the balance pan. Use a weighing bottle, beaker, watch glass, etc.

(7) After you have completed weighing, check the following:

①You have recorded your results correctly;

②The balance pan is clean;

③There are no objects left on the pan.

(8) Corrosive liquids and solids are always placed in a vapor tight, pre-weighed container before weighing on an analytical balance.

(9) Report and record anything unusual.

2. The Electronic Analytical Balance

The single pan electronic analytical balance (Figure 1 – 27) is fast and convenient.

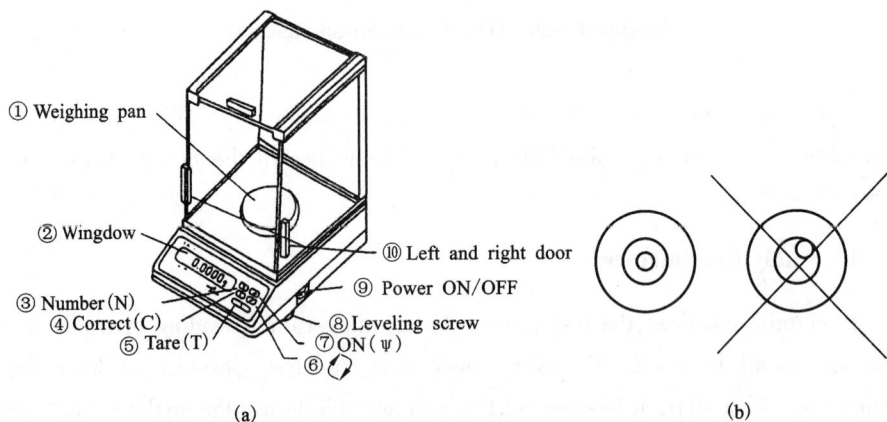

Figure 1 – 27　Electronic analytica balance

(a) The structure of electronic analytical balance; (b) Horizontal-levelbubble gauge

Its rapid digital readout obscures the subtlety of its operation. It is the device with the highest precision that you will use in the laboratory. It is both delicate and expensive. Some simple rules must be followed in its use.

(1) Powdered and liquid materials must never be placed directly on the balance pan.

(2) Spills should be avoided by not transferring material into containers while they are on the balance.

(3) If they do occur, they should be cleaned up at once (carefully).

(4) Do not try to calibrate an analytical balance. If in doubt, ask an instructor

to check it.

An analytical balance is one which weighs to ± 0. 1 mg and has a maximum capacity of 200 g. A number of companies produce reliable, reproducible analytical balances. The balances in the laboratory are produced by the Changshu Bailing Company. They are accurate to ±0. 1 mg and have a maximum capacity of 100 g.

3. Procedure for Obtaining the Mass of An Object

Using analytical instrumentation correctly requires practice. This exercise will give you the opportunity to develop confidence in yourweighing skills.

(1) Check the zero load before and after each weighing.

(2) Determine the weight of a clean, dry weighing bottle (without lid) to within +0. 1 mg.

CAUTION: Do not handle the glassware with your fingers (or thumbs or toes) unless otherwise instructed.

(3) Determine the weight of the weighing bottle lid to within +0. 1 mg.

(4) Determine the total weight of the weighing bottle plus lid to within +0. 1 mg.

(5) Repeat Steps (2) – (4) at least two more times to demonstrate the reproducibility of these measurements.

(6) Make sure you have recorded all your data in your laboratory notebook. Compare the sum of the weights of the weighing bottle and liddetermined separately with the weight of the lid and bottle determined together. Explain any discrepancies between the weights that you find. Suggest using a value that is most likely to be closest to the "real true weight" in your data for this series of experiments.

4. Weighing Out Samples and Using Weighing

Weighing bottles are small glass bottles with ground glass tops. Weighing bottles are to be used only for drying, storing, and weighing solid standards and unknowns. Weighing bottles should be numbered in pencil on the ground glass surface. Samples to be dried are placed in the weighing bottles without the stopper and placed in a beaker with a watch glass cover and a piece of paper with your name. This entire apparatus is then placed in the oven for the specified time. Upon removal from the oven, the weighing bottle is allowed to cool until it can be easily handled and then transferred to the desiccator. The weighing bottle should not be inserted until the bottle has come to room temperature in the desiccator.

The best way to manipulate the weighing bottle is to use a band of dry paper pulled firmly around the bottle (Figure 1 – 28). Do not use your fingers directly on the weighing bottle as the moisture from your fingers will affect the weight. If the weighing bottle stands for several hours in the desiccator before taking the next sample, its weight should be rechecked.

Figure 1 – 28   The use of weighing bottle

5. Weighing by Addition

Samples can be added to a clean, dry container, often a weighing boat, which has been previously weighed or tared on the balance. Extreme care must be taken when samples are transferred from this container to insure that no material is lost. Normally, the solvent should be used to wash any residue from the boat into the container at the end of the transfer. Steps in weighing by addition are as follows.

(1) Check to ensure that the horizontal position of the balance is level. Each balance is equipped with a level indicator. If not level, ask an instructor to adjust it.

(2) Clean the balance pan with a brush.

(3) Place a weighing boat, small beaker may also be used, on the center of the balance pan and be sure to close the balance doors.

(4) Tare the balance and wait until it reads 0. 0000g.

(5) Remove the weighing boat from the balance. Gently add solid sample to the weighing boat. Never add reagent while the weighing boat is in the balance!

(6) Return the weighing boat to the balance and close the balance doors. Record the weight to the nearest 0. 1 mg (i. e. four places after the decimal).

(7) Remove the boat and sample. Transfer the sample to the appropriate container.

6. Weighing by Difference

The preferred method is known as weighing by difference. To do this first determine the amount of solid you wish to weigh. This is called the target value. Your task is to hit this target value within about 10%, then record the mass to the nearest 0. 0001 g. With practice you can learn to "tap" the solid from your weighing bottle into the flask or beaker outside the balance (Figure 1 –26). Start by tapping less than the amount you will need. Reweigh the weighing bottle and calculate the mass transferred by subtracting the previous mass. Estimate how much more solid you will need to add to reach the target value. Tap in a bit less than that amount, and calculate the new mass transferred. If necessary, do a third tap. With practice you can hit your value (plus or minus 10%) in three tries. Don't waste time trying to hit your target value exactly.

This technique is useful when several duplicate masses of a material must be weighed out. Steps in weighing by difference are as follows.

(1) Check to ensure that the horizontal position of the balance is level. Each balance is equipped with a level indicator. If not level, ask an instructor to adjust it.

(2) Clean the balance pan with a brush.

(3) Place a weighing bottle on the center of the balance pan and be sure to close the balance doors.

(4) Tare the balance and wait until it reads 0. 000 0 g.

(5) Remove the weighing bottle from the balance. Place 1 – 4 g of the material to be measured into the weighing bottle. Never add reagent while the weighing bottle is in the balance!

(6) Return the weighing bottle to the balance and close the balance doors. Record the weight to the nearest 0. 1 mg (i. e. four places after the decimal).

(7) Carefully remove some material from the weighing bottle and place it in an appropriate container. Reweigh the weighing bottle. The difference between the original weight and the final weight is equal to the weight of sample taken. Repeat as necessary.

Chapter 4   Basic Operation of Volumetric Glassware

The apparatus for the precise measurement of volume are pipets, burets and volumetric flasks.

I . Buret

A buret is a common piece of equipment used to deliver known volumes of liquids. It is a slender glass tube of uniform bore with graduationmarks along its length. A 50 mL buret usually has the milliliter marks numbered, with unnumbered 0. 1 mL subdivisions in between. To obtain maximum precision, volumes are estimated to 0.01 mL.

Two kinds of burets are illustrated in Figure 1 – 29. Geiser buret(or acidic buret) for use with stopcock is useful for acid solutions or oxidation solutions, and Mohr buret (or basic buret) for use with pinchcock is useful for basic solutions.

Figure 1 – 29   Buret
(a) Geiser buret;
(b) Mohr buret for use with pinchcock

1. Cleaning a Buret

A scrupulously cleaned buret is one that drains smoothly to leave an unbroken invisible film of liquid on its walls, droplets indicate dirty walls and a buret unfit for accurate work. Cleaning is best accomplished by the use of liquid detergent solution and scrubbing with a buret brush with a long wire handle.

Either chromic acid or alkaline permanganate can be used; the latter reagent is generally preferred, but it must be used to be hot (near the boiling point) and repeatedly. To clean a buret, turn it upside down and immerse it in the boiling alkaline permanganate in a beaker. Suck up the permanganate into the buret by placing a rubber pipet bulb over the tip. To retain the solution in the buret simply

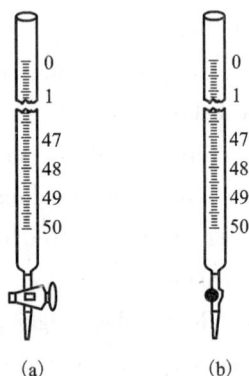

close the stopcock (Figure 1 – 30). Let the solution stand until cool, drain and wash the buret thoroughly with tap water and then with distilled water.

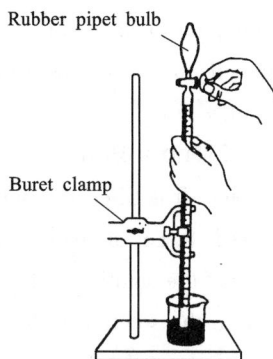

In cleaning a Mohr buret, remove the glass bead inside the rubber tube, replace the rubber tube, and suck the cleaning solution into the buret by placing a rubber pipet bulb over the tip of the buretwhich has been and immersed in the hot alkaline permanganate. To retain the solution in the buret, place a pinch cock on the rubber tubing. Do not suck the cleaning solution up into the rubber tubing.

Figure 1 – 30   Cleaning buret
with cleaning solution

2. Cleaning and Lubricating a Stopcock

A buretstopcock must be regreased after the buret is cleaned. The technique for cleaning and lubricating a stopcock is demonstrated in Figure 1 – 31.

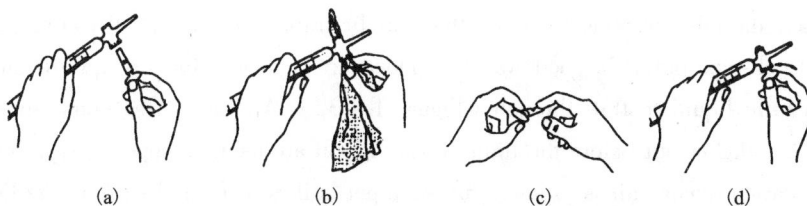

(a)   (b)   (c)   (d)

Fogire 1 – 31   Cleaning and lubricating a stopcock

(a) Removing stopcock; (b) Cleaning plug; (c) Lubricating stopcock; (d) Oscillating plug

(1) Remove stopcock.

(2) Remove all old grease. Clean and dry the plug also.

(3) Apply a thin film ofstopcock grease to the plug on each side of the bore.

(4) Distribute the grease by oscillating the plug slightly, making sure that there is a uniform film of grease over all the stopcock.

If there is grease in the beret tip, fill the buret with water, open the valve and lower the tip into the beaker of hot water. Do not use a shortcut method of heating the tip with a match.

3. Checking a Buret for Leakage

To check your buret for leakage, clamp your beret on a buret stand and fill it to above the "0" mark with distilled water. After eliminating any air bubbles from the tip, lower the meniscus to approximately "0" or a bit below. Let the buret stand for 5 min. If the meniscus drops, the stopcock leaks and it must be adjusted or replaced.

4. Filling a Buret

Before it is filled the buret should always be rinsed three or more times with small portions of the solution to be filled. Rinse the buret with the solution to be used by taking a portion of about 5 mL from a beaker or flask (not a large reagent bottle), pouring it into the buret, tilting the buret horizontal and rocking and rotating it until all its walls have been washed. Then turn the buret upright

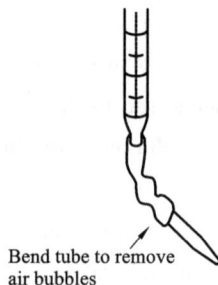

Bend tube to remove air bubbles

Fignre 1 – 32 Removing air bubbles

and let the solution drain through the tip (not the open end). Repeat rinsing, then fill the buret with solution above the "0"mark. Last, open the stopcock or pinchcock and shake the buret to remove any air that may be trapped in the tip. If you are using a pinchcock-type buret, a good way to remove air is to aim the tip upward and to allow a little liquid to flow through (Figure 1 – 32). Air being less dense that the liquid should flow out before the liquid does. If you are using a stopcock-type buret, a good way to remove air is to use a rubber pipet bulb to exert pressure on the liquid and to increase the rate of flow. Adjust the solution level until it is at or below the "0" mark. Clamp the buret in a buret clamp.

5. Reading a Buret

Water will wet a clean glass surface, the top surface of a water column in a glass cylinder is a concave-upward meniscus. Two readings are necessary to measure a dispensed volume. If both readings are made in consistent fashion, volumes can be measured to ±0.02 mL with a 50 mL buret. To make a reading, place a white paper with a black band just below the curved meniscus (A match box or even a finger can serve this purpose). Sight along a line horizontal with the meniscus, you should see the white paper directly behind the meniscus and the black surface reflected in the curved meniscus itself. In this way the meniscus is clearly outlined and the liquid level can be read along its flat bottom surface (Figure 1 – 33). Repeat this technique every time you read a beret, always remember to sight along a horizontal line. Always record readings to ±0.01 mL, if the meniscus is at 7.00 mL, record it as 7. 00 mL, not 7 mL.

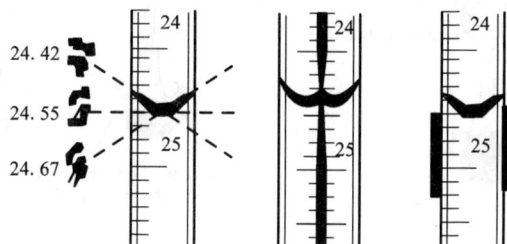

Figure 1 – 33   Reading a buret

The buret tip should be handled in the same way as the pipet tip. It should be wiped with a tissue; then, after a bit of liquid is withdrawn, it should be touched to some wet glass surface before the initial readingis taken or before a measured volume is withdrawn. After the desired volume is withdrawn, the tip should again be touched to the wet receiver wall before the final buret reading is taken.

6. Titration

The most satisfactory method of manipulating the rate of delivery from a buret is to use the left hand to turn the stopcock (this is assuming you are right-handed), the right hand is used to swirl the titration flask. With the scale of the buret facing the operator, the handle of the stopcock is on the operator's right. With the base of the left hand to the left hand to the left of the buret, the thumb and first two fingers encircle the buret to control the handle of the plug, the last two fingers against the left of the tip. This braced position of the hand leads to maximum control of the stopcock. It also makes it possible to keep constant pull on the plug into a secure position in the seat (This technique was developed for glass stopcocks to keep the stopcock from being pulled out). This is essential with glass stopcocks to avoid leakage. Although this technique may seem awkward at first, a little practice will remove the strangeness and you will find you can titrate more rapidly than other procedures. The technique for titrating with a Geiser buret is shown in Figure 1 –34(a).

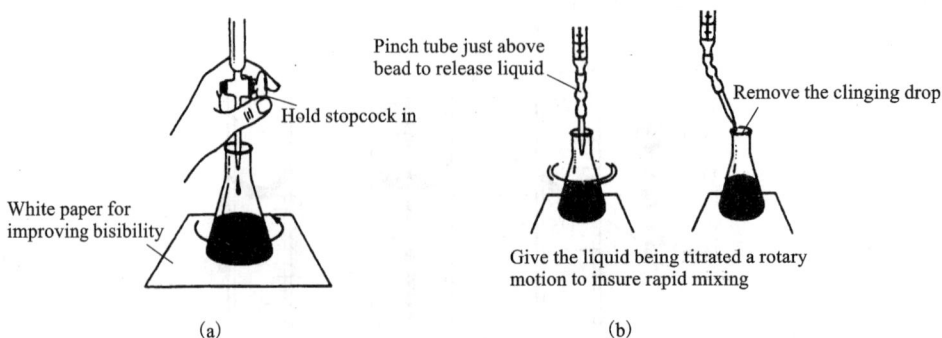

Figure 1 – 34　Titration

(a)Titrating with a Geiseer buret; (b)Titrating with a Mohr buret

In the following discussion we shall assume that you have followed the instructions in the preceding section for rinsing, filling, and preparing the buret for use. After recording the initial buret reading to the nearest 0. 01 mL, the titrant may be added rapidly at first to the titration vessel (usually an Erlenmeyer flask). Remember a buret cannot be drained too rapidly or too much liquid will adhere to the walls-rate of 0. 2 mL/s is satisfactory. The technique for titrating with a Mohr buret is shown in Figure 1 – 34(b). Using the right hand to swirl the flask and the left hand to control the buret, pinch tube just above bead to release liquid, add liquid at a rapid and uniform rate. Reaction in the localized region of mixing produces an indicator change. The addition of the titrant is periodically stopped and the rapidity with which the indicator returns to its color in the first solution is observed. Using this as a guide, the addition of the titrant is continued at a gradually decreasing rate. The tip of the buret and the walls of the flask are washed down with a small volume of distilled water from the wash bottle. The process of addition and rinsing is continued until the end has been located within a drop or within a partial drop. After a suitable drainage period the buret is read.

Near the end of the titration the rate of addition should be decreased until the titrant is being added drop by drop. Very near the end point partial drops of titrant can be added in several ways (A typical drop is 0. 05 mL).

(1) Let part of a drop form on the tip of the buret and wash it into the flask with water from your wash bottle.

(2) Let a partial drop form on the tip and detach it by touching the tip to the

inside surface of the flask. Then rinse the sides of the flask with the wash bottle.

(3) Very rapidly flip the buret stopcock through 180°. This will deliver a partial drop directly into the flask. This technique should be practised before the end point is reached, for the speed of the flip determines the volume of titrant added.

When the end point is observed, the walls of the flask should be rinsed with a stream of distilled water from a wash bottle. If necessary, additional partial drops of the titrant should then be added.

II . Pipet

The transfer and accurate measurement of relatively small volumes of liquids are often carried out by means of pipets. Two kinds of pipets are illustrated in Figure 1 – 35. The ungraduated form is called a "volumetric" or "transfer" pipet, and the other is called a "graduated" pipet or Mohr measuring pipet. The use of a volumetric pipet will be described. A volumetric pipet has a single calibration mark and delivers the volume printed on the bulb of the pipet (A graduated pipet has calibrations along the length of the pipet). Volumes can be measured more accurately

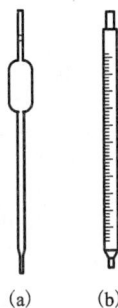

(a)   (b)

Figure 1 – 35   Pipet

(a) Transfer pipet;

(b) Mohr measuring pipet

with a volumetric pipet than with a graduated pipet. The techniques of handling them may be summarized as follows.

The pipet should be clean and rinsed three times with the liquid to be pipeted. The liquid to be pipeted is sucked into a clean pipet by means of a rubber pipet bulb, not by mouth. A right-handed person should hold the upper end of the pipet in the right hand between the thumb and fingers, leaving the index finger free (Figure 1 – 36(a)). The rubber pipet bulb should be held in the left hand and compressed before connecting it to the upper end of the pipet, and the connection should be just secure enough so as to avoid air leakage (Figure 1 – 36(b)). The tip of the pipet should be placed below the surface of the liquid to be dispensed. As soon as the liquid level rises to a bit above the desired graduation mark, the bulb is slipped off to one side and the end of the index finger is quickly placed over the end of the pipet (Figure 1 – 36(c)).

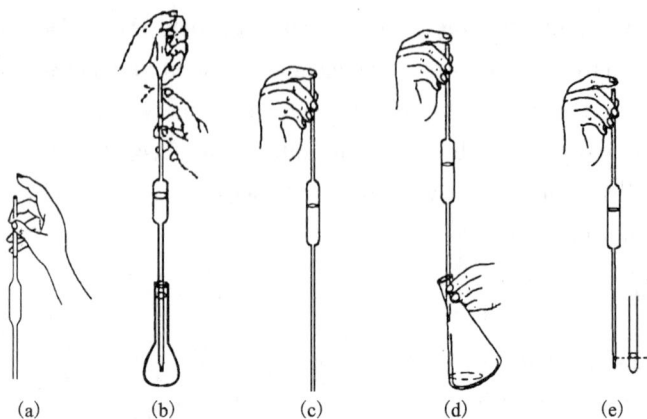

(a)   (b)   (c)   (d)   (e)

Figure 1 – 36   Operation of a pipet

The tip of the pipet is withdrawn from the supply of liquid and the outside of the lower stem is wiped dry with a cleaning tissue. Pressure from the index finger is relieved slightly to permit the liquid meniscus to descend to the desired graduation mark, then increased again so as to hold the liquid at this position. Persons with very dry skin may have a difficult time holding the liquid meniscus at a fixed position unless they first moisten the index finger. When the meniscus is at the mark the pipet tip is touched to a wet glass wall of something like a beaker in order to detach the drop or partial drop of liquid held there. The filled pipet with its carefully adjusted meniscus must not be subjected to sudden, jerky motions that may result in loss of some of the liquid from the tip. When using the volumetric pipet, the full contents of the pipet are allowed to flow into the desired container. After waiting a few seconds for draining, the pipet tip is touched to the wet container wall to remove the drop or partial drop held there and in the capillary end of the pipet (Figure 1 – 36(d)). Do not blow out the remaining liquid, the small amount that stays behind is reproducible and has been allowed for in the calibration of the pipet (Figure 1 – 36(e)). A left-handed person should hold the pipet in the left hand and the bulb in the right. Because it seriously hampers control of the pipet you should not control the flow of liquid from the pipet by means of your thumb. When the liquid drains from the walls of the pipet, there should be left a continuous invisible film of liquid, not droplets. If droplets appear, the pipet must be cleaned before use with liquid detergent solution followed by copious rinsing, first with tap water and then with distilled water.

When using the graduated pipet, the handling. techniques are the same as for the volumetric pipet except that instead of draining the entire contents, the meniscus is allowed to drop only to the desired level. The partial drop hanging at the tip is also removed as before. For reproducibly accurate work, the pipet must not be drained too fast; minimums of 20 s for a 10 mL pipet and 30 s for a 50 mL pipet are recommended. Pipets which meet the standards set by the National Bureau of Standards will have errors that do not exceed 0. 02 mL for a 10 mL pipet or 0. 03 mL for a 25 mL pipet.

III. Volumetric Flask

A volumetric flask is commonly used in the preparation of chemical solutions. Because it has a narrow, clearly-marked neck, it is possible to fill it in a reproducible way and minimize the error of measurement. When the flask is filled, the bottom of the meniscus (the curved line of the water surface) should appear to lie in the plane that passes through the circular line etched on the neck of the flask. The eyes of the observer must be in this same plane. The error that results when the eyes are above or below this plane is called parallax error. The close-fitting stopper prevents evaporation and permits the vigorous shaking mixing in the preparation of solutions.

As with other volumetric ware the walls must be scrupulously cleaned so that no drops will cling to the neck above the mark. Care must also be taken to insure that no significant amount of liquid is trapped by the stopper. During the filling process the flask should be held only by the neck above the mark, otherwise the heat of your hand will raise the temperature of the liquid which will expand and fill the flask with less liquid than is required at room temperature. Volumetric flasks should never be placed on a flame or hot plate to dissolve a solute. This can cause permanent changes in the volume of the flask.

If a solute is difficult to dissolve, carry out the dissolution in a beaker or flask and then quantitatively transfer the solution to the volumetric flask by the stirring rod (Figure 1 – 37). Solvent is added until the bulb of the flask is about three-quarters filled and a uniform solution is obtained by swinging the flask in a small circle to promote swirling of the liquid without bringing it into the neck. Mixing at this time allows volume changes which accompany dilution to take place before the solution is made up to volume. Solvent is now added to bring the solution to the calibration mark. The last few drops may be added with a medicine dropper. After the final dilution, remember to mix your solution thoroughly, by inverting the flask and shaking. Since a uniform solution has been prepared, solution that is now removed on the stopper in no way changes the concentration of the solution.

According to the minimum standards set by the National Institute of Standards and Technology, a first-class 50 mL flask should be in error by no more than 0. 05 mL; a 250 mL flask by no more than 0. 12 mL. For very accurate work, the volumetric

Figure 1 – 37   Transferring the solution

flask should be calibrated so that a suitable correction can be made. In any case, it is important to realize that the volumetric flask will contain the volume for which it is marked, not deliver it.

Chapter 5   Instruction for pH Meter

I. Principles of Operation

One of the most common instruments in any laboratory is the pH meter. The pH meter consists of a voltaic cell and a device to measure the EMF (electromotive force) derived from the cell. The EMF is dependent upon the concentration of hydrogen ions in the solution. Therefore the pH meter allows us to measure the hydrogen ion concentration indirectly by measuring the EMF produced by the pH-sensitive cell.

The cell is a combination of two half-cells as shown in Figure 1 – 38. The overall cell can be represented as

Ag,AgCl | Cl⁻ | glass | H⁺ (variable) ‖ Cl⁻ (saturated) | Hg_2Cl_2(s),Hg,Pt

One half-cell consists of a reference electrode designed to give a constant potential regardless of the hydrogen ion concentration. The other half-cell consists of the measurine electrode and the sample solution into which it dips. The EMF of the measuring electrode is dependent upon the pH. Electrical contact between two cells is provided by a fiber plug (this serves the same purpose as a salt bridge) in the reference electrode.

A calomel electro de is commonly used as the reference electrode (Figure 1 – 38 (b)). The half-cell reaction is

$$Hg_2Cl_2(s) + 2e^- \rightleftharpoons 2Hg(1) + 2Cl^-$$

(Constant concentration present from saturated KCl solution)

$$\varphi_{SCE} = \varphi^\theta - 0.05916 \lg[Cl^-] = 0.2412 \text{ V}$$

Hg_2Cl_2 and mercury metal are placed in the inner tube of the electrode, the inner tube is in contact with the surrounding solution of chloride ions [Cl⁻]. A constant [Cl⁻] is maintained by filling the electrode with a saturated solution of KC1. A few crystals of solid KC1 keep the solution saturated. Because the concentrations of all species involved in the half-cell reaction are constant (provided the temperature does not change) ,the EMF value of the half-cell remains constant.

The pH-measuring electrode is called a glass electrode (Figure 1 – 38 (a)).

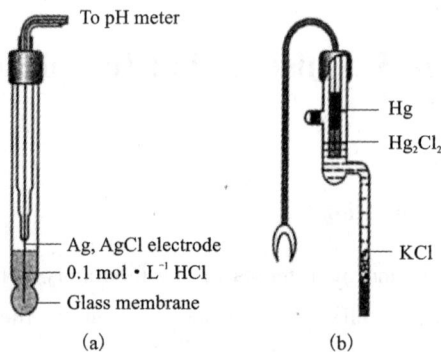

Figure 1 – 38　Electrodes for a pH meter

(a) Glass measuring electrode; (b) Calomel reference electrode

At the end of this electrode is a thin bulb made of special glass membrane which develops a potential if the [H +] is different in solutions contacting the two sides of the glass. The mechanism responsible for this potential is complicated and we don't need concern here. The application of the Nernst Equation to the half-reaction has the form:

$$\varphi_G = \varphi_G^\theta + \frac{2.303RT}{F} \lg \frac{[H^+]_{outside}(\text{variable})}{[H^+]_{inside}(\text{constant})} = \varphi_G^\theta - 0.05916 pH_x$$

If the bulb of the glass electrode is filled with a solution of HCl, the [H^+] inside, cannot change. The internal electrode of the glass electrode assembly is a silver—silver chloride electrode whose potential likewise does not change. The overall cell voltage then depends only on the [H +] in the unknown solution:

$$\varphi_{obs} = \varphi_{SCE} - \varphi_G = \varphi_{SCE} - \varphi_G^\theta + \frac{2.303RT}{F} pH_x = E^* + \frac{2.303RT}{F} pH_x$$

In this form of the Nernst Equation we have used the symbol E^* to indicate a constant term, the value of which will depend upon the constant conditions of the particular electrode assembly. The equation shows that E_{obs} depends directly on the pH of the solution. The scale of the meter can be marked in volts or directly in pH units. Calibration of the meter is accomplished by dipping the electrode assembly into a standard solution of known pH and adjusting the circuit to make the meter reading agree with the known pH of the solution.

II. Details of Operation

1. pH Meter PB – 10 Setup

The setup of pH meter PB – 10 is shown in Figure 1 – 39.

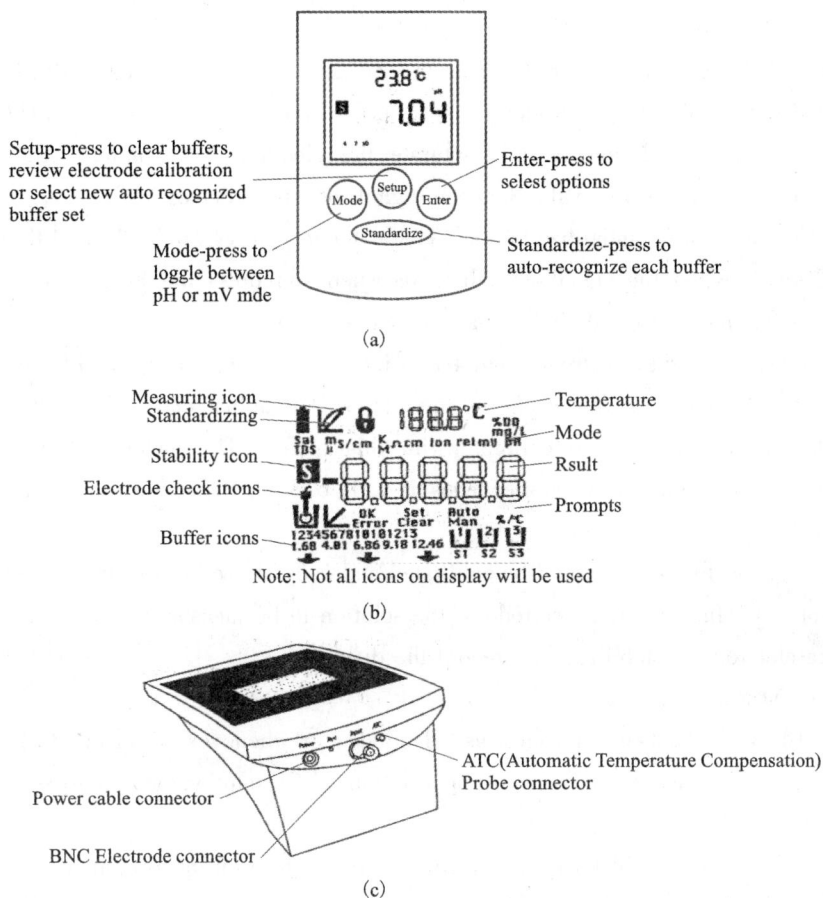

Setup-press to clear buffers, review electrode calibration or select new auto recognized buffer set

Enter-press to selest options

Mode-press to loggle between pH or mV mde

Standardize-press to auto-recognize each buffer

(a)

Measuring icon
Standardizing
Stability icon
Electrode check inons
Buffer icons

Temperature
Mode
Rsult
Prompts

Note: Not all icons on display will be used

(b)

ATC(Automatic Temperature Compensation) Probe connector

Power cable connector

BNC Electrode connector

(c)

Figure 1 – 39   pH meter PB – 10

(a) Front panel controls; (b) Digital display; (c) Rear panel connectors

2. Digital pH Meter Operating Instructions

(1) Connect power cable to meter power jack and to AC power source.

(2) Remove the shorting cap on the BNC connector. Install the combination glass pH/ATC electrode byplugging it into the input connection (push on and twist to

lock) and the ATC connector into the ATC jack.

(3) Press the Mode button to select pH.

(4) Press the Setup button, then press the Enter button to clear the existing standardization.

(5) Press the Setup button, then press the Enter button to select a new set of buffers "1.68 4.01 6.86 9.18 12.46".

(6) Remove the electrode from the bottle of storage solution. Rinse with distilled water. Blot dry electrode (do not wipe). Immerse the electrode in pH 6.86 buffer. Swirl the solution to fully saturate the electrode with buffer. Allow the electrode to reach a stable value and the digital display indicates "S".

(7) Press the Standardize button. The meter recognizes the buffer and flashes a buffer icon. When the signal is stable, or when you press the Enter button, the buffer is entered and the digital display indicates "6.86".

(8) Remove the electrode from the pH 6.86 buffer, rinse and blot dry the electrode. Immerse the electrode in the pH 4.01 buffer and swirl. Press the Standardize button again to calibrate with this buffer. The meter will display a calibration slope and the two buffer icons, "4.01 6.86", then return to the measure screen.

(9) Now the meter should be calibrated andready to use for measuring the pH of any solution. Immerse the electrode in the solution to be measured, stir, allow time for the electrode to stabilize, and record the display.

3. Notes

(1) Before first use of your glass electrode, or whenever the electrode is dry, soak over night in an electrode filling solution, KCl solution or electrode storage solution.

(2) Rinse and blot dry the electrode between each measurement (do not wipe). Rinse the electrode with distilled water or deionized water, or part of the next solution to be measured.

(3) The pH meter allows automatic standardization using up to three buffers. When you enter a fourth buffer, the buffer farthest away is replaced by the pH of new buffer.

(4) The pH meter performs automatic temperature compensation. If an ATC probe is used, the meter continually adjusts for temperature. Therefore, buffers may vary slightly from the nominal values because of temperature. Default temperature is 25℃.

Chapter 6   Instructions for 7200 Spectrophotometer

I . Principles of Operation

Incident radiation of radiant Power I passes through a solution of an absorbing species at concentration c and path length b, and the transmitted radiation has radiant Power I. This radiant power is the quantity measured by spectrometric detectors. Bouguer in 1729 and Lambert in 1760 recognized that when electromagnetic energy is absorbed, the power of the transmitted energy decreases geometrically (exponentially). Since the fraction of radiant energy transmitted decays exponentially with path length, we can write it in exponential form:

$$T = \frac{I}{I_o} = 10^{-abc}$$

It is more convenient to omit the negative sign on the right-hand side of the equation and to define a new term:

$$A = -\lg T = \lg \frac{I_o}{I} = abc$$

Where A is the absorbance. This is the common form of Beer's law. Note that it is the absorbance that is directly proportional to the concentration. The path length b is expressed in centimeters and the concentration c in grams per liter. The constant a is called the absorptivity and is dependent on the wavelength and the nature of the absorbing material. In an absorption spectrum, the absorbance varies with wavelength in direct proportion to a (b and c are constant). The product of the absorptivity and the molecular weight of the absorbing species is called the molar absorptivity s. Thus,

$$A = \varepsilon bc$$

Where c is now in moles per liter.

II . Details of Operation

1. WFJ 7200 Spectrophotometer Setup

WFJ 7200 Spectrophotometer (Figure 1 – 40) measures the amount of visible light absorbed by a colored solution. It boasts a 325 nm to 1000 nm visible range, a 5nm bandwidth. absorbance. transmittance, concentration, and factor modes an easy-to-read LED.

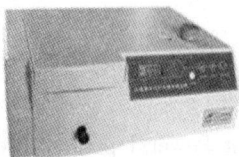

Figure 1 – 40　　Spectrophotometer

2. WFJ 7200 Spectrophotometer Operating Instructions

(1) Check that the instrument is turned on. Allow 20 min for warming up.

(2) Press the Mode button to select absorbance mode.

(3) Set the wavelength to the desired value using the knob on the top.

(4) Obtain a cuvette. Make sure it is clean inside and out. Rinse the cavette in, with small amounts of the blank. Then fill the cuvette about three-quarters full. Dry the outside of the cuvette carefully, and insert it into the sample compartment with its transparent sides aligned with the light path. Close the cover.

(5) Press 0A/100% T to set the absorbance of the blank to zero. After a few moments the BLA will be displayed. The absorbance should be 0. 000.

(6) To analyze your sample, insert sample cuvette and read the absorbance value on the LED.

(7) Turn off the instrument when finished.

3. Notes

(1) Sample cuvettes have two opposite sides that are cloudy and two opposite sides that are transparent, for the light to pass through. Place your cuvette in the sample compartment, with the transparent sides aligned with the light path.

(2) Never touch the transparent sides of the cuvette, since this will disrupt the path of the light beam.

(3) Rinse the cuvette thoroughly with distilled water and then with the liquid to be filled.

Part Two   Experiments in Basic Chemistry

Experiment 1   Common Ion Effect and Principle of Solubility Product

I . Learning Objects

(1) Observe the common ion effect on solubility.

(2) Understand the precipitation equilibrium and its shift.

(3) Judge the formation, dissolution, and transformation of the precipitates.

II . Principles

The shift in equilibrium caused by the addition of a compound having an ion in common with the dissolved substance is called the common-ion effect. For example:

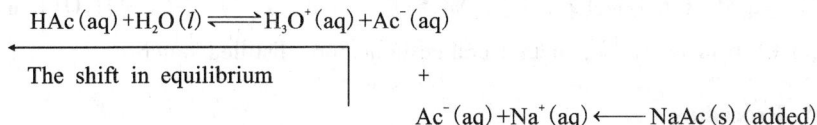

$$HAc\,(aq) + H_2O\,(l) \Longleftrightarrow H_3O^+\,(aq) + Ac^-\,(aq)$$

| The shift in equilibrium | + |

$$Ac^-\,(aq) + Na^+\,(aq) \longleftarrow NaAc\,(s)\ (added)$$

Because NaAc is a strong electrolyte, it provides Ac^- ion, which is present on the right side of the equation for acetic acid ionization. According to Le Ch ? telier's principle, the equilibrium composition should shift to the left. The equilibrium was disturbed by the addition of Ac^-, a shift in the equilibrium will occur that will reduce the concentration of Ac^-. The degree of ionization of acetic acid is decreased by the addition of NaAc. Consequently, the $[H^+]$ of the solution is reduced. The NaAc has the acetate ion in common with HAc, so the influence is known as the common-ion effect.

A general precipitation equilibrium can be expressed as follow,

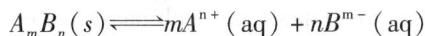

$$A_mB_n\,(s) \Longleftrightarrow mA^{n+}\,(aq) + nB^{m-}\,(aq)$$

At any moment, the ion product(IP) of the reaction is

$$IP(A_mB_n) \Longleftrightarrow c(A^{n+})^m \times c(B^{m-})^n$$

The ion product is the product of the molar concentrations of the ions raised to the power of their stoichiometric coefficients. The ion product has the same form as the solubility product expression, but the concentrations of substances are not necessarily equilibrium values.

At equilibrium, the ion product (IP) of the reactionbecomes equilibrium constant expression,

$$K_{sp}^{\ominus}(A_m B_n) = [A^{n+}]^m \times [B^{m-}]^n$$

Here, $K_{sp}^{\ominus}$ is called solubility product constant.

There are three possible relationships between $K_{sp}^{\ominus}$ and IP for a given compound.

IP $< K_{sp}^{\ominus}$ unsaturated solution, no precipitate occurs

IP $= K_{sp}^{\ominus}$ saturated solution

IP $> K_{sp}^{\ominus}$ supersaturated solution, precipitate occurs until reach a new equilibrium

III. Apparatus and Reagents

Test tubes, centrifuge tubes(10 mL), centrifuge(4000 r/min), stirring rod, medicine spoon, graduated cylinder.

HAc($0.2 \ mol \cdot L^{-1}$), NaAc($0.2 \ mol \cdot L^{-1}$), Pb(NO_3)$_2$($0.1 \ mol \cdot L^{-1}$), saturated PbI_2 solution, NaCl($1 \ mol \cdot L^{-1}$), K_2CrO_4($0.1 \ mol \cdot L^{-1}$), KI($0.1 \ mol \cdot L^{-1}$), AgNO$_3$($0.1 \ mol \cdot L^{-1}$), Na$_2$S($0.1 \ mol \cdot L^{-1}$), NH$_3 \cdot$ H$_2$O($6 \ mol \cdot L^{-1}$), HCl($6 \ mol \cdot L^{-1}$), methyl orange indicator, distilled water.

IV. Procedure

1. Common ion effect

(1)Add 1 mL of $0.2 \ mol \cdot L^{-1}$ HAc solution to a test tube, then add one drop methyl orange solution, mix well and observe solution color, continue to add $0.2 \ mol \cdot L^{-1}$ NaAc solution dropwsie, what color appear? Explain the reason of change.

(2)Add 1 mL of saturated PbI_2 solution to a test tube, then add 5 drops of $0.1 \ mol \cdot L^{-1}$ KI solution, what happens? Explain the reason of change.

2. The application of the principle of solubility product

(1) Formation of the precipitates and multiphase ion equilibrium

Add 10 drops of $0.1 \ mol \cdot L^{-1}$ Pb(NO_3)$_2$ solution into a centrifuge tube, then add 5 drops of $1 \ mol \cdot L^{-1}$ NaCl solution, shake the tube slightly. When the precipitation reaction completes, Centrifuge any precipitate that forms, decant and save the supernatant (that is the liquid), To the supernatant add 5 drops of

0.1 mol $\cdot$ L^{-1} K$_2$ CrO$_4$, observe if there are precipitates forming, Explain the change.

(2) Transformation of precipitates

Add 2 drops of 1 mol $\cdot$ L^{-1} NaCl solution into a test tube, then add 1 drop of 0.1 mol $\cdot$ L^{-1} AgNO$_3$ and shake the tube thoroughly, what do you observe? Continue to add 3 drops of 0.1 mol $\cdot$ L^{-1} KI solution, stir the solution with glass rod, record the color change of the precipitate, explain the change and write the ion reaction equation concerned.

(3) Fractional precipitates

Add 5 drops of 0.1 mol $\cdot$ L^{-1} Na$_2$S solution and 2 drops of 0.1 mol $\cdot$ L^{-1} K$_2$ CrO$_4$ solution into a test tube, dilute the mixture to 5 mL with distilled water, then dropwisely add 0.1 mol $\cdot$ L^{-1} Pb(NO$_3$)$_2$ solution (~ 3 drops), shake the mixture thoroughly and boil it for several minutes, let the solution cool down and stand for several minutes, observe the color the precipitate first formed. After the precipitate subsides, continue to add Pb(NO$_3$)$_2$ solution, observe the color change of the precipitate, Interpret your results in terms of solubility product.

(4) Dissolution of the precipitates

① Take a small amount of CaCO$_3$ solid into a test tube, add 2 mL water, observe if it dissolves. Then add several dorps of 6 mol $\cdot$ L^{-1} HCl, observe and explain what happens, write the ion reaction equation.

② Add 10 drops of 0.1 mol $\cdot$ L^{-1} AgNO$_3$ solution into a test tube and then add 3 drops of 1 mol $\cdot$ L^{-1} NaCl, what happens?

Dropwisely add 6 mol $\cdot$ L^{-1} NH$_3$ $\cdot$ H$_2$O into the solution mentioned above, what happens? Explain and write the reaction equation.

V. Questions

(1) What is the difference between common ion effect and salt effect? Do they coexist?

(2) When CaCl$_2$ is added into CaSO$_4$ saturated solution, what happens? Explain the phenomenon in terms of Le Chatelier's principle.

(3) How to judge the formation, dissolution, and transformation of the precipitates.

Experiment 2　Preparation and Properties of Buffer Solutions

I. Learning Objects

(1) Learn the properties of buffer solutions

(2) Learn the preparation of buffer solutions

(3) Grasp the operation methods of pH meter and Mohr measuring pipet.

II. Principles

A buffer solution, called buffer system (or a buffer pair) contains both an acid species and a base species in equilibrium (Table 1 – 2).

Table 1 – 2　Selected Buffer Systems

Buffer system	Conjugate acid	Conjugate base	Proton – transfer equilibrium	pK_a (at 25℃)
HAc – NaAc	HAc	Ac^-	$HAc + H_2O \rightleftharpoons Ac^- + H_3O^+$	4. 76
Tris · HCl – Tris	Tris · H^+	Tris	Tris · $H^+ + H_2O \rightleftharpoons$ Tris $+ H_3O^+$	7. 85
$NH_4Cl – NH_3$	NH_4^+	NH_3	$NH_4^+ + H_2O \rightleftharpoons NH_3 + H_3O^+$	9. 25
$NaH_2PO_4 – Na_2HPO_4$	$H_2PO_4^-$	HPO_4^{2-}	$H_2PO_4^- + H_2O$	
$HPO_4^{2-} + H_3O^+$	7. 21			

An example would be a solution containing HAc and the salt NaAc. Thus this buffer solution contains a significant concentration of HAc molecules and Ac^- ions. The equilibrium is illustrated as follows:

$$HAc(aq) + H_2O(l) \rightleftharpoons H_3O^+(aq) + Ac^-(aq)$$

　　　　　　↑　　　　　　　　　　　　　　　↑

large concentration　　large concentration due to NaAc

Here the acetate ion (the conjugate base) plays a role in resisting the change in pH when a strong acid is added, and is called as an antiacid component.

(added acid)　$Ac^-(aq) + H_3O^+(aq) \rightleftharpoons HAc(aq) + H_2O(l)$

HAc (the conjugate acid) can resist a change in pH when the addition of a strong base, and is called as an antibase component.

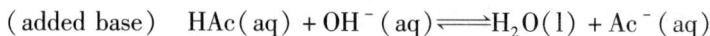

(added base) $HAc(aq) + OH^-(aq) \rightleftharpoons H_2O(l) + Ac^-(aq)$

The amounts of weak acid and weak base in the buffer must be significantly larger than the amounts of H_3O^+ or OH^- that will be added, otherwise, the pH cannot remain approximately constant.

The pH of a buffer can be calculated by an equationknown as the Henderson-Hasselbalch equation. This equation expresses the pH in terms of the pK_a for the acid and the ratio of concentrations of HA and A^-.

$$pH = pKa + \log[c(B^-)/c(HB)]$$

The buffer capacity is defined as the amount of strong acid or base needed to change the pH of one liter of buffer by 1 unit.

$$\beta = \frac{dn_{a(b)}}{V|dpH|}$$

Where β is the buffer capacity and has units of moles per liter per pH ($mol \cdot L^{-1} \cdot pH^{-1}$); $dn_{a\,(or\,b)}$ stands for moles of strong acid or strong base which are added to a buffer solution to cause the change in pH, dpH.

The magnitude of β indicates the relative strength of buffer capacity. The larger the value of β, the greater is the capacity of the buffer to resist changes in pH.

Buffer capacity is largely affected by two factors: buffer ratio and total concentration of buffer.

III. Apparatus and Reagents

Geiser buret (50 mL), Mohr measuring pipet (10 mL, 1 mL), pH meter, beakers, rubber pipet bulb, test tubes, graduated cylinder.

$Na_2HPO_4(0.2\ mol \cdot L^{-1})$, $KH_2PO_4(0.2\ mol \cdot L^{-1}, 2\ mol \cdot L^{-1})$, $NaOH(2\ mol \cdot L^{-1})$, $HCl(1\ mol \cdot L^{-1})$, $NaOH(1\ mol \cdot L^{-1})$, $NaCl(0.9\%, W/V)$, distilled water.

IV. Procedure

1. Preparation of buffer solution

(1) Calculate the volumes (mL) of $0.2\ mol \cdot L^{-1}Na_2HPO_4$ and $0.2\ mol \cdot L^{-1}$ KH_2PO_4 required for preparing 80 mL buffer with pH = 7.40. ($pKa_2(H_3PO_4) = 7.21$)

(2) Based on the results of calculation, take theNa$_2$HPO$_4$ and KH$_2$PO$_4$ solution

with graduated cylinder into a 150 mL beaker and mix the solution well. Measure the pH of the buffer you prepared with a pH meter.

(3) Adjust the pH of the buffer to 7. 4 with 2 mol $\cdot$ L^{-1} NaOH and 2 mol $\cdot$ L^{-1} KH$_2$PO$_4$. Then the buffer is ready for next step.

2. Properties of buffer solution

(1) To two 50 mL beakers add 0. 9% (W/V) NaCl 25. 00 mL with pipet and measure their pH with pH meter. Then, add 1 mol $\cdot$ L^{-1} NaOH 0. 25 mL with measuring pipet to one beaker and add 1 mol $\cdot$ L^{-1} HCl 0. 25 mL with measuring pipet to another, measure their pH respectively, record the pH and make explanation.

(2) To two beakers add buffer solution you prepared 25. 00 mL with pipet and measure their pH with pH meter; Then, add 0. 25 mL of 1 mol $\cdot$ L^{-1} NaOH with measuring pipet to one beaker and add 0. 25 mL of 1 mol $\cdot$ L^{-1} HCl with measuring pipet to another, measure their pH respectively, record the pH, calculate buffer capacity, β, and make explanation (Table1 − 3).

Table1 − 3   Effect of acid and base on pH of different solutions

No.	Volumes of solution (mL)	pH	Volumes of acid or base added (mL)	pH	ΔpH	β
1	0.9% W/V NaCl 25.00 mL		1 mol $\cdot$ L^{-1} HCl 0.25 mL			
2	0.9% W/V NaCl 25.00 mL		1 mol $\cdot$ L^{-1} NaOH 0.25 mL			
3	Buffer 25.00 mL (pH = 7.40)		1 mol $\cdot$ L^{-1} HCl 0.25 mL			
4	Buffer 25.00 mL (pH = 7.40)		1 mol $\cdot$ L^{-1} NaOH 0.25 mL			

V. Questions

(1) How to prepare pH = 7.40 buffer solution?

(2) In a Na$_2$HPO$_4$ and KH$_2$PO$_4$ buffer system, what are antiacid component and antibase component? Explain the mechanism of this buffer.

(3) When a buffer is diluted, does it remain the same buffer capacity?

Experiment 3   Determination of Ionization Constant of Acetic Acid

I . Learning Objects

(1) Learn how to use basic burette.

(2) Consolidate the ability to use volumetric flask, pipet, and pH meter.

(3) Learn and grasp how to determine the ionization constant of weak electrolytes.

II . Principles

In aqueous solution, acetic acid(CH_3COOH, HAc) is a weak electrolyte, which ionizes and reach an equilibrium at a given temperature in water as below

$$HAc + H_2O \rightleftharpoons H_3O^+ + Ac^-$$

The equilibrium constant (ionization constant of HAc, $K_{a(HAc)}$) of this reaction is expressed as

$$K_{a(HAc)} = \frac{[H_3O^+][Ac^-]}{[HAc]} \tag{1}$$

Taking negative logarithm of both sides of equation (1), we have

$$-\log K_{a(HAc)} = -\log([H_3O^+][Ac^-]/[HAc])$$

Mathematically, we can write above equation as below

$$-\log K_{a(HAc)} = -\log[H_3O^+] - \log[Ac^-]/[HAc]$$

Because $-\log[H_3O^+] = p[H_3O^+] = pH$ and $-\log K_{a(HAc)} = pK_{a(HAc)} = pK_a$, we have

$$pK_a = pH - \log[Ac^-]/[HAc] \tag{2}$$

Equation (2) can also be written as

$$pK_a = pH + \lg \frac{[HAc]}{[Ac^-]} \tag{3}$$

Obviously, when $[HAc] = [Ac^-]$, $\log[HAc]/[Ac^-] = \log 1 = 0$, so, $pK_a = pH$.

When we titrate HAc with NaOH, the following reaction occurs

$$HAc + NaOH^- \rightleftharpoons NaAc + H_2O$$

According to this reaction, when one half of the total amount of HAc is consumed by NaOH, in the final mixture, the concentration of HAc is approximately equal to that of NaAc. In other word, $[HAc] = [Ac^-]$. If we measure the pH of this mixture, we can easily calculate the value of pK_a since $pK_a = pH$ at this point.

Ⅲ. Apparatus and Reagents

pH meter, volumetric flask(100 mL), Mohr buret(50 mL), Erlenmeyer flask (250 mL), beakers(50 mL, 100 mL), measuring pipet(5 mL, 10 mL), transfer pipet(25 mL, 50 mL), rubber bulb NaOHstandard solution(about 0.2 mol $\cdot$ L^{-1}), HAc solution (0.2 mol $\cdot$ L^{-1}), standard buffer solution (pH = 4.00), Phenolphthalein indicator.

Ⅳ. Procedure

1. Pipet 25.00 mL of 0.1 mol $\cdot$ L^{-1} HAc solution into a 250 mL Erlenmeyer flask, add 2 drops of Phenolphthalein indicator and stir, then titrate the HAc solution with NaOH standard solution until a weak pink color remains for about 30s. Stop the titration and record the buret reading. Repeat the titration two more times, the difference between any two volumes of NaOH consumed should be less than 0.10 mL. Calculate the average volume of NaOH consumed by three titrations.

2. Pipet 25.00 mL of 0.1 mol $\cdot$ L^{-1} HAc solution into a 100 mL beaker, accurately add NaOH standard solution(using basic buret) by one half of the average volume calculated in Step 1, thoroughly mix the solution, finally measure and record the pH of the solution.

3. Repeat step 2 two more times.

V. Data and calculations

1. Titration of acetic acid solution

Table 1 – 4   Determination of Volume of NaOH

	1	2	3
V_{HAc}/mL	25.00	25.00	25.00
V_{NaOH} consumed /mL			
$\overline{V_{NaOH}}$ consumed / mL			
$\frac{1}{2}\overline{V_{NaOH}}$ consumed /mL			

Table1 – 5   Determination of ionization constant of HAc Temperature _____ ℃

	1	2	3
V_{HAc} in beaker / mL	25.00	25.00	25.00
$\frac{1}{2}\overline{V_{NaOH}}$ added /mL			
pH			
K_a			
Average of K_a			

VI. Questions

(1) How do the ionization constant, K, and ionization degree, α, of acetic acid change with temperature and concentration?

(2) Why was phenolphthalein selected? What is the role of an indicator?

(3) Do you know any other method to determine ionization constant of acetic acid?

Experiment 4   Preparation and Standardization
of Acid and Base Standard Solution

Ⅰ. Learning Objects

(1) Learn the procedure for cleaning volumetric apparatus.

(2) Understand the principles of acid-base titration.

(3) Learn the method of preparation of acid and base standard solutions

(4) Learn how to use the buret and judge the endpoints of the titration.

Ⅱ. Principles

In an acid-base titration the standard solutions cannot be prepared directly because common acids and bases contain some impurities and are unstable. At first, prepare solutions of the approximate concentrations required, and then standardize these solutions.

1. The standardization of HCl solution

HCl solution is generally standardized by primary standard substance, Na_2CO_3。

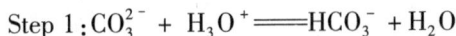

$$Step\ 1: CO_3^{2-} + H_3O^+ \rightleftharpoons HCO_3^- + H_2O$$

$$Step\ 2: HCO_3^- + H_3O^+ \rightleftharpoons H_2CO_3 + H_2O$$

The solution at equivalence point is acidic, so, methyl orange is selected as indicator. At the endpoint, the color of the solution is changed from yellow to orange (endpoint color).

The concentration of HCl can be calculated using the following expression:

$$C(HCl) \cdot V(HCl) = \frac{W(Na_2CO_3)}{M(\frac{1}{2}Na_2CO_3)} \times 1000$$

$W(Na_2CO_3)$——mass of Na_2CO_3 used in each titration(g)

$M(\frac{1}{2}Na_2CO_3)$——$\frac{1}{2}Na_2CO_3$(g/mol)

$V(HCl)$——volume of HCl solution used in each titration(mL)

$C(HCl)$——the exact concentration of HCl(mol/L)

＊The standardization of NaOH solution

NaOH   solution   is   generally   standardized   by   primary   standard substance, $KHC_8H_4O_4$ or $H_2C_2O_4 \cdot 2H_2O$。

2. The determination of NaOH concentration using HCl standard solution

$$H_3O^+ + OH^- = 2H_2O$$

At equivalence point $n(HCl) = n(NaOH)$

The concentration of NaOH can be calculated using the following expression,

$$C_{(HCl)} \times V_{(HCl)} = C_{(NaOH)} \times V_{(NaOH)}$$

III. Apparatus and Reagents

Geiser buret (50 mL), Mohr buret (50 mL), transfer pipet (25 mL), Erlenmeyer flask (250 mL), buret stand, wash bottle, rubber pipet bulb, alcohol burner, graduated cylinder, volumetric flask, electronic balance, beakers (50 mL, 100 mL, 500 mL), medicine dropper.

Standardized HCl solution, NaOH sample solution, anhydrous Na2CO3, methyl orange indicator, Phenolphthalein indicator.

IV. Procedure

1. Preparation of approximate 0.1 mol $\cdot$ L^{-1} HCl and 0.1 mol $\cdot$ L^{-1} NaOH solutions

(1) Measure out about 25 mL of 2 mol $\cdot$ L^{-1} HCl solution and add about 475 mL of distilled water, stir thoroughly and finally labelthe bottle.

(2) Measure out about 25 mL of 2 mol $\cdot$ L^{-1} NaOH solution and add about 475 mL of distilled water, stir thoroughly and finally labelthe bottle.

2. Preparation of Na_2CO_3 standard solution

Precisely weigh out 0.48 ~ 0.52g of primary standard substance, Na_2CO_3, place it in a clean 50 mL beaker, dissolve it with 30 mL distilled water, stir the solution with glass rod and make Na_2CO_3 dissolved completely. Then, transfer the solution with stirring rod into a 100 mL volumetric flask carefully, wash the inner wall of the beaker with a small amount of distilled water 3 times, transfer the washing water of each time into the volumetric flask, this is to ensure that the Na_2CO_3 weighed is quantitatively transferred into the flask. Then, add distilled water to bring the solution to the calibration mark (the last few drops should be added with a medicine dropper), mix the solution thoroughly by inverting the flask and shaking.

3. Standardization of $0.1 \ mol \cdot L^{-1}$ HCl solution

(1) Clean the buret and check the tip for leakage. If no leakage is observed, rinse your cleaned buret oncewith distilled water and two or three times with the prepared $0.1 \ mol \cdot L^{-1}$ HCl solution(about 10 mL each time). Add the acid solution into the buret and work air bubbles out.

(2) Adjust the meniscus surface of the solution to zero mark or a little lower and remain for one minute. Record the initial reading(the volume is read to the nearest 0.01 mL).

(3) Suck out 25.00 mL Na_2CO_3 solution from volumetric flask with transfer pipet and put it into a clean 250 mL Erlenmeyer flask(but not necessarily dry), add 2 drops of methyl orange, place a sheet of white paper under the flask to aid in the detection of any color change. The solution shows yellow.

(4) Titratethe Na_2CO_3 solution in the flask with the acid solution from the buret, note the solution color as the drops of HCl solution hit theNa$_2$CO$_3$ solution. Swirl the liquid in the flask gently and continuously as you add the HCl solution to drive out CO_2(for more accurate determination, it is necessary to heat the solution to expel CO_2). When the orange color begins to change, slow down the rate of addition of HCl solution. Rinse down the inside of the flask with a jet of distilled water from your wash bottle just before the termination of the titration. In the final stages of the titration add the HCl solution drop by drop, stirring between dorps, until the orange color persists for about 30s, carefully record the final reading on the HCl buret.

(5) Repeat the titration two more times. The deviation between any two volumes of HCl should be less than 0.05 mL. Calculate the accurate concentration(keep 4 significant figures) of HCl solution.

4. Determination of the accurate concentration of $0.1 \ mol \cdot L^{-1}$ NaOH solution

(1) Treatthe Mohr buret with $0.1 \ mol \cdot L^{-1}$ NaOH solution like the manner used in treating acid buret with0.1 $mol \cdot L^{-1}$ HCl solution . add the NaOH solution into the buret.

(2) Adjust the meniscus surface of the solution to zero mark or a little lower and remain for one minute. Record the initial reading(the volume is read to the nearest 0.01 mL).

(3) Suck out 25.00 mL standardizedHCl solution into a clean 250 mL Erlenmeyer flask(but not necessarily dry), add 2 drops ofPhenolphthalein, place a

sheet of white paper under the flask to aid in the detection of any color change. The solution shows colorless.

(4) Titratethe HCl solution in the flask with the NaOH solution form the buret, note the solution color as the drops of NaOH solution hit the HCl solution. Swirl the liquid in the flask gently and continuously as you add the NaOH solution. When the pink color begins to change, slow down the rate of addition of NaOH solution. Rinse down the inside of the flask with a jet of distilled water from your wash bottle just before the termination of the titration. In the final stages of the titration add the NaOH solution drop by drop, stirring between dorps, until the pink color persists for about 30s. Carefully record the final reading on the buret.

(5) Repeat the titration two more times. The deviation between any two volumes of NaOH should be less than 0.05 mL. Calculate the accurate concentration(keep 4 significant figures) of NaOH solution.

V. Data and calculations

Table 1 −6   Standardization of 0.1 mol · L^{-1} HCl solution

No. items			1	2	3
V(Na$_2$CO$_3$)/mL			25.00	25.00	25.00
V(HCl)		Initial buret reading/mL			
		Final buret reading/mL			
		V(HCl) consumed/mL			
Accurate concentration of HCl/mol · L^{-1}					
Average ofAccurate concentration of HCl/mol · L^{-1}					
Relative average deviation, $R\overline{d}$, (%)					

Table 1 -7   Determination of the accurate concentration of 0.1 mol · L^{-1} NaOH solution

items		No.	1	2	3
$V(\text{HCl})/\text{mL}$			25.00	25.00	25.00
NaOH	Initial buret reading/mL				
	Final buret reading/mL				
	$V(\text{NaOH})$ consumed/mL				
Accurate concentration of NaOH/mol · L^{-1}					
Average ofAccurate concentration of NaOH/mol · L^{-1}					
Relative average deviation, $R\bar{d}$, (%)					

VI. Questions

(1) The initial reading should be set at or veryclose to the zero mark in each titration, why?

(2) TheErlenmeyer flask cannot be rinsed by the solution to be titrated, why?

(3) The buret must be rinsed by the solution to be filled, why?

(4) Methyl orange was used as the indicator in this experiment. What is the role when selecting indicator? Why was methyl orange used and not phenolphthalein indicator?

Experiment 5   Redox Reaction and Electrode Potential

I . Learning Objects

(1) Qualitatively compare the electrode potential of redox couple.

(2) Understand the relationship between redox reaction and electrode potential.

(3) Understand the effect of concentration and acidity on redox reations.

II . Principles

$$pOx + ne - = qRed$$

The reduction state in a redox couple with lower φ^θ is stronger reducing agent. The oxidation state in a redox couple with higher φ^θ is stronger oxidizing agent.

When writing a spontaneous redox reaction, the left side (reactants) must contain the stronger oxidizing and reducing agents to ensure the cell potential of the voltaic cell constructed is more than zero(Table 1 -8).

Table 1 -8   Selected Standard electrode Potential

electrodes	φ^θ/V
$\varphi^\theta(Fe^{3+}/Fe^{2+})$	0.771
$\varphi^\theta(I_2/I^-)$	0.536
$\varphi^\theta(Br_2/Br^-)$	1.066
$\varphi^\theta(MnO_4^-/Mn^{2+})$	1.507
$\varphi^\theta(MnO_2/Mn^{2+})$	1.224
$\varphi^\theta(Cu^{2+}/Cu)$	0.324
$\varphi^\theta(Cl_2/Cl^-)$	1.358
$\varphi^\theta(H_2O_2/H_2O)$	1.776
$\varphi^\theta(O_2/H_2O_2)$	0.695
$\varphi^\theta(Zn^{2+}/Zn)$	-0.762

For the redox couple, the Nernst equation is

$$\varphi_{(Ox/Red)} = \varphi^{\theta}_{(Ox/Red)} + \frac{0.05916V}{n} \lg \frac{(c_{Ox})^p}{(c_{Red})^q}$$

Based on the equation, we can deduce that:

(1) The value of the standard electrode potential, $\varphi^{\theta}(Ox/Red)$, is related to the nature of oxidant and reductant, but not to the concentration.

(2) The electrode potential, $\varphi_{Ox/Red}$, depends on the nature of oxidant and reductant, the concentrations of the species involved in half-cells, and the temperature.

(3) At a given temperature, the value of $\varphi_{(Ox/Red)}$ increases with increasing the concentrations of oxidized species or decreasing the concentrations of reduced species. Conversely, $\varphi_{(Ox/Red)}$ decreases with decreasing the concentrations of oxidized species or increasing the concentrations of reduced species.

III. Apparatus and Reagents

Voltmeter(or pH meter), test tubes, alcohol burner, beakers, test tube racks, dropping bottles, salt bridge.

$ZnSO_4$(1 mol $\cdot$ L^{-1}), $CuSO_4$(1 mol $\cdot$ L^{-1}, 0.001 mol $\cdot$ L^{-1}), $FeSO_4$ (1 mol $\cdot$ L^{-1}), KI(0.1 mol $\cdot$ L^{-1}), $FeCl_3$(0.1 mol $\cdot$ L^{-1}), CCl_4, K_3[Fe (CN)$_6$] (0.1 mol $\cdot$ L^{-1}), KBr(0.1 mol $\cdot$ L^{-1}), $FeSO_4$(0.1 mol $\cdot$ L^{-1}), Br_2 water(saturated solution), KSCN(0.1 mol $\cdot$ L^{-1}), I_2 saturated solution, HNO_3(12 mol $\cdot$ L^{-1}, 2 mol $\cdot$ L^{-1}), small pellet of zinc, H_2SO_4(1 mol $\cdot$ L^{-1}), KIO_3(0.1 mol $\cdot$ L^{-1}), NaOH(6 mol $\cdot$ L^{-1}), $KMnO_4$(0.01 mol $\cdot$ L^{-1}), Na_2SO_3(0.1 mol $\cdot$ L^{-1}), $H_2C_2O_4$(2 mol $\cdot$ L^{-1}), $MnSO_4$(0.2 mol $\cdot$ L^{-1}), NH_4F(3 mol $\cdot$ L^{-1}), $H_2 O_2$(0.1 mol $\cdot$ L^{-1}), $NaBiO_3$(solid), $Na_2 S_2 O_3$(0.05 mol $\cdot$ L^{-1}), Cl_2 water (saturated solution), distilled water. copper strip, zinc strip.

IV. Procedure

1. Measure cell potentials and compare the strength of electrode potentials

(1) Based on the following figure and table information, assemblethe apparatus to construct a primary cell.

Table 1 −9   Construction of primary cell

A beaker (50 mL)	B beaker(50 mL)
1 、Zn I ZnSO$_4$ (1 mol · L^{-1})	CuSO$_4$ (1 mol · L^{-1}) I Cu
2 、Zn I ZnSO$_4$ (1 mol · L^{-1})	CuSO$_4$ (0.001 mol · L^{-1}) I Cu
3 、Cu I CuSO$_4$ (0.001 mol · L^{-1})	CuSO$_4$ (1 mol · L^{-1}) I Cu

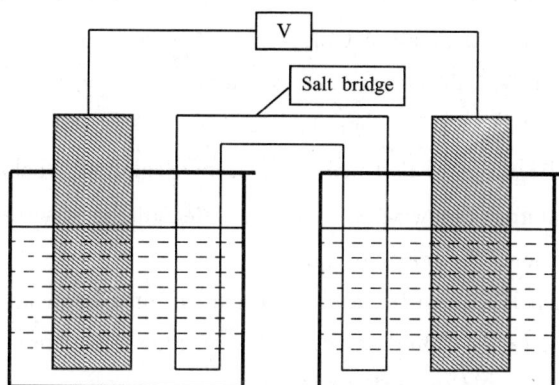

Flgure 2 −1   primary cell Piagram

(2) Recordthe cell potential for each of the cells, list the strength of the electrode potential from the lowest to the highest.

(3) Explain why is the electrode potential of "CuSO$_4$ (1 mol · L^{-1}) I Cu" higher than that of   "CuSO$_4$ (0.001 mol · L^{-1}) I Cu".

2. Relationship between electrode potential and redox reaction

(1) Qualitatively compare with electrode potential:

① Take two test tubes, add 2 drops of 0.1 mol · L^{-1} FeCl$_3$ solution and 5 drops of CCl$_4$ to each tube. Then add 10 drops of 0.1 mol · L^{-1} KI solution to one tube and 10 drops of 0.1 mol · L^{-1} KBr solution to another tube, shake the test tubes sufficiently. Observe the color change in CCl$_4$ layer and explain the phenomenon. Write down their reaction equations.

② Take two test tubes, add 10 drops of bromine water to one tube, add 10 drops of iodine water to another one. Then, add 10 drops of 0.1 mol · L^{-1} FeSO$_4$ solution in each tube, shake, finally add 5 drops of 0.1 mol · L^{-1} KSCN solution.

Observe the color change and explain the phenomenon. Write down their reaction equations.

③ Compare with the electrode potentials of I_2/I^-, Br_2/Br^-, and Fe^{3+}/Fe^{2+}. Point out the strongest oxidant and reductant among these redox couples. Deduce the relationship between electrode potential and redox reaction.

(2) Relativity of redox:

① Oxidation of H_2O_2: To a test tube add 10 drops of $0.1\ mol \cdot L^{-1}$ KI solution, then add 3 drops of $1.0\ mol \cdot L^{-1}$ H_2SO_4 solution, 2 drops of $0.1\ mol \cdot L^{-1}$ H_2O_2 solution and 20 drops of CCl_4, shake the tube. Record the phenomenon and explain the change with the reaction equations.

② Reduction of H_2O_2: To a test tube add 5 drops of $0.01\ mol \cdot L^{-1}$ $KMnO_4$ solution, then add 3 drops of $1.0\ mol \cdot L^{-1}$ H_2SO_4 solution, shake. Now add $0.1\ mol \cdot L^{-1}$ H_2O_2 solution dropwise with stirring. Record the observation and explain the change, write down the reaction equation.

Conclude when H_2O_2 can be used as an oxidant or a reductant and indicate its oxidized product or reduced product with redox reaction equations.

3. The effect of pH on redox reaction

(1) Take two test tubes, add 10 drops of $0.1\ mol \cdot L^{-1}$ KBr solution and 5 drops of CCl_4 to each tube. Then add 5 drops of $1.0\ mol \cdot L^{-1}$ H_2SO_4 solution to one tube and 5 drops of $6.0\ mol \cdot L^{-1}$ HAc solution to another tube, shake the test tubes sufficiently. Then to each tube add 1 drop of $0.01\ mol \cdot L^{-1}$ $KMnO_4$ solution. Compare the reaction rate and explain it.

(2) To three test tubes add 10 drops of $0.1\ mol \cdot L^{-1}$ Na_2SO_3 solution. Then to one test tube add 10 drops of $1.0\ mol \cdot L^{-1}$ H_2SO_4 solution, to the second tube add 10 drops of distilled water, and to the third tube add 10 drops of $6.0\ mol \cdot L^{-1}$ NaOH solution, shake them all sufficiently. Then add 5 drops of $0.01\ mol \cdot L^{-1}$ $KMnO_4$ solution into each tube. Record and explain the observation, write down the reaction equations.

V. Questions

(1) What factors will affect the electrode potential?

(2) What does the salt bridge function in a primary cell?

(3) Can we get the conclusion that the larger the cell potential, the higher

thereaction rate?

(4) Does a positive value for the standard cell potential, E^{θ}_{cell}, indicate a spontaneous reaction?

Experiment 6　Determination of the Composition of Base Mixture

I. Learning Objects

(1) Learn how to select a proper indictor for an acid-base titration.

(2) Learn how to determine the composition of base mixture.

(3) Understandthe "double endpoint method" for the determination of the composition of base mixture.

II. Principles

The base mixture is possibly constituted from either Na_2CO_3 and $NaHCO_3$, or Na_2CO_3 and NaOH, or only one compound, such as Na_2CO_3, $NaHCO_3$, and NaOH. In this experiment, HCl standard solution (standardized by primary standard substance) is used to titrate the base mixture. Phenolphthalein or Methyl orange or both(for "double endpoint method") are selected as indicator to show the endpoint of titration. We designate that V_1 is the volume of HCl standard solution consumed by the base mixture when Phenolphthalein is used as indicator, and that V_2 is the volume of HCl standard solution consumed by the base mixture when Methyl orange is used as indicator. Based on the size of V_1 and V_2, we can deduce what the composition and concentration are in the base mixture.

(1) If the base mixture is composed of Na_2CO_3 and $NaHCO_3$, when the HCl standard solution is added into the mixture, it reacts with Na_2CO_3 first to form HCO_3^-,

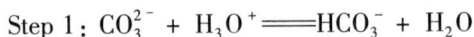

$$\text{Step 1: } CO_3^{2-} + H_3O^+ = HCO_3^- + H_2O$$

Since the pH at endpoint is approximate 8.34, Phenolphthalein, whose pH range of color change is 9.6 ~ 8.0, from red to colorless (the endpoint color is colorless), is a possible indicator for Step 1 titration. When the color of the base mixture just changes from red to colorless, we stop titration and record buret reading. The volume of HCl standard solution consumed in this step of titration is signified as V_1.

When we continue to add HCl standard solution into the base mixture(now it is colorless), HCl will react with HCO_3^- to form H_2CO_3 or CO_2. The HCO_3^- in the mixture comes from two parts, one is from $NaHCO_3$ which is originally existed in the sample, the other is from step 1 titration, the product of step 1 reaction. So,

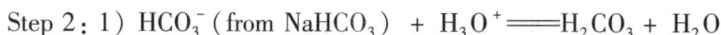

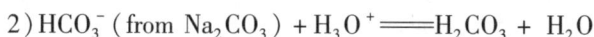

Step 2: 1) HCO_3^- (from $NaHCO_3$) $+ H_3O^+ \Longrightarrow H_2CO_3 + H_2O$

2) HCO_3^- (from Na_2CO_3) $+ H_3O^+ \Longrightarrow H_2CO_3 + H_2O$

Since the pH at endpoint is approximate 3.87, Methyl orange, whose pH range of color change is 4.4 ~ 3.1, from yellow to red (the endpoint color is orange), is a possible indicator for Step 2 titration. When the color of the base mixture just changes from yellow to orange, we stop titration and record buret reading. The volume of HCl standard solution consumed by the second titration is signified as V_2.

Obviously in this case, $V_2 > V_1$. In other word, if $V_2 > V_1$, we can deduce that the composition of base mixture is Na_2CO_3 and $NaHCO_3$.

Based on the information stated above, the concentration of Na_2CO_3 and $NaHCO_3$ can be calculated as following,

$$\rho(NaHCO_3) = \frac{(V_2 - V_1) \times c(HCl) \times M(NaHCO_3)}{V_{sample}} (mg/mL)$$

$$\rho(Na_2CO_3) = \frac{V_1 \times c(HCl) \times M(Na_2CO_3)}{V_{sample}} (mg/mL)$$

(2) If the base mixture is composed of Na_2CO_3 and $NaOH$, when the HCl standard solution is added into the mixture, it reacts with both Na_2CO_3 and $NaOH$ first to form HCO_3^- and H_2O,

Step 1: 1) $CO_3^{2-} + H_3O^+ \Longrightarrow HCO_3^- + H_2O$

2) $OH^- + H_3O^+ \Longrightarrow H_2O + H_2O$

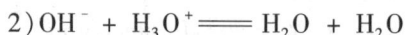

Likewise, the pH at endpoint is still approximate 8.34, Phenolphthalein is again selected as the indicator for Step 1 titration. When the color of the base mixture just changes from red to colorless, we stop titration and record buret reading. The volume of HCl standard solution consumed is signified as V_1.

We continue to add HCl standard solution into the base mixture (now it is colorless), HCl will react with HCO_3^- to form H_2CO_3 or CO_2. In this case, the HCO_3^- in the mixture only comes from reaction 1) in step 1, So,

Step 2: HCO_3^- (only from Na_2CO_3) $+ H_3O^+ \Longrightarrow H_2CO_3 + H_2O$

Since the pH at endpoint is approximate 3.87, Methyl orange again is selected as the indicator for step 2 titration. When the color the base mixture just changes from yellow to orange, stop titration and record buret reading. The volume of HCl standard solution consumed by the second titration is signified as V_2.

Obviously, in this case, $V_2 < V_1$. In other word, if $V_2 < V_1$, we can deduce that the composition of base mixture is Na_2CO_3 and NaOH.

Based on the information stated above, the concentration of Na_2CO_3 and NaOH can be calculated as following,

$$\rho(NaOH) = \frac{(V_1 - V_2) \times c(HCl) \times M(NaOH)}{V_{sample}}(mg/mL)$$

$$\rho(Na_2CO_3) = \frac{V_2 \times c(HCl) \times M(Na_2CO_3)}{V_{sample}}(mg/mL)$$

(3) If the base mixture is composed only of Na_2CO_3, when the HCl standard solution is added into the solution, it reacts with Na_2CO_3,

$$\text{Step 1: 1)} \quad CO_3^{2-} + H_3O^+ \Longrightarrow HCO_3^- + H_2O$$

Since the pH at endpoint is still approximate 8.34, Phenolphthalein is again selected as the indicator for Step 1 titration. Like that in case 1 and 2, record the volume of HCl standard solution consumed, signified as V_1 for step 1 titration. Continue to add HCl standard solution into the base mixture(now it is colorless), HCl will react with HCO_3^- to form H_2CO_3 or CO_2.

$$\text{Step 2:} \quad HCO_3^- (\text{only from } Na_2CO_3) + H_3O^+ \Longrightarrow H_2CO_3 + H_2O$$

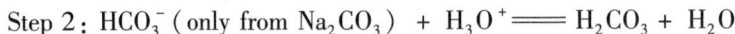

The pH at endpoint is approximate 3.87, Methyl orange again is selected as the indicator for Step 2 titration. When the color the base mixture just changes from yellow to orange, stop titration and record buret reading. The volume of HCl standard solution consumed by the Step2 titration is signified as V_2.

Obviously, in this case, $V_2 = V_1$. In other word, if $V_2 = V_1$, we can deduce that the composition of base mixture is Na_2CO_3 alone.

Based on the information stated above, the concentration of Na_2CO_3 can be calculated as following,

$$\rho(Na_2CO_3) = \frac{V_1(\text{or } V_2) \times c(HCl) \times M(Na_2CO_3)}{V_{sample}}(mg/mL)$$

In Case 1, 2 and 3, since there are two pH jumps in each case, we use two indicators to show the two endpoints of the titration respectively. This method is called as "double endpoint method" or "double indicator method".

(4) If the base mixture is composed only of $NaHCO_3$, when the HCl standard solution is added into the solution, it reacts only with $NaHCO_3$,

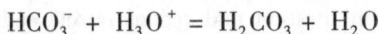

$$HCO_3^- + H_3O^+ = H_2CO_3 + H_2O$$

Since the pH at endpoint is approximate 3.87, Methyl orange again is selected as the indicator for the titration. When the color of the base mixture just changes from yellow to orange, we stop titration and record buret reading. The volume of HCl standard solution consumed by this step of titration is still signified as V_2.

Obviously, unlike that in the case 1, 2, and 3, reaction, $CO_3^{2-} + H_3O^+ = HCO_3^- + H_2O$, does not exist in this case, suggesting that there is only one pH jump and Phenolphthalein is not used, so, $V_1 = 0$. In other word, if $V_1 = 0$, $V_2 > 0$, we can deduce that the composition of base mixture is $NaHCO_3$ alone.

The concentration of $NaHCO_3$ can be calculated as following,

$$\rho(NaHCO_3) = \frac{V_2 \times c(HCl) \times M(NaHCO_3)}{V_{sample}}(mg/mL)$$

(5) If the base mixture is composed only of NaOH, when the HCl standard solution is added into the solution, it reacts with NaOH to form H_2O,

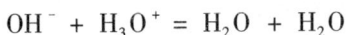

$$OH^- + H_3O^+ = H_2O + H_2O$$

The pH at endpoint is approximate 7.0, if phenolphthalein is again selected as the indicator for the titration, the color the base mixture changes from red to colorless at the endpoint. Once the color of the mixture becomes colorless, we stop titration and record buret reading. The volume of HCl standard solution consumed is signified as V_1.

Like that in case 4, only one pH jump exists in this case and Methyl orange is not used, indicating $V_2 = 0$. Obviously, if $V_1 > 0$, $V_2 = 0$, we can deduce that the composition of base mixture is NaOH alone.

The concentration of NaOH can be calculated as following,

$$\rho(NaOH) = \frac{V_1 \times c(HCl) \times M(NaOH)}{V_{sample}}(mg/mL)$$

In summary, according to the analysis described above, we can determine the composition and concentration of a base mixture composed of Na_2CO_3, $NaHCO_3$, and NaOH based on the size of volume of HCl standard solution consumed by each endpoint of titration.

III. Apparatus and Reagents

Geiser buret(50 mL), transfer pipet (25 mL), Erlenmeyer flask (250 mL), buret stand, wash bottle, rubber pipet bulb.

Sample solution A of Base mixture, Sample solution B of Base mixture, HCl standard solution, Phenolphthalein indicator, methyl orange indicator.

IV. Procedure

1. Determination of Sample Solution A

(1) Pipet 25.00 mL of the sample A solution into a 250 mL Erlenmeyer flask, add 2 drops of Phenolphthalein indicator, and titrate with HCl standard solution until the color just changes from red to colorless. Stop titration and record the buret reading and volume of HCl standard solution consumed (V_1), Then add 2 drops of methyl orange indicator and continue the titration until the color just changes from yellow to orange. Stop titration and record the buret reading and volume of HCl standard solution consumed only by the second titration (V_2).

(2) Repeat Step (1) at least two more times in order that you have three sets of data among which the biggest deviation should be less than 0.05 mL.

(3) Data and calculation

Table 1 – 10　Data of determination of sample solution A

NO. items	1	2	3
$V_{\text{sample A}}$ (mL)	25.00	25.00	25.00
V_{HCl} (initial) (mL)			
V_{HCl} (final – 1) (mL)			
V_1 (mL) $= V_{\text{HCl}}$ (final – 1) $- V_{\text{HCl}}$ (initial)			
V_{HCl} (final – 2) (mL)			
V_2 (mL) $= V_{\text{HCl}}$ (final – 2) $- V_{\text{HCl}}$ (final – 1)			
$\rho(\text{NaHCO}_3)$ (mg · mL^{-1}) Or, $\rho(\text{NaOH})$ (mg · mL^{-1})			
Average of $\rho(\text{NaHCO}_3)$ (mg · mL^{-1}) Or, Average of $\rho(\text{NaOH})$ (mg · mL^{-1})			
Relative average deviation, $R\bar{d}$, (%)			
$\rho(\text{Na}_2\text{CO}_3)$ (mg · mL^{-1})			
Average of $\rho(\text{Na}_2\text{CO}_3)$ (mg · mL^{-1})			
Relative average deviation, $R\bar{d}$, (%)			

2. Determination of sample solution B

(1) Pipet 25.00 mL of the sample B solution into a 250 mL Erlenmeyer flask, add 2 drops of Phenolphthalein indicator, and titrate with HCl standard solution until the color just changes from red to colorless. Stop titration and record the buret reading and volume of HCl standard solution consumed (V_1), Then add 2 drops of methyl orange indicator and continue the titration until the color just changes from yellow to orange. Stop titration and record the buret reading and volume of HCl standard solution consumed only by the second titration (V_2).

(2) Repeat step:

①at least two more times in order that you have three sets of data among which the biggest deviation should be less than 0.05 mL.

②Data and calculation.

Table 1 – 11   Data of determination of sample solution B

NO. / items	1	2	3
$V_{\text{sample A}}$ (mL)	25.00	25.00	25.00
V_{HCl} (initial) (mL)			
V_{HCl} (final – 1) (mL)			
V_1 (mL) = V_{HCl} (final – 1) – V_{HCl} (initial)			
V_{HCl} (final – 2) (mL)			
V_2 (mL) = V_{HCl} (final – 2) – V_{HCl} (final – 1)			
$\rho(\text{NaHCO}_3)$ (mg · mL^{-1}) Or, $\rho(\text{NaOH})$ (mg · mL^{-1})			
Average of $\rho(\text{NaHCO}_3)$ (mg · mL^{-1}) Or, Average of $\rho(\text{NaOH})$ (mg · mL^{-1})			
Relative average deviation, $R\bar{d}$, (%)			
$\rho(\text{Na}_2\text{CO}_3)$ (mg · mL^{-1})			
Average of $\rho(\text{Na}_2\text{CO}_3)$ (mg · mL^{-1})			
Relative average deviation, $R\bar{d}$, (%)			

V. Question

(1) Explain why the addition of too much indicator could lead to an inaccurate result.

(2) When sodium hydroxide, NaOH (in the flask) is to be titrated with hydrochloric acid (in the burette), which indicator should you select, why? (Suppose that only methyl orange and phenolphthalein indicator are available).

(3) If the base mixture is placed in theErlenmeyer flask too long a time before titration, does the composition of the base mixture change? Why?

Experiment 7   Determination of Concentration of Hydrogen Peroxide

I . Learning Objects

(1) Learn the method of preparation and standardization of $KMnO_4$ solution.

(2) Grasp the principles of determining H_2O_2 by $KMnO_4$.

(3) Practice and grasp the operation methods of transfer pipet, Erlenmeyer flask, and Mohr buret.

II . Principles

Since the oxidation number of oxygen in H_2O_2 is -1, hydrogen peroxide can function as either a oxidant or a reductant. In clinic, hydrogen peroxide is widely used as disinfector. In this experiment, we will analyze an unknown solution of hydrogen peroxide by titrating it with $KMnO_4$ solution. In a dilute sulfuric acid and at room temperature, H_2O_2 can be quantitatively oxidized by $KMnO_4$ to form O_2 and colorless Mn^{2+}. The reaction is:

$$5H_2O_2 + 2MnO_4^- + 6H^+ =\!=\!= 2Mn^{2+} + 5O_2 \uparrow \ + 8H_2O$$

Thus, when $KMnO_4$ solution is dropwisely added to an acidified H_2O_2 solution, each drop is decolorized until all the H_2O_2 is used up. The next drop added will remain the color of $KMnO_4$ (pink) which is the endpoint color. Based on the volume and the concentration of $KMnO_4$ solution (standard solution here), we will be able to calculate the number of moles of H_2O_2 oxidized.

$KMnO_4$ is a strong oxidant which is easily reduced by dust and other organic matter, it is not available as a primary standard. Therefore, the standard solution of $KMnO_4$ is prepared approximately and then standardized by a primary standard reagent. In neutral solution, the reduction product of $KMnO_4$ is insoluble MnO_2. So, before standardization $KMnO_4$ solution is boiled to react with any impurities and then filtered to remove MnO_2.

Sodium oxalate ($Na_2C_2O_4$) is often used as one of the primary standards to standardize $KMnO_4$ solution. The reaction is:

$$5C_2O_4^{2-} + 2MnO_4^- + 16H^+ =\!=\!= 2Mn^{2+} + 10CO_2 \uparrow \ + 8H_2O$$

This reaction proceeds somewhat slowly at the beginning. However, the Mn^{2+}

produced in the reaction can function as a catalyst to the reaction, therefore, with the increase of the amount of Mn^{2+} produced, the reaction is speeded up and goes to completion quickly.

According to the reaction stoichiometry, it can be derived that: $5n(KMnO_4)$ = $2n(Na_2C_2O_4)$, then, the concentration of $KMnO_4$ standard solution can be calculated easily.

III. Apparatus and Reagents

Geiser buret(50 mL), transfer pipet(25 mL), Erlenmeyer flask(250 mL), buret stand, wash bottle, rubber pipet bulb, volumetric flask(1000 mL, 250 mL).

30% H_2O_2 solution (sample), 0.05 mol·L^{-1} $KMnO_4$ standard solution, 3 mol·L^{-1} H_2SO_4 solution.

IV. Procedure

1. Preparation of H_2O_2 solution

(1) Pipet 25.00 mL of 30% H_2O_2 solution into a 1000 mL volumetric flask, and dilute it to the calibration mark with distilled water(the last few drops should be added with a medicine dropper), mix the solution thoroughly by inverting the flask and shaking(this solution can also be prepared by laboratory technician).

Pipet 25.00 mL of the solution prepared above from the 1000 mL volumetric flask into a 250 mL volumetric flask, then, add distilled water to bring the solution to the calibration mark(the last few drops should be added with a medicine dropper), mix the solution thoroughly by inverting the flask and shaking.

2. Determination of the concentration of H_2O_2 solution

(1) Suck out 25.00 mL of H_2O_2 solution diluted by Step 1 with transfer pipet and transfer it into a clean 250 mL Erlenmeyer flask(but not necessarily dry), add 5 mL of 3 mol·L^{-1} H_2SO_4 solution, swirl to mix. Titrate the H_2O_2 solution in the flask with $KMnO_4$ standard solution from the buret, note the solution color as the drops of $KMnO_4$ standard solution hit the H_2O_2 solution. Swirl the liquid in the flask gently and continuously as you add $KMnO_4$ standard solution. When the color of the solution begins to show pink, slow down the rate of addition of $KMnO_4$ standard solution. Rinse down the inside of the flask with a jet of distilled water from your wash bottle just before the termination of the titration. In the final stages of the titration add

KMnO$_4$ standard solution drop by drop, stirring between dorps, until the pink color persists for about 30s, carefully record the final reading on the acidic buret and the volume of KMnO$_4$ standard solution used.

(2) Repeat Step (1) at least two more times in order that you have three sets of data among which the biggest deviation should be less than 0.10 mL.

The concentration of H$_2$O$_2$ solution can be calculated by following method,

$$\rho(\text{H}_2\text{O}_2) = \frac{5}{2} \times \frac{c(\text{KMnO}_4) \times V(\text{KMnO}_4) \times M(\text{H}_2\text{O}_2)}{25.00 \times \dfrac{25.00}{250.0} \times \dfrac{25.00}{1000.0}} (\text{mg/mL})$$

V. Data and calculations

Table 1-12   Determination of H$_2$O Canlentration

items	Titration number		
	1	2	3
$V(\text{H}_2\text{O}_2)$ (mL)	25.00	25.00	25.00
$V(\text{KMnO}_4)$ (initial) (mL)			
$V(\text{KMnO}_4)$ (final) (mL)			
$V(\text{KMnO}_4)$ (used) (mL)			
ρ (H$_2$O$_2$) (mg/mL)			
Average of $\rho(\text{H}_2\text{O}_2)$ (mg/mL)			
Relative average deviation, $R\bar{d}$, (%)			

VI. Questions

(1) In this experiment, can we use HCl solution instead of 3 mol $\cdot$ L^{-1} H$_2$SO$_4$ solution to acidify the reaction system? Explain.

(2) Why does the color of the first drop ofKMnO$_4$ standard solution added into H$_2$O$_2$ solution disappear much slower than that of the latter drops?

(3) List andexplain the applications of H$_2$O$_2$ and KMnO$_4$ in clinic.

Experiment 8　Preparation and Properties of Coordination Compounds

I . Learning Objects

(1) Prepare coordination compounds.

(2) understand the difference between coordination compounds and simple compounds.

(3) Know the stability of coordination compounds and the factors that will influence the stability of coordination compounds.

(4) Learn the chelating effects.

II. Principles

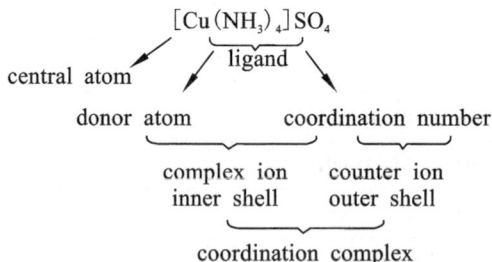

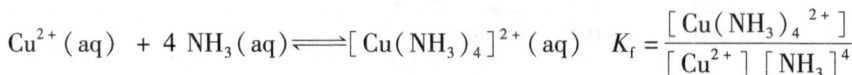

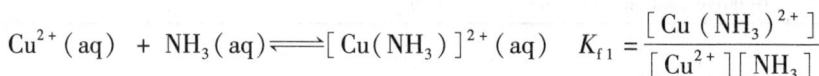

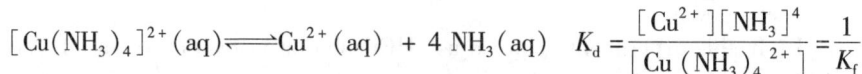

$$\underset{\substack{\text{central atom} \\ \text{donor atom}}}{\underbrace{[Cu}(\underbrace{NH_3)_4}]SO_4}_{\substack{\text{ligand} \\ \text{coordination number}}}$$

$$\underbrace{\underbrace{\text{complex ion}}_{\substack{\text{inner shell}}} \quad \underbrace{\text{counter ion}}_{\text{outer shell}}}_{\text{coordination complex}}$$

$$Cu^{2+}(aq) + 4\,NH_3(aq) \Longrightarrow [Cu(NH_3)_4]^{2+}(aq) \quad K_f = \frac{[Cu(NH_3)_4^{\,2+}]}{[Cu^{2+}][NH_3]^4}$$

$$Cu^{2+}(aq) + NH_3(aq) \Longrightarrow [Cu(NH_3)]^{2+}(aq) \quad K_{f1} = \frac{[Cu(NH_3)^{2+}]}{[Cu^{2+}][NH_3]}$$

$$Cu^{2+}(aq) + 4\,NH_3(aq) \Longrightarrow [Cu(NH_3)_4]^{2+}(aq) \quad K_f = K_{f1} \times K_{f2} \times K_{f3} \times K_{f4}$$

$$[Cu(NH_3)_4]^{2+}(aq) \Longrightarrow Cu^{2+}(aq) + 4\,NH_3(aq) \quad K_d = \frac{[Cu^{2+}][NH_3]^4}{[Cu(NH_3)_4^{\,2+}]} = \frac{1}{K_f}$$

The larger the K_f, the more stable the complex ion for ions of the same coordination number. For the different coordination number, compare the stability of the complex by calculation.

III. Apparatus and Reagents

Test tubes, alcohol burner, beakers, test tube racks, dropping bottles.

$H_2SO_4(1\ mol \cdot L^{-1})$, $NaOH(0.1\ mol \cdot L^{-1},\ 2\ mol \cdot L^{-1})$, $NH_3 \cdot H_2O$

$(2 \ mol \cdot L^{-1})$, NaCl (0.1 mol $\cdot$ L^{-1}), $BaCl_2$ (0.1 mol $\cdot$ L^{-1}), $FeCl_3$ (0.1 mol $\cdot$ L^{-1}), $AgNO_3$(0.1 mol $\cdot$ L^{-1}), $CuSO_4$ (0.1 mol $\cdot$ L^{-1}), $NiSO_4$(0.1 mol $\cdot$ L^{-1}, 0.5 mol $\cdot$ L^{-1}), $Na_2S_2O_3$ (0.1 mol $\cdot$ L^{-1}), KSCN(0.1 mol $\cdot$ L^{-1}), NH_4F(1 mol $\cdot$ L^{-1}), KBr(0.1 mol $\cdot$ L^{-1}), KI(0.1 mol $\cdot$ L^{-1}), $K_3[Fe(CN)_6]$ (0.1 mol $\cdot$ L^{-1}), I_2 saturated solution, CCl_4, $K_4P_2O_7$ (2 mol $\cdot$ L^{-1}), $(NH_4)_2Fe$ $(SO_4)_2$(solid), dimethylglyoxime(1%).

IV. Procedure

1. Formation and composition of complex ions

(1) To a test tube add 20 drops of 0.1 mol $\cdot$ L^{-1} $CuSO_4$ solution, then add 2.0 mol $\cdot$ L^{-1} $NH_3 \cdot H_2O$

solution drop by drop until a precipitate forms, what is the color of the precipitate? Continue to add more 2.0 mol $\cdot$ L^{-1} $NH_3 \cdot H_2O$ solution to the tube until the precipitate formed is dissolved. What is the product? Keep the resulting solution for next experiment.

(2) Equally separate the resulting solution of step (1) to two test tubes, then, add 0.1 mol $\cdot$ L^{-1} $BaCl_2$ solution to one tube and 0.1 mol $\cdot$ L^{-1} NaOH solution to the other, respectively. What happens? Explain and write the reaction equations.

2. Difference between simple ions andcomplex ions

To one test tube add 5 drops of 0.1 mol $\cdot$ L^{-1} $FeCl_3$ solution, to another test tube add 5 drops of 0.1 mol $\cdot$ L^{-1} $K_3[Fe(CN)_6]$ solution, then, to both tubes add 0.1 mol $\cdot$ L^{-1} KSCN solution. What happens? Explain and write the reaction equations.

3. Coordination equilibrium and redox reaction

Add 5 drops of 0.1 mol $\cdot$ L^{-1} $FeCl_3$ solution to one test tube, add 5 drops of $0.$ 1 mol $\cdot$ L^{-1} $K_3[Fe(CN)_6]$ solution to another tube, then, to both tubes add 1 drop of 0.1 mol $\cdot$ L^{-1} KI solution, shake the tube sufficiently. Observe the color change in the solution, explain the difference between two tubes, and write the reaction equations.

4. Coordination equilibrium and precipitate reaction

To a test tube add 1 drop of 0.1 mol $\cdot$ L^{-1} $AgNO_3$ solution and several drops of 0.1 mol $\cdot$ L^{-1} NaCl solution, observe the phenomenon, then, add 2.0 mol $\cdot$ L^{-1} $NH_3 \cdot H_2O$ solution dropwise until the precipitate formed is dissolved (what is the products), continue to add $0.1 \, mol \cdot L^{-1}$ KBr solution, observe the formation of precipitate. To the precipitate add $0.1 \, mol \cdot L^{-1}$ $Na_2 S_2 O_3$ solution until the precipitate formed is dissolved(what is the products). Finally add $0.1 \, mol \cdot L^{-1}$ KI solution, what happens? Explain and write the reaction equations.

Based on the results above, compare the solubility of AgCl, AgBr, AgI and the stability of

$$[Ag(NH_3)_2]^+ \text{ and } [Ag(S_2O_3)_2]^{3-}.$$

5. Transformation between coordination compounds

To a test tube add 5 drop of $0.1 \, mol \cdot L^{-1}$ $FeCl_3$ solution and several drops of $0.1 \, mol \cdot L^{-1}$ KSCN solution, observe the phenomenon, then, add $1 \, mol \cdot L^{-1}$ NH_4 F in a dropwise manner, observe the color change of the solution. Explain and write the reaction equations.

6. Effect of acidity of the solution on coordination equilibrium

To a test tube add 10 drops of $0.1 \, mol \cdot L^{-1}$ $NiSO_4$ solution, then, add $2.0 \, mol \cdot L^{-1}$ $NH_3 \cdot H_2O$ solution drop by drop until theprecipitate(cloudy) formed is dissolved, observe the phenomenon, what is the products? Equally separate the resultant solution to two test tubes, then, slowly add $1 \, mol \cdot L^{-1}$ $H_2 SO_4$ solution to one tube and $0.1 \, mol \cdot L^{-1}$ NaOH solution to the other respectively. What happens? Explain and write the reaction equations.

7. Formation of chelate compounds

To a test tube add10 drops of $0.1 \, mol \cdot L^{-1}$ $CuSO_4$ solution, then, add $2.0 \, mol$ $\cdot L^{-1}$ $K_4P_2O_7$ solutiondrop by drop, observe the phenomenon. Continue to add 2.0 $mol \cdot L^{-1}$ $K_4P_2O_7$ solution in a dropwise manner. What happens? Explain and write the reaction equations.

V. Questions

(1)What is a coordination compound? List some examples.

(2)How to judge the relative stability of various coordination compounds? What factors will influence the stability of coordination compounds?

Experiment 9 Preparation and Properties of the Colloid

I . Learning Objects

(1) Learn to prepare a colloid and how to purify a colloid.

(2) Grasp the factors affecting the stability of colloid system.

(3) Understand optical properties and electric properties of a colloid.

II . Principles

A colloid may be prepared using condensation method which usually employs chemical reactions. For instance, a dark red colloidal suspension of iron (III) hydroxide($Fe(OH)_3$) can be prepared by mixing a concentrated solution of iron (III) chloride with hot water. A colloidal suspension of antimony(III) sulfide($Sb_2 S_3$) is produce by the reaction of hydrogen sulfide with antimony potassium tartrate dissolved in water. Dialysis is the separation of a solution from a colloid(protein) by means of a semipermeable membrane. Such a membrane is a dialyzing membrane. In this experiment, we will separate and sublimate the colloid by means of a dialyzing membrane.

Colloidal dispersions involve particles whose size is larger than those found in solutions but smaller than those in suspensions. When a strong beam of light passes through a colloid, the Tyndall effect is exhibited as the large-size colloidal particles reflect and scatter the light while smaller-size particles in solution, such as ions, do not show this effect. The Tyndall effect is one property that distinguishes colloidal dispersions from solutions.

One of the most important properties of dispersed colloidal particles is that they are usually electrically charged. When an iron(III) hydroxide solution is placed in electrolytic cell, the dispersed particles move to the negative electrode. This is good evidence that the iron(III) hydroxide particles are positively charged. The charges on colloidal particles result from the adsorption of ions that exist in the dispersion medium. Most hydroxides of metals have positive charges, while most sulfides of metals form negatively charged colloidal dispersions. These electrically charged crystals repel one another, so aggregation to large particles is prevented.

These very small crystals will aggregate into larger crystals when adding ions of opposite charge. The iron(Ⅲ) hydroxide solution can be made to aggregate by the addition of an ionic solution, particularly if the solution contains anions with multiple charge(such as phosphate ions, PO_4^{3-}). The iron(Ⅲ) hydroxide colloidal particles are positively charged, so the more the negative charge on anions, the more effective is the coagulation effect. The order of efficiency in coagulating positively charged iron (Ⅲ) hydroxide colloidal particles would be: $Na_3PO_4 > MgSO_4 > NaCl$.

The additionof large amount of a macromolecule such as gelatin, albumin, etc. to a colloid frequently renders the latter less sensitive to the precipitating effect of electrolytes; this is an illustration of the protective effect of macromolecule to colloid system. On the other hand, the addition of small amount of a macromolecule to a colloid frequently renders the latter more sensitive to the precipitating effect of electrolytes; this is an illustration of the phenomenon of sensitive action. It appears that the macromolecule absorbs the dispersed particles and let them precipitate.

Ⅲ. Apparatus and Reagents

U-tube, instrument for Tyndall effect(or laser pointer), beakers, test tubes, alcohol burner, test tube racks, medicine droppers, wire gauze, electrophoresis apparatus, stirring rod, dropping bottles.

$FeCl_3(0.1 \text{ mol} \cdot L^{-1})$, antimony potassium tartrate$(0.4\%, w/v)$, $KSCN(0.1 \text{ mol} \cdot L^{-1})$, $HCl(0.004 \text{ mol} \cdot L^{-1})$, $NaCl(0.005 \text{ mol} \cdot L^{-1})$, $CaCl_2(0.005 \text{ mol} \cdot L^{-1})$, $AlCl_3(0.005 \text{ mol} \cdot L^{-1})$, gelatine$(1\%)$, H_2S(saturated solution), NaCl $(5\%, w/v)$, $AgNO_3(0.1 \text{ mol} \cdot L^{-1})$, distilled water.

Ⅳ. Procedure

1. Preparation of the colloid

(1)　Preparation of iron(Ⅲ) hydroxide colloidal solution. Boil 120 mL distilled water in a 400 mL beaker. While the water is boiling, add 12 mL of 0.1 mol $\cdot L^{-1}$ $FeCl_3$ solution dropwise with stirring, continue boiling for $1 \sim 2$ min, dark red colloidal iron(Ⅲ) hydroxide forms. Keep it for next use. Write the structure of iron (Ⅲ) hydroxide colloidal group.

(2) Preparation of the antimony(Ⅲ) sulfide(Sb_2S_3) colloidal solution. Measure out about 20 mL of antimony potassium tartrate into a 100 mL beaker. Add saturated H_2S solution drop by drop with stirring until the solution changes orange red. Write the structure of antimony(Ⅲ) sulfide colloidal group. Keep it for next use.

2. Properties ofcolloidal system.

(1) Optical Property-Tyndall effect. Take three test tubes and fill each one third full with one of the following liquids: the colloidal iron (III) hydroxide, the colloidal antimony (III) sulfide, NaCl solution. Shine a light with laser pointer through each test tube. Observe the Tyndall effect. Which liquids did not give the Tyndall effect? Why?

(2) Electrical properties-electrophoresis. Place the colloidal iron(III) hydroxide solution in U-tube and in an electrophoresis device. Observe the dispersed particles and the color change in both arms of the U-tube. To which electrode did the iron (III) hydroxide colloidal particle move? Why?

(3) Coagulation. Add 1 mL of Sb_2S_3 colloidal solution into a clean test tube, add 0.005 mol $\cdot$ L^{-1} NaCl solution dropwise. Swirl the tube gently and continuously as you add the NaCl solution. When the precipitation begins to persist, record the number of drops you add. How about 0.005 mol $\cdot$ L^{-1} CaCl$_2$ solution and 0.005 mol $\cdot$ L^{-1} AlCl$_3$ solution instead of 0.005 mol $\cdot$ L^{-1} NaCl solution? Record and compare the number of drops for the precipitation appearing in three cases. Explain.

Add 2 mL of iron(III) hydroxide colloidal solution into a small test tube, then add 2 mL of Sb_2S_3 colloidal solution, swirl the mixture. Record your observation and explain.

Place 2 mL of the Sb_2S_3 colloidal solution into a small test tube. Heat the tube. Record your observation and explain.

3. Protectiveeffect and sensitive effect of macromolecule on the colloid

(1) Protective effect. Transfer 2 mL of 1% gelatine and 2 mL of distilled water, respectively, into two test tubes with a measuring cup. Add 2 mL of Sb_2S_3 colloidal solution into both tubes, carefully shake the tubes for 3min. Then, add 5% (w/v) NaCl solution to both tubes dropwise with shaking until precipitates begin to appear in both tubes. What happens in both tubes. Explain the difference.

(2) Sensitive effect. To two test tubes, add 2 mL of Sb_2S_3 colloidal solution into each of them. Then, to one tube add 5% (w/v) NaCl solution dropwise with shaking until precipitates begin to appear, stop the addition and record the number of drops of 5% (w/v) NaCl solution added. Slightly tilt the other tube, add 1 drop of 1% gelatine on the inner wall of the tube and let it follow slowly into the Sb_2S_3 colloidal solution, focus on the change in the place where both solutions mix together. What happens in both tubes. Explain the difference.

V. Questions

(1) What are the main stabilizing factors for a colloidal solution?

(2) How to precipitate a colloidal solution? List at least three methods.

(3) Why is a macromolecule system more stable than a colloid system?

(4) How does a colloidal particle get electrically charged in a colloidal solution?

Experiment 10   Determination of General Hardness of Water

I. Learning Objects

(1) Learn and grasp complexometric titration method.

(2) Learn how to determine water hardness by EDTA.

II. Principles

Water hardness is the term for the calcium or magnesium carbonate dissolved in water as Ca^{2+}, Mg^{2+}, and HCO_3^- (bicarbonate) ions. There are two measures of water hardness, hardness and alkalinity. Hardness measures the amount of positive calcium and magnesium ions; alkalinity the negative bicarbonate ions. Usually, alkalinity is less than hardness, although some mineral waters and ion exchange softened waters rich in sodium or potassium may have higher levels of alkalinity.

Sometimes alkalinity caused by the presence of $Ca(HCO_3)_2$ and $Mg(HCO_3)_2$ in water is called "temporary" or "carbonate" hardness because they can be removed by just boiling water to form $CaCO_3$ and $MgCO_3$ precipitate.

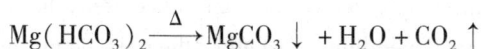

$$Ca(HCO_3)_2 \xrightarrow{\Delta} CaCO_3 \downarrow + H_2O + CO_2 \uparrow$$

$$Mg(HCO_3)_2 \xrightarrow{\Delta} MgCO_3 \downarrow + H_2O + CO_2 \uparrow$$

The hardness caused chloridesor sulphates of calcium and magnesium is called "permanent" hardness because they can not be removed by just boiling water. Ion exchange method is widely used to soften the permanent hardness of water.

General hardness (or total hardness) of water is defined as the sum of both "temporary hardness" and "permanent hardness".

In nature, water is hardened by the passage of rainwater containing dissolved carbon dioxide through layers of stone such as chalk, gypsum, or limestone. Hard water contains multiply charged ions such as calcium, magnesium, and heavy metal ions, which replace sodium and potassium ions in soaps and detergents to form precipitates. These precipitates interfere with cleaning action and leave bathtub rings and scum. Calcium carbonate ($CaCO_3$) is the most common precipitate. It is water insoluble and is the main component of the scale that clogs pipes.

Total water hardness is usually expressed as the milligrams of $CaCO_3$ equivalent to the total amount of calcium and magnesium present in one liter of water (mg/liter, i. e. ppm). Water hardness may range from zero to hundreds of ppm, depending on the source. The classification of degree of water hardness according to the US Geological Survey is as follows:

Soft $0 - 60$ mg $\cdot$ L^{-1} (ppm) CaCO$_3$ equivalents

Moderately hard $61 - 120$ mg $\cdot$ L^{-1} (ppm) CaCO$_3$ equivalents

Hard $121 - 180$ mg $\cdot$ L^{-1} (ppm) CaCO$_3$ equivalents

Very hard > 181 mg $\cdot$ L^{-1} (ppm) CaCO$_3$ equivalents

Complexometric titration is based on the formation of a complex ion between titrant and metal ions. Ethlyenediaminetetraacetic acid(EDTA or H$_4$Y, where Y = C$_{10}$H$_{12}$N$_2$O$_8$) is a complexing agent designed to bind metal ions quantitatively, forming stable, water soluble complexes with a 1:1 stoichiometry for most metal ions (i. e. 1 EDTA binds to 1 metal ion). EDTA binds to both calcium and magnesium, but binds more tightly to calcium, thus:

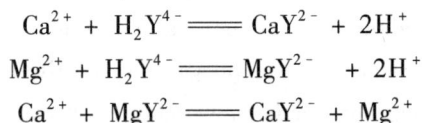

$$Ca^{2+} + H_2Y^{4-} \Longrightarrow CaY^{2-} + 2H^+$$
$$Mg^{2+} + H_2Y^{4-} \Longrightarrow MgY^{2-} + 2H^+$$
$$Ca^{2+} + MgY^{2-} \Longrightarrow CaY^{2-} + Mg^{2+}$$

As a sample is titrated with EDTA, the calcium ions in the sample are preferentially complexed by the EDTA, while magnesium complexes with the indicator. EDTA-metal complexes are generally uncolored; however, metallochromic indicators change colour depending on whether they are bound or unbound. After all the free calcium and magnesium are bound by EDTA, additional EDTA extracts the magnesium ions from the Eriochrome Black T indicator, restoring it to its uncomplexed blue colour, and an endpoint is observed.

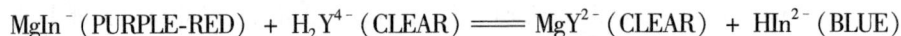

MgIn$^-$(PURPLE-RED) + H$_2$Y^{4-}(CLEAR) $\Longrightarrow$ MgY^{2-}(CLEAR) + HIn^{2-}(BLUE)

Eriochrome Black T does not give a sharp colour change for water containing calcium, but no magnesium. To make sure that there is some magnesium present in your sample, we add a small amount of Mg-EDTA (Mg already complexed with EDTA). Since we add the same amount of EDTA as Mg^{2+}, the addition of Mg-EDTA has no net effect on the subsequent titration.

Both EDTA and the metallochromic indicators are weak acids and their actions are very pH dependent; thus we use a pH 10 buffer to hold the solutions at an appropriate pH for both the EDTA and the Eriochrome Black T indicator to work well. Some metal ions interfere with this titration by causing indistinct endpoints, or by complexing with EDTA and/or the indicator more strongly than the metals of interest. We will look at an example of a chemical masking agent that is used to counteract such occurrences.

In this experiment, we will use EDTA complexometric titration to determine the general hardness of a sample of tap water in laboratory or water brought from your home. The total hardness of water will be measured. You are encouraged to bring your own water sample to study.

The general hardness of water can be calculated as follow,

$$\text{General hardness of water(ppm)} = \frac{c(\text{EDTA}) \times V(\text{EDTA}) \times M(\text{CaCO}_3)}{V(\text{sample})} \times 10^3 \, (\text{mg/L})$$

III. Apparatus and Reagents

Geiser buret (50 mL), transfer pipets (50 mL), Mohr measuring pipets (1.0 mL, 5.0 mL, 10.0 mL), Erlenmeyer flask (250 mL), buret stand, wash bottle, rubber pipet bulb, beakers(1000 mL, 250 mL).

HCl (6 mol $\cdot$ L^{-1}), Triethanolamine (1.5 mol $\cdot$ L^{-1}), Na$_2$ S (0.25 mol $\cdot$ L^{-1}), NH$_3$ – NH$_4$Cl buffer(pH = 10),

Eriochrome Black T indicator, standardized 0.01M EDTA solution, and water sample.

IV. Procedure

1. Clean your buret. Rinse it out with a small aliquot of EDTA solution to remove any metals from the glass. Shake your EDTA solution well, and then fill the burette with EDTA solution.

2. Transfer a 100.00 mL of your water sample into a 250 mL Erlenmeyer flask. Add 2 drops of 6 mol $\cdot$ L^{-1} HCl, shake the solution well to remove carbon dioxide. Add 5 mL of 1.5 mol $\cdot$ L^{-1} Triethanolamine, 10 mL of NH$_3$ – NH$_4$Cl buffer(pH = 10), 1.0 mL of 0.25 mol $\cdot$ L^{-1} Na$_2$ S, 3 drops of Eriochrome Black T indicator, shake the solution well. Then, titrate carefully with EDTA standard solution until the color changes to the clear blue. The reaction is slow at the endpoint, so add your titrant slowly (3 ~ 5 seconds between drops near the end) and swirl your mixture thoroughly. The color first changes from purple-red to purple. One more drop will change it to clear blue.

3. Repeat Step 2 two more times. Calculate the total hardness of your water

sample.

V. Data and calculations

Table 1 – 13　Determination of Havdness of Water

NO.　　　　　Items	1	2	3
Volume of water sample(mL)	100.0	100.0	100.0
Volume of EDTA consumed (mL)			
General water hardness (mg/L)			
Average of water General hardness (mg/L)			
Relative average deviation, $R\bar{d}$, (%)			

VI. Questions

(1) In this experiment, why is $NH_3 \cdot NH_4Cl$ buffer used? Can we use HAc · NaAc buffer?

(2) List the application of EDTA in medicine.

附　录

附录1　不同温度下水的饱和蒸气压

（ $\times 10^2$ Pa，273.2 ~ 313.2 K）

温度/K	0.0	0.2	0.4	0.6	0.8
273	–	6.105	6.195	6.286	6.379
274	6.473	6.567	6.663	6.759	6.858
275	6.958	7.058	7.159	7.262	7.366
276	7.473	7.579	7.687	7.797	7.907
277	8.019	8.134	8.249	8.365	8.483
278	8.603	8.723	8.846	8.970	9.095
279	9.222	9.350	9.481	9.611	9.745
280	9.881	10.017	10.155	10.295	10.436
281	10.580	10.726	10.872	11.022	11.172
282	11.324	11.478	11.635	11.792	11.952
283	12.114	12.278	12.443	12.610	12.779
284	12.951	13.124	13.300	13.478	13.658
285	13.839	14.023	14.210	14.397	14.587
286	14.779	14.973	15.171	15.369	15.572
287	15.776	15.981	16.191	16.401	16.615
288	16.831	17.049	17.260	17.493	17.719
289	17.947	18.177	18.410	18.648	18.886
290	19.128	19.372	19.618	19.869	20.121
291	20.377	20.634	20.896	21.160	21.426
292	21.694	21.968	22.245	22.523	22.805
293	23.090	23.378	23.669	23.963	24.261

续上表

温度/K	0.0	0.2	0.4	0.6	0.8
294	24.561	24.865	25.171	25.482	25.797
295	26.114	26.434	26.758	27.086	27.418
296	27.751	28.088	28.430	28.775	29.124
297	29.478	29.834	30.195	30.560	30.928
298	31.299	31.672	32.049	32.432	32.820
299	33.213	33.609	34.009	34.413	34.820
300	35.232	35.649	36.070	36.496	36.925
301	37.358	37.796	38.237	38.683	39.135
302	39.593	40.054	40.519	40.990	41.466
303	41.945	42.429	42.918	43.411	43.908
304	44.412	44.923	45.439	45.958	46.482
305	47.011	47.547	48.087	48.632	49.184
306	49.740	50.301	50.869	51.441	52.020
307	52.605	53.193	53.788	54.390	54.997
308	55.609	56.229	56.854	57.485	58.122
309	58.766	59.412	60.067	60.727	61.395
310	62.070	62.751	63.437	64.131	64.831
311	65.537	66.251	66.969	67.693	68.425
312	69.166	69.917	70.673	71.434	72.202
313	72.977	73.759	—	—	—

附录2　钠钙玻璃量器容积校正用表

钠钙玻璃量器容积校正用表($\beta = 2.6 \times 10^{-5}$ K^{-1})

标称容量/mL	温度/K															
	283	284	285	286	287	288	289	290	291	292	293	294	295	296	297	298
1000	998.39	998.32	998.24	998.14	998.04	997.92	997.79	997.64	997.49	997.32	997.15	996.96	996.77	996.56	996.35	996.12
	1.61	1.68	1.76	1.86	1.96	2.08	2.21	2.36	2.51	2.68	2.85	3.04	3.23	3.44	3.65	3.88
500	499.19	499.16	499.12	499.07	499.02	498.96	498.89	498.82	498.74	498.66	498.57	498.48	498.38	498.28	498.17	498.06
	0.81	0.84	0.88	0.93	0.98	1.04	1.11	1.18	1.26	1.34	1.43	1.52	1.62	1.72	1.83	1.94
250	249.60	249.58	249.56	249.54	249.51	249.48	249.45	249.41	249.37	249.33	249.29	249.24	249.19	249.14	249.09	249.03
	0.40	0.42	0.44	0.46	0.49	0.52	0.55	0.59	0.63	0.67	0.71	0.76	0.81	0.86	0.91	0.97
200	199.678	199.664	199.648	199.628	199.608	199.58	199.56	199.53	199.50	199.46	199.43	199.39	199.35	199.31	199.27	199.22
	0.322	0.336	0.352	0.372	0.392	0.42	0.44	0.47	0.50	0.54	0.57	0.61	0.65	0.69	0.73	0.78
125	124.80	124.79	124.78	124.77	124.76	124.74	124.72	124.71	124.69	124.67	124.64	124.62	124.60	124.57	124.54	124.52
	0.20	0.21	0.22	0.23	0.24	0.26	0.28	0.29	0.31	0.33	0.36	0.38	0.40	0.43	0.46	0.48
100	99.839	99.832	99.824	99.814	99.804	99.792	99.779	99.764	99.749	99.732	99.715	99.696	99.677	99.656	99.635	99.612
	0.161	0.168	0.176	0.186	0.196	0.208	0.221	0.236	0.251	0.268	0.285	0.304	0.323	0.344	0.365	0.388
50	49.919	49.916	49.912	49.907	49.902	49.896	49.889	49.882	49.874	49.866	49.857	49.848	49.838	49.828	49.817	49.806
	0.081	0.084	0.088	0.093	0.098	0.104	0.111	0.118	0.126	0.134	0.143	0.152	0.162	0.172	0.183	0.194
30	29.952	29.950	29.947	29.944	29.941	29.937	29.934	29.929	29.925	29.920	29.914	29.909	29.903	29.897	29.890	29.884
	0.048	0.050	0.053	0.056	0.059	0.063	0.066	0.071	0.075	0.080	0.086	0.091	0.097	0.103	0.110	0.116
15	14.976	14.975	14.974	14.972	14.971	14.969	14.967	14.965	14.962	14.960	14.957	14.954	14.952	14.948	14.945	14.942
	0.024	0.025	0.026	0.028	0.029	0.031	0.033	0.035	0.038	0.040	0.043	0.046	0.048	0.052	0.055	0.058

注：上行为质量值(g)，下行为差值。25 mL可查250 mL一行，将小数点向左移一位。

附录 3　弱电解质的解离常数

（近似浓度 $0.01 \sim 0.003 \ \text{mol} \cdot \text{L}^{-1}$，温度 298 K）

名称	化学式	解离常数，K	pK
醋酸	HAc	1.76×10^{-5}	4.75
碳酸	H_2CO_3	$K_1 = 4.30 \times 10^{-7}$	6.37
		$K_2 = 5.61 \times 10^{-11}$	10.25
草酸	$H_2C_2O_4$	$K_1 = 5.90 \times 10^{-2}$	1.23
		$K_2 = 6.40 \times 10^{-5}$	4.19
亚硝酸	HNO_2	$4.6 \times 10^{-4}(285.5 \ \text{K})$	3.37
磷酸	H_3PO_4	$K_1 = 7.52 \times 10^{-3}$	2.12
		$K_2 = 6.23 \times 10^{-8}$	7.21
		$K_3 = 2.2 \times 10^{-13}(291 \ \text{K})$	12.67
亚硫酸	H_2SO_3	$K_1 = 1.54 \times 10^{-2}(291 \ \text{K})$	1.81
		$K_2 = 1.02 \times 10^{-7}$	6.91
硫酸	H_2SO_4	$K_2 = 1.20 \times 10^{-2}$	1.92
硫化氢	H_2S	$K_1 = 9.1 \times 10^{-8}(291 \ \text{K})$	7.04
		$K2 = 1.1 \times 10^{-12}$	11.96
氢氰酸	HCN	4.93×10^{-10}	9.31
铬酸	H_2CrO_4	$K_1 = 1.8 \times 10^{-1}$	0.74
		$K_2 = 3.20 \times 10^{-7}$	6.49
*硼酸	H_3BO_3	5.8×10^{-10}	9.24
氢氟酸	HF	3.53×10^{-4}	3.45
过氧化氢	H_2O_2	2.4×10^{-12}	11.62
次氯酸	HClO	$2.95 \times 10^{-5}(291 \ \text{K})$	4.53
次溴酸	HBrO	2.06×10^{-9}	8.69
次碘酸	HIO	2.3×10^{-11}	10.64
碘酸	HIO_3	1.69×10^{-1}	0.77
砷酸	H_3AsO_4	$K_1 = 5.62 \times 10^{-3}(291 \ \text{K})$	2.25
		$K_2 = 1.70 \times 10^{-7}$	6.77
		$K_3 = 3.95 \times 10^{-12}$	11.40
亚砷酸	$HAsO_2$	6×10^{-10}	9.22
铵离子	NH_4^+	5.56×10^{-10}	9.25
氨水	$NH_3 \cdot H_2O$	1.79×10^{-5}	4.75
联胺	N_2H_4	8.91×10^{-7}	6.05

续上表

名称	化学式	解离常数，K	pK
羟氨	NH_2OH	9.12×10^{-9}	8.04
氢氧化铅	$Pb(OH)_2$	9.6×10^{-4}	3.02
氢氧化锂	$LiOH$	6.31×10^{-1}	0.2
氢氧化铍	$Be(OH)_2$	1.78×10^{-6}	5.75
	$BeOH^+$	2.51×10^{-9}	8.6
氢氧化铝	$Al(OH)_3$	5.01×10^{-9}	8.3
	$Al(OH)_2^+$	1.99×10^{-10}	9.7
氢氧化锌	$Zn(OH)_2$	7.94×10^{-7}	6.1
氢氧化镉	$Cd(OH)_2$	5.01×10^{-11}	10.3
*乙二胺	$H_2NC_2H_4NH_2$	$K_1 = 8.5 \times 10^{-5}$	4.07
		$K_2 = 7.1 \times 10^{-8}$	7.15
*六亚甲基四胺	$(CH_2)_6N_4$	1.35×10^{-9}	8.87
*尿素	$CO(NH_2)_2$	1.3×10^{-14}	13.89
*质子化六亚甲基四胺	$(CH_2)_6N_4H^+$	7.1×10^{-6}	5.15
甲酸	$HCOOH$	$1.77 \times 10^{-4}(293\ K)$	3.75
氯乙酸	$ClCH_2COOH$	1.40×10^{-3}	2.85
氨基乙酸	NH_2CH_2COOH	1.67×10^{-10}	9.78
*邻苯二甲酸	$C_6H_4(COOH)_2$	$K_1 = 1.12 \times 10^{-3}$	2.95
		$K_2 = 3.91 \times 10^{-6}$	5.41
柠檬酸	$(HOOCCH_2)_2C(OH)COOH$	$K_1 = 7.1 \times 10^{-4}$	3.14
		$K_2 = 1.68 \times 10^{-5}(293\ K)$	4.77
		$K_3 = 4.1 \times 10^{-7}$	6.39
酒石酸	$(CH(OH)COOH)_2$	$K_1 = 1.04 \times 10^{-3}$	2.98
		$K_2 = 4.55 \times 10^{-5}$	4.34
*8－羟基喹啉	C_9H_6NOH	$K_1 = 8 \times 10^{-6}$	5.1
		$K_2 = 1 \times 10^{-9}$	9.0
苯酚	C_6H_5OH	$1.28 \times 10^{-10}(293\ K)$	9.89
*对氨基苯磺酸	$H_2NC_6H_4SO_3H$	$K_1 = 2.6 \times 10^{-1}$	0.58
		$K_2 = 7.6 \times 10^{-4}$	3.12
*乙二胺四乙酸	$(CH_2COOH)_2NH^+$	$K_5 = 5.4 \times 10^{-7}$	6.27
（EDTA）	$CH_2CH_2NH^+(CH_2COOH)_2$	$K_6 = 1.12 \times 10^{-11}$	10.95

注：摘自 R. C. Weast, Handbook of Chemistry and Physics D－165, 70th. edition , 1989—1990

*摘自其他参考书。

附录4　实验室常用酸、碱的浓度

试剂名称	密度/(g·mL^{-1})	质量分数/%	物质的量浓度/(mol·L^{-1})
浓硫酸	1.84	98	18
稀硫酸	1.06	9	1
浓盐酸	1.19	38	12
稀盐酸	1.03	7	2
浓硝酸	1.40	67	15
稀硝酸	1.07	12	2
浓磷酸	1.70	85	15
稀磷酸	1.05	9	1
浓高氯酸	1.67	70	11.6
稀高氯酸	1.12	19	2
浓氢氟酸	1.13	40	23
氢溴酸	1.38	40	7
氢碘酸	1.70	57	7.5
冰醋酸	1.05	99~100	17.5
稀醋酸	1.04	30	5
稀醋酸	1.02	12	2
浓氢氧化钠	1.44	40	14.4
稀氢氧化钠	1.09	8	2
浓氨水	0.91	28	14.8
稀氨水	0.96	11	6
稀氨水	0.98	3.5	2

附录5　常用酸碱指示剂与试纸

1. 几种常用酸碱指示剂及其溶液的配制

指示剂名称	变色范围(pH)及颜色	溶液的配制
甲基橙	3.1~4.4　红~黄	0.1%的水溶液
甲基红	4.4~6.2　红~黄	0.1 g指示剂溶于100 mL 60%乙醇溶液中
石蕊	5.0~8.0　红~蓝	0.5%的水溶液
酚酞	8.0~10.0　无色~红	0.1 g指示剂溶于100 mL 60%乙醇溶液中
溴百里酚蓝	6.0~7.6　黄~蓝	0.05 g指示剂溶于100 mL 20%乙醇溶液中
	1.2~2.8　红~黄	0.1 g指示剂溶于100 mL 20%乙醇溶液中
百里酚蓝	8.0~9.6　黄~蓝	0.1 g指示剂溶于100 mL 水中 (内含4.3 mL 0.05 mol·L^{-1})
中性红	6.8~8.0　红~黄橙	0.1 g指示剂溶于100 mL 60%乙醇中
苯酚红	6.8~8.4　黄~红	0.1 g指示剂溶于100 mL 60%乙醇中

2. 几种常用试纸的制备

试纸名称	配制方法	用　途
广泛pH试纸 (黄)	将滤纸浸入混合指示剂中,取出晾干	测定溶液的近似pH
石蕊试纸 (红及蓝)	用热的乙醇处理市售石蕊除去夹带的红色素。残渣1份与6份水浸并不断振荡,滤去不溶物,将滤液分成两份,一份加稀H_3PO_4或H_2SO_4至变红,另一份加稀NaOH至变蓝,然后以这样的溶液浸湿白滤纸,在蔽光的无酸碱蒸气的房间中晾干	指示溶液的酸碱性:红色在碱性溶液中变蓝,蓝色在酸性溶液中变红
淀粉碘化钾试纸(白色)	将3 g淀粉与25 mL水搅匀,倾入225 mL沸水中,加入1 g KI和1 g固体Na_2CO_3,用水稀至500 mL。将滤纸浸入,取出后晾干	用以检验氧化剂(特别是游离的卤素),作用时变蓝
醋酸铅试纸 (白色)	将滤纸浸入3%的醋酸铅溶液中,取出后在无H_2S的房间中晾干	用以检验痕迹的H_2S,作用时变黑

注:一般化学实验用的淀粉碘化钾试纸及醋酸铅试纸由学生实验时制备,即在小滤纸条上滴1滴1%淀粉溶液和1滴0.1 mol·L^{-1} KI溶液或1滴0.1 mol·L^{-1} Pb(Ac)$_2$溶液。

附录6　常见配离子的稳定常数(298.15 K)

配离子	K_f	配离子	K_f
$AgCl_2^-$	1.84×10^5	$Al(OH)_4^-$	3.31×10^{33}
$AgBr_2^-$	1.93×10^7	AlF_6^{3-}	(6.9×10^{19})
AgI_2^-	4.80×10^{10}	$Al(EDTA)^-$	(1.3×10^{16})
$Ag(NH_3)^+$	2.07×10^3	$Ba(EDTA)^{2-}$	(6.0×10^7)
$Ag(NH_3)_2^+$	1.67×10^7	$Be(EDTA)^{2-}$	(2×10^9)
$Ag(CN)_2^-$	2.48×10^{20}	$BiCl_4^-$	7.96×10^6
$Ag(SCN)_2^-$	2.04×10^8	$BiCl_6^{3-}$	2.45×10^7
$Ag(S_2O_3)_2^{3-}$	(2.9×10^{13})	$BiBr_4^-$	5.92×10^7
$Ag(en)_2^+$	(5.0×10^7)	BiI_4^-	8.88×10^{14}
$Ag(EDTA)^{3-}$	(2.1×10^7)	$Bi(EDTA)^-$	(6.3×10^{22})
$Ca(EDTA)^{2-}$	(1×10^{11})	$Fe(C_2O_4)_3^{4-}$	1.7×10^5
$Cd(NH_3)_4^{2+}$	2.78×10^7	$Fe(EDTA)^{2-}$	(2.1×10^{14})
$Cd(CN)_4^{2-}$	1.95×10^{18}	$Fe(EDTA)^-$	(1.7×10^{24})
$Cd(OH)_4^{2-}$	1.20×10^9	$HgCl^+$	5.73×10^6
$CdBr_4^{2-}$	(5.0×10^3)	$HgCl_2$	1.46×10^{13}
$CdCl_4^{2-}$	(6.3×10^2)	$HgCl_3^-$	9.6×10^{13}
CdI_4^{2-}	4.05×10^5	$HgCl_4^-$	1.31×10^{15}
$Cd(en)_3^{2+}$	(1.2×10^{12})	$HgBr_4^{2-}$	9.22×10^{20}
$Cd(EDTA)^{2-}$	(2.5×10^{16})	HgI_4^{2-}	5.66×10^{29}
$Co(NH_3)_4^{2+}$	1.16×10^5	HgS_2^{2-}	3.36×10^{51}
$Co(NH_3)_6^{2+}$	1.3×10^5	$Hg(NH_3)_4^{2+}$	19.5×10^{19}
$CO(NH_3)_6^{3+}$	(1.6×10^{35})	$Hg(CN)_4^{2-}$	1.82×10^{41}
$Co(NCS)_4^{2-}$	(1.0×10^3)	$Hg(CNS)_4^{2-}$	4.98×10^{21}
$Co(EDTA)^{2-}$	(2.0×10^{16})	$Hg(EDTA)^{2-}$	(6.3×10^{21})
$Co(EDTA)^-$	(1×10^{36})	$Ni(NH_3)_6^{2+}$	8.97×10^8
$Cr(OH)_4^-$	(7.8×10^{29})	$Ni(CN)_4^{2-}$	1.31×10^{30}
$Cr(EDTA)^-$	(1.0×10^{23})	$Ni(N_2H_4)_6^{2+}$	1.04×10^{12}
$CuCl_2^-$	6.91×10^4	$Ni(en)_3^{2+}$	2.1×10^{18}
$CuCl_3^{2-}$	4.55×10^5	$Ni(EDTA)^{2-}$	(3.6×10^{18})

续上表

配离子	K_f	配离子	K_f
CuI_2^-	(7.1×10^8)	$Pb(OH)_3^-$	8.27×10^{13}
$Cu(SO_3)_2^{3-}$	4.13×10^8	$PbCl_3^-$	27.2
$Cu(NH_3)_4^{2+}$	2.30×10^{12}	$PbBr_3^-$	15.5
$Cu(P_2O_7)_2^{6-}$	8.24×10^8	PbI_3^-	2.67×10^3
$Cu(C_2O_4)_2^{2-}$	2.35×10^9	PbI_4^{2-}	1.66×10^4
$Cu(CN)_2^-$	9.98×10^{23}	$Pb(CH_3CO_2)^+$	152.4
$Cu(CN)_3^{2-}$	4.21×10^{28}	$Pb(CH_3CO_2)_2$	826.3
$Cu(CN)_4^{3-}$	2.03×10^{30}	$Pb(EDTA)^{2-}$	(2×10^{18})
$Cu(CNS)_4^{3-}$	8.66×10^9	$PdCl_3^-$	2.10×10^{10}
$Cu(EDTA)^{2-}$	(5.0×10^{18})	$PdBr_4^{2-}$	6.05×10^{13}
FeF^{2+}	7.1×10^6	PdI_4^{2-}	4.36×10^{22}
FeF_2^+	3.8×10^{11}	$Pd(NH_3)_4^{2+}$	3.10×10^{25}
$Fe(CN)_6^{3-}$	4.1×10^{52}	$Pd(CN)_4^{2-}$	5.20×10^{41}
$Fe(CN)_6^{4-}$	4.2×10^{45}	$Pd(CNS)_4^{2-}$	9.43×10^{23}
$Fe(NCS)^{2+}$	9.1×10^2	$Pd(EDTA)^{2-}$	(3.2×10^{18})
$FeBr^{2+}$	4.17	$PtCl_4^{2-}$	9.86×10^{15}
$FeCl^{2+}$	24.9	$PtBr_4^{2-}$	6.47×10^{17}
$Fe(C_2O_4)_3^{3-}$	(1.6×10^{20})	$Pt(NH_3)_4^{2+}$	2.18×10^{35}
$Sc(EDTA)^-$	1.3×10^{23}	$Zn(CN)_4^{2-}$	5.71×10^{16}
$Zn(OH)_3^-$	1.64×10^{13}	$Zn(CNS)_4^{2-}$	19.6
$Zn(OH)_4^{2-}$	2.83×10^{14}	$Zn(C_2O_4)_2^{2-}$	2.96×10^7
$Zn(NH_3)_4^{2+}$	3.60×10^8	$Zn(EDTA)^{2-}$	(2.5×10^{16})

附录7　某些试剂的配制

名　　称	浓度/($\text{mol} \cdot \text{L}^{-1}$)	配　制　方　法
氨水		往水中通入直至饱和
溴水		往水中滴入液溴直至饱和
碘水	0.01	溶解 2.5 g 碘和 3 gKI 于尽可能少量的水中，加水稀释至 1 L
淀粉溶液	1%	将 1 g 淀粉和少量冷水调成糊状，倒入 100 mL 沸水中，煮沸后，冷却
盐桥		将 2 g 琼胶和 30 g KCl 加入 100 mL 水中，在不断搅拌下，加热溶解，煮沸几分钟，趁热倒入 U 形管中，冷却后即可使用
氯化亚锡($SnCl_2$)	0.1	溶解 2.6 g $SnCl_2 \cdot 2H_2O$ 于 160 mL 浓盐酸中，加水稀释至 1 L，加入数粒纯锡，以防止氧化。
三氯化铋($BiCl_3$)	0.1	溶解 31.6 g $BiCl_3$ 于 160 mL 浓盐酸中，加水稀释至 1 L
三氯化锑($SbCl_3$)	0.1	溶解 22.8 g $BiCl_3$ 于 160 mL 浓盐酸中，加水稀释至 1 L
氯化氧钒(VO_2Cl)	0.1	将 1 g 偏钒酸铵固体加入 20 mL 6 $\text{mol} \cdot \text{L}^{-1}$ HCi 和 10 mL 水中
硝酸汞($Hg(NO_3)_2$)	0.1	溶解 33.4 g $Hg(NO_3)_2 \cdot \frac{1}{2}H_2O$ 于 1 L 0.6 $\text{mol} \cdot \text{L}^{-1}$ HNO_3 中
硝酸亚汞($Hg_2(NO_3)_2$)	0.1	溶解 56.1 g $Hg(NO_3)_2 \cdot 2H_2O$ 于 1 L 0.6 $\text{mol} \cdot \text{L}^{-1}$ HNO_3 中，并加入少量汞
硫酸氧钛 $TiOSO_4$		在烧杯中加入 5 g TiO_2 固体和 9 mL 浓 H_2SO_4，加热半小时，使 TiO_2 溶解。冷后将溶液慢慢倒入 50 mL 水中，澄清备用
二苯胺基脲 ($C_{13}H_{14}ON_4$)		取 0.1 g 二苯胺基脲加入 50 mL 95% 乙醇中，溶解后再加入 200 mL 1:9H_2SO_4。此试剂应为无色液体。易变质，应贮于冰箱中，变色后不能使用(最好现配现用。)
铬黑 T		将铬黑 T 和烘干和 NaCl 按 1:50 的比例研细、混匀，贮于冰箱中，变色后不能使用 (最好现配现用)
洗液		将 5 g 研细的重铬酸钾固体加入 100 mL 热的浓硫酸中，搅匀，得红棕色液体。冷却后倒入带磨口的瓶中备用。洗液变绿即失效。

附录8　国际原子量表

符号	名称	原子量	符号	名称	原子量	符号	名称	原子量
Ac	锕	[227]	H	氢	1.0079	Pt	铂	195.08
Ag	银	107.8682	He	氦	4.00260	Ra	镭	226.0254
Al	铝	26.98154	Hf	铪	178.49	Rb	铷	85.4678
Am	镅	[243]	Hg	汞	164.9304	Re	铼	186.207
Ar	氩	39.948	Ho	钬	126.9045	Rh	铑	102.9055
As	砷	74.9216	I	碘	114.82	Rn	氡	[222]
At	砹	[210]	In	铟	192.22	Ru	钌	101.07
Au	金	196.9665	Ir	铱	39.0983	S	硫	32.06
B	硼	10.81	K	钾	83.80	Sb	锑	121.75
Ba	钡	137.34	Kr	氪	138.9055	Sc	钪	44.9559
Be	铍	9.01218	La	镧	6.941	Se	硒	78.96
Bi	铋	208.9804	Li	锂	174.967	Si	硅	28.0855
Br	溴	79.904	Lu	镥	24.305	Sm	钐	150.36
C	碳	12.011	Mg	镁	54.9380	Sn	锡	118.69
Ca	钙	40.08	Mn	锰	95.94	Sr	锶	87.62
Cd	镉	112.41	Mo	钼	14.0067	Ta	钽	180.9479
Ce	铈	140.12	N	氮	22.98977	Tb	铽	158.9254
Cl	氯	35.453	Na	钠	92.9064	Tc	锝	[98]
Co	钴	58.9332	Nb	铌	144.24	Te	碲	127.60
Cr	铬	132.9054	Nd	钕	20.179	Th	钍	232.0381
Cs	铯	132.9054	Ne	氖	58.69	TI	钛	47.88
Cu	铜	63.546	Ni	镍	15.9994	Tl	铊	204.383
Dy	镝	162.50	O	氧	190.2	Tu	铥	168.9342
Er	铒	167.26	Os	锇	30.97376	U	铀	238.0289
Eu	铕	151.96	P	磷	231.0359	V	钒	50.9415
F	氟	18.998403	Pa	镤	207.2	W	钨	183.85
Fe	铁	55.847	Pb	铅	106.42	Xe	氙	131.29
Fr	钫	[223]	Pd	钯	[145]	Y	钇	88.9059
Ga	镓	69.72	Pm	钷	[145]	Yb	镱	173.04
Gd	钆	157.25	Po	钋	140.9077	Zn	锌	65.38
Ge	锗	72.59	Pr	镨		Zr	锆	91.22

附录 9　常见有机化合物的物理常数

化合物	沸点/℃	熔点/℃	相对密度 $\rho \dfrac{20}{4}$	水中溶解度
环乙烷	80.7	6.5	0.779	
苯	80.1	5.5	0.8765	0.08
甲苯	110.6	-95	0.8669	
二甲苯(o-m-p-)	140			不溶
1-氯丁烷	78.44		0.8862	
1-溴丁烷	101.6		1.2758	
氯化苄	179		1.10	
溴苯	155	-31	1.499	不溶
氯仿	61.7		1.4832	
甲醇	64.96	-93.9	0.7914	∞
乙醇	78.5	-114.1	0.7893	∞
正丁醇	117	-80	0.810	∞
甘油	290(分解)	17	1.2656	∞
三苯甲醇	>360	162.5	1.188	难溶
苯酚	181	43		9
乙醚	34.5	-116.3	0.713	8
二氧六环	101(170mm)	11.8	1.0337	
甲醛	-21	-92	0.815	易溶
乙醛	21	-123.5	0.7834	∞
苯甲醛	179	-26	1.0509	—
丙酮	56.1	-94.8	0.7899	∞
环乙酮	155~156	-45	0.9478	溶
二苯甲酮	306(升华)	49	1.0976	不溶
甲酸	100.7	8.4	1.22	∞
乙酸	117.9	16.6	1.0498	∞
草酸	189.5			溶
乳酸	122(15 mm)	16.8	1.249(15℃)	∞
正己酸	205	-150	0.929	0.30
苯甲酸	250	122.4	1.266(15℃)	
乙酰氯	15~52	-112	1.1051	遇水剧烈分解
乙酸酐	139	-73	1.082	遇水分解
乙酸乙酯	77.06	-84	0.901	8.6
乙酰乙酸乙酯	180(分解)	<-45	1.025	14.3
硝基苯	210.8	5.7	1.203	0.9
苯胺	184.4	-6	1.022	3.4

附录 10　金属离子指示剂

指示剂名称	解离平衡和颜色变化	溶液配制方法
铬黑 T （EBT）	$H_2In^- \xrightleftharpoons[]{pK_{a2}=6.3} HIn^{2-} \xrightleftharpoons[]{pK_{a3}=11.5} In^{3-}$ 　紫红　　　　　　蓝　　　　　　橙	1. 5 g·L^{-1}水溶液 2. 与 NaCl 按 1:100 质量比混合
二甲酚橙 （XO）	$H_2In^{4-} \xrightleftharpoons[]{pK_a=6.3} HIn^{5-}$ 　　　黄　　　　　　红	2 g·L^{-1}水溶液
K－B 指示剂	$H_2In \xrightleftharpoons[]{pK_{a1}=8} HIn^- \xrightleftharpoons[]{pK_{a2}=13} In^{2-}$ 　红　　　　　蓝　　　　　紫红 （酸性铬蓝 K）	0. 2 g 酸性铬蓝 K 与 0. 34 g 萘酚绿 B 溶于 100 mL 水中。配制后需调节 K－B 的比例，使终点变化明显
钙指示剂	$H_2In^- \xrightleftharpoons[]{pK_{a2}=7.4} HIn^{2-} \xrightleftharpoons[]{pK_{a3}=13.5} In^{3-}$ 　酒红　　　　　蓝　　　　　酒红	5 g·L^{-1}的乙醇溶液
吡啶偶氮萘酚 （PAN）	$H_2In \xrightleftharpoons[]{pK_{0.1}=1.9} HIn \xrightleftharpoons[]{pK_{a2}=12.2} In^-$ 　黄绿　　　　　黄　　　　　淡红	1 g·L^{-1}或 3 g·L^{-1}的乙醇溶液
Cu－PAN （CuY－PAN 溶液）	$\dfrac{CuY + PAN}{\text{浅绿}} + \dfrac{M^{n+}}{\text{无色}} = MY + \dfrac{Cu-PAN}{\text{红色}}$	取 0. 05 mol·L^{-1} Cu^{2+} 溶液 10 mL，加 pH 为 5~6 的 HAc 缓冲溶液 5 mL，1 滴 PAN 指示剂，加热至 333 K 左右，用 EDTA 滴至绿色，得到约 0. 025 mol·L^{-1} 的 CuY 溶液。使用时取 2~3 mL 于试液中，再加数滴 PAN 溶液
磺基水杨酸	$H_2In \xrightleftharpoons[]{pK_{a2}=2.7} HIn^- \xrightleftharpoons[]{pK_{a3}=13.1} In^{2-}$ 　　　　　　（无色）	10 g·L^{-1}或 100 g·L^{-1}的水溶液
钙镁试剂 （Calmagnite）	$H_2In \xrightleftharpoons[]{pK_{a2}=8.1} HIn^{2-} \xrightleftharpoons[]{pK_{a3}=12.4} In^{3-}$ 　红　　　　　蓝　　　　　红橙	5 g·L^{-1}水溶液
紫脲酸铵	$H_4In \xrightleftharpoons[]{pK_{a2}=9.2} H_3In^{2-} \xrightleftharpoons[]{pK_{a3}=10.9} H_2In^{3-}$ 　红紫　　　　　紫　　　　　蓝	与 NaCl 按 1:100 质量比混合

注：EBT、钙指示剂、K－B 指示剂等在水溶液中稳定性较差，可以配成指示剂与 NaCl 之比为 1:100 或 1:200 的固体粉末。

附录 11　沉淀滴定吸附指示剂

指示剂	被测离子	滴定剂	滴定条件	溶液配制方法
荧光黄	Cl^-	Ag^+	pH 7～10(一般 7～8)	2 g·L^{-1}乙醇溶液
二氯荧光黄	Cl^-	Ag^+	pH 4～10(一般 5～8)	1 g·L^{-1}水溶液
曙红	Br^-, I^-, SCN^-	Ag^+	pH 2～10(一般 3～8)	5 g·L^{-1}水溶液
溴甲酚绿	SCN^-	Ag^+	pH 4～5	1 g·L^{-1}水溶液
甲基紫	Ag^+	Cl^-	酸性溶液	1 g·L^{-1}水溶液
罗丹明 6G	Ag^+	Br^-	酸性溶液	1 g·L^{-1}水溶液
钍试剂	SO_4^{2-}	Ba^{2+}	pH 1.5～3.5	5 g·L^{-1}水溶液
溴酚蓝	Hg_2^{2+}	Cl^-, Br^-	酸性溶液	1 g·L^{-1}水溶液

附录12　化合物的溶度积常数表

化合物	溶度积	化合物	溶度积	化合物	溶度积
醋酸盐		氢氧化物		* CdS	8.0×10^{-27}
* * AgAc	1.94×10^{-3}	* AgOH	2.0×10^{-8}	* CoS(α-型)	4.0×10^{-21}
卤化物		* Al(OH)$_3$(无定形)	1.3×10^{-33}	* CoS(β-型)	2.0×10^{-25}
* AgBr	5.0×10^{-13}	* Be(OH)$_2$(无定形)	1.6×10^{-22}	* Cu$_2$S	2.5×10^{-48}
* AgCl	1.8×10^{-10}	* Ca(OH)$_2$	5.5×10^{-6}	* CuS	6.3×10^{-36}
* AgI	8.3×10^{-17}	* Cd(OH)$_2$	5.27×10^{-15}	* FeS	6.3×10^{-18}
BaF$_2$	1.84×10^{-7}	* * Co(OH)$_2$(粉红色)	1.09×10^{-15}	* HgS(黑色)	1.6×10^{-52}
* CaF$_2$	5.3×10^{-9}	* * Co(OH)$_2$(蓝色)	5.92×10^{-15}	* HgS(红色)	4×10^{-53}
* CuBr	5.3×10^{-9}	* Co(OH)$_3$	1.6×10^{-44}	* MnS(晶形)	2.5×10^{-13}
* CuCl	1.2×10^{-6}	* Cr(OH)$_2$	2×10^{-16}	* * NiS	1.07×10^{-21}
* CuI	1.1×10^{-12}	* Cr(OH)$_3$	6.3×10^{-31}	* PbS	8.0×10^{-28}
* Hg$_2$Cl$_2$	1.3×10^{-18}	* Cu(OH)$_2$	2.2×10^{-20}	* SnS	1×10^{-25}
* Hg$_2$I$_2$	4.5×10^{-29}	* Fe(OH)$_2$	8.0×10^{-16}	* * SnS$_2$	2×10^{-27}
HgI$_2$	2.9×10^{-29}	* Fe(OH)$_3$	4×10^{-38}	* * ZnS	2.93×10^{-25}
PbBr$_2$	6.60×10^{-6}	* Mg(OH)$_2$	1.8×10^{-11}	磷酸盐	
* PbCl$_2$	1.6×10^{-5}	* Mn(OH)$_2$	1.9×10^{-13}	* Ag$_3$PO$_4$	1.4×10^{-16}
PbF$_2$	3.3×10^{-8}	* Ni(OH)$_2$(新制备)	2.0×10^{-15}	* AlPO$_4$	6.3×10^{-19}
* PbI$_2$	7.1×10^{-9}	* Pb(OH)$_2$	1.2×10^{-15}	* CaHPO$_4$	1×10^{-7}
SrF$_2$	4.33×10^{-9}	* Sn(OH)$_2$	1.4×10^{-28}	* Ca$_3$(PO$_4$)$_2$	2.0×10^{-29}
碳酸盐		* Sr(OH)$_2$	9×10^{-4}	* * Cd$_3$(PO$_4$)$_2$	2.53×10^{-33}
Ag$_2$CO$_3$	8.45×10^{-12}	* Zn(OH)$_2$	1.2×10^{-17}	Cu$_3$(PO$_4$)$_2$	1.40×10^{-37}
* BaCO$_3$	5.1×10^{-9}	草酸盐		FePO$_4$·2H$_2$O	9.91×10^{-16}
CaCO$_3$	3.36×10^{-9}	Ag$_2$C$_2$O$_4$	5.4×10^{-12}	* MgNH$_4$PO$_4$	2.5×10^{-13}
CdCO$_3$	1.0×10^{-12}	* BaC$_2$O$_4$	1.6×10^{-7}	Mg$_3$(PO$_4$)$_2$	1.04×10^{-24}
* CuCO$_3$	1.4×10^{-10}	* CaC$_2$O$_4$·H$_2$O	4×10^{-9}	* Pb$_3$(PO$_4$)$_2$	8.0×10^{-43}
FeCO$_3$	3.13×10^{-11}	CuC$_2$O$_4$	4.43×10^{-10}	* Zn$_3$(PO$_4$)$_2$	9.0×10^{-33}
Hg$_2$CO$_3$	3.6×10^{-17}	* FeC$_2$O$_4$·2H$_2$O	3.2×10^{-7}	其他盐	
MgCO$_3$	6.82×10^{-6}	Hg$_2$C$_2$O$_4$	1.75×10^{-13}	* [Ag$^+$][Ag(CN)$_2^-$]	7.2×10^{-11}
MnCO$_3$	2.24×10^{-11}	MgC$_2$O$_4$·2H$_2$O	4.83×10^{-6}	* Ag$_4$[Fe(CN)$_6$]	1.6×10^{-41}
NiCO$_3$	1.42×10^{-7}	MnC$_2$O$_4$·2H$_2$O	1.70×10^{-7}	* Cu$_2$[Fe(CN)$_6$]	1.3×10^{-16}
* PbCO$_3$	7.4×10^{-14}	* * PbC$_2$O$_4$	8.51×10^{-10}	AgSCN	1.03×10^{-12}
SrCO$_3$	5.6×10^{-10}	* SrC$_2$O$_4$·H$_2$O	1.6×10^{-7}	CuSCN	4.8×10^{-15}
ZnCO$_3$	1.46×10^{-10}	ZnC$_2$O$_4$·2H$_2$O	1.38×10^{-9}	* AgBrO$_3$	5.3×10^{-5}

续上表

化合物	溶度积	化合物	溶度积	化合物	溶度积
铬酸盐		硫酸盐		* $AgIO_3$	3.0×10^{-8}
Ag_2CrO_4	1.12×10^{-12}	* Ag_2SO_4	1.4×10^{-5}	$Cu(IO_3)_2 \cdot H_2O$	7.4×10^{-8}
* $Ag_2Cr_2O_7$	2.0×10^{-7}	* $BaSO_4$	1.1×10^{-10}	* * $KHC_4H_4O_6$(酒石酸氢钾)	3×10^{-4}
* $BaCrO_4$	1.2×10^{-10}	* $CaSO_4$	9.1×10^{-6}	* * $Al(8-羟基喹啉)_3$	5×10^{-33}
* $CaCrO_4$	7.1×10^{-4}	Hg_2SO_4	6.5×10^{-7}	* $K_2Na[Co(NO_2)_6] \cdot H_2O$	2.2×10^{-11}
* $CuCrO_4$	3.6×10^{-6}	* $PbSO_4$	1.6×10^{-8}	* $Na(NH_4)_2[Co(NO_2)_6]$	4×10^{-12}
* Hg_2CrO_4	2.0×10^{-9}	* $SrSO_4$	3.2×10^{-7}	* * $Ni(丁二酮肟)_2$	4×10^{-24}
* $PbCrO_4$	2.8×10^{-13}	硫化物		* * $Mg(8-羟基喹啉)_2$	4×10^{-16}
* $SrCrO_4$	2.2×10^{-5}	* Ag_2S	6.3×10^{-50}	* * $Zn(8-羟基喹啉)_2$	5×10^{-25}

摘自 David R. Lide, Handbook of Chemistry and Physics, 78th. edition, 1997—1998

* 摘自 J. A. Dean Ed. Lange's Handbook of Chemistry, 13th. edition 1985

* * 摘自其他参考书。

附录 13　标准电极电势表

1. 在酸性溶液中(298 K)

电对	方程式	$\varphi^{\ominus}/V$
Li(I) – (0)	$Li^+ + e^- = Li$	-3.0401
Cs(I) – (0)	$Cs^+ + e^- = Cs$	-3.026
Rb(I) – (0)	$Rb^+ + e^- = Rb$	-2.98
K(I) – (0)	$K^+ + e^- = K$	-2.931
Ba(II) – (0)	$Ba^{2+} + 2e^- = Ba$	-2.912
Sr(II) – (0)	$Sr^{2+} + 2e^- = Sr$	-2.89
Ca(II) – (0)	$Ca^{2+} + 2e^- = Ca$	-2.868
Na(I) – (0)	$Na^+ + e^- = Na$	-2.71
La(III) – (0)	$La^{3+} + 3e^- = La$	-2.379
Mg(II) – (0)	$Mg^{2+} + 2e^- = Mg$	-2.372
Ce(III) – (0)	$Ce^{3+} + 3e^- = Ce$	-2.336
H(0) – (– I)	$H_2(g) + 2e^- = 2H^-$	-2.23
Al(III) – (0)	$AlF_6^{3-} + 3e^- = Al + 6F^-$	-2.069
Th(IV) – (0)	$Th^{4+} + 4e^- = Th$	-1.899
Be(II) – (0)	$Be^{2+} + 2e^- = Be$	-1.847
U(III) – (0)	$U^{3+} + 3e^- = U$	-1.798
Hf(IV) – (0)	$HfO^{2+} + 2H^+ + 4e^- = Hf + H_2O$	-1.724
Al(III) – (0)	$Al^{3+} + 3e^- = Al$	-1.662
Ti(II) – (0)	$Ti^{2+} + 2e^- = Ti$	-1.630
Zr(IV) – (0)	$ZrO_2 + 4H^+ + 4e^- = Zr + 2H_2O$	-1.553
Si(IV) – (0)	$[SiF_6]^{2-} + 4e^- = Si + 6F^-$	-1.24
Mn(II) – (0)	$Mn^{2+} + 2e^- = Mn$	-1.185
Cr(II) – (0)	$Cr^{2+} + 2e^- = Cr$	-0.913
Ti(III) – (II)	$Ti^{3+} + e^- = Ti^{2+}$	-0.9
B(III) – (0)	$H_3BO_3 + 3H^+ + 3e^- = B + 3H_2O$	-0.8698
* Ti(IV) – (0)	$TiO_2 + 4H^+ + 4e^- = Ti + 2H_2O$	-0.86
Te(0) – (– II)	$Te + 2H^+ + 2e^- = H_2Te$	-0.793
Zn(II) – (0)	$Zn^{2+} + 2e^- = Zn$	-0.7618
Ta(V) – (0)	$Ta_2O_5 + 10H^+ + 10e^- = 2Ta + 5H_2O$	-0.750
Cr(III) – (0)	$Cr^{3+} + 3e^- = Cr$	-0.744
Nb(V) – (0)	$Nb_2O_5 + 10H^+ + 10e^- = 2Nb + 5H_2O$	-0.644
As(0) – (– III)	$As + 3H^+ + 3e^- = AsH_3$	-0.608

续上表

电对	方程式	$\varphi^\ominus/V$
U(Ⅳ) – (Ⅲ)	$U^{4+} + e^- = U^{3+}$	-0.607
Ga(Ⅲ) – (0)	$Ga^{3+} + 3e^- = Ga$	-0.549
P(Ⅰ) – (0)	$H_3PO_2 + H^+ + e^- = P + 2H_2O$	-0.508
P(Ⅲ) – (Ⅰ)	$H_3PO_3 + 2H^+ + 2e^- = H_3PO_2 + H_2O$	-0.499
*C(Ⅳ) – (Ⅲ)	$2CO_2 + 2H^+ + 2e^- = H_2C_2O_4$	-0.49
Fe(Ⅱ) – (0)	$Fe^{2+} + 2e^- = Fe$	-0.447
Cr(Ⅲ) – (Ⅱ)	$Cr^{3+} + e^- = Cr^{2+}$	-0.407
Cd(Ⅱ) – (0)	$Cd^{2+} + 2e^- = Cd$	-0.4030
Se(0) – (–Ⅱ)	$Se + 2H^+ + 2e^- = H_2Se(aq)$	-0.399
Pb(Ⅱ) – (0)	$PbI_2 + 2e^- = Pb + 2I^-$	-0.365
Eu(Ⅲ) – (Ⅱ)	$Eu^{3+} + e^- = Eu^{2+}$	-0.36
Pb(Ⅱ) – (0)	$PbSO_4 + 2e^- = Pb + SO_4^{2-}$	-0.3588
In(Ⅲ) – (0)	$In^{3+} + 3e^- = In$	-0.3382
Tl(Ⅰ) – (0)	$Tl^+ + e^- = Tl$	-0.336
Co(Ⅱ) – (0)	$Co^{2+} + 2e^- = Co$	-0.28
P(Ⅴ) – (Ⅲ)	$H_3PO_4 + 2H^+ + 2e^- = H_3PO_3 + H_2O$	-0.276
Pb(Ⅱ) – (0)	$PbCl_2 + 2e^- = Pb + 2Cl^-$	-0.2675
Ni(Ⅱ) – (0)	$Ni^{2+} + 2e^- = Ni$	-0.257
V(Ⅲ) – (Ⅱ)	$V^{3+} + e^- = V^{2+}$	-0.255
Ge(Ⅳ) – (0)	$H_2GeO_3 + 4H^+ + 4e^- = Ge + 3H_2O$	-0.182
Ag(Ⅰ) – (0)	$AgI + e^- = Ag + I^-$	-0.15224
Sn(Ⅱ) – (0)	$Sn^{2+} + 2e^- = Sn$	-0.1375
Pb(Ⅱ) – (0)	$Pb^{2+} + 2e^- = Pb$	-0.1262
*C(Ⅳ) – (Ⅱ)	$CO_2(g) + 2H^+ + 2e^- = CO + H_2O$	-0.12
P(0) – (–Ⅲ)	$P(white) + 3H^+ + 3e^- = PH_3(g)$	-0.063
Hg(Ⅰ) – (0)	$Hg_2I_2 + 2e^- = 2Hg + 2I^-$	-0.0405
Fe(Ⅲ) – (0)	$Fe^{3+} + 3e^- = Fe$	-0.037
H(Ⅰ) – (0)	$2H^+ + 2e^- = H_2$	0.0000
Ag(Ⅰ) – (0)	$AgBr + e^- = Ag + Br^-$	0.07133
S(Ⅱ.Ⅴ) – (Ⅱ)	$S_4O_6^{2-} + 2e^- = 2S_2O_3^{2-}$	0.08
*Ti(Ⅳ) – (Ⅲ)	$TiO^{2+} + 2H^+ + e^- = Ti^{3+} + H_2O$	0.1
S(0) – (–Ⅱ)	$S + 2H^+ + 2e^- = H_2S(aq)$	0.142
Sn(Ⅳ) – (Ⅱ)	$Sn^{4+} + 2e^- = Sn^{2+}$	0.151

续上表

电对	方程式	$\varphi^{\ominus}/V$
Sb(Ⅲ) – (0)	$Sb_2O_3 + 6H^+ + 6e^- = 2Sb + 3H_2O$	0.152
Cu(Ⅱ) – (Ⅰ)	$Cu^{2+} + e^- = Cu^+$	0.153
Bi(Ⅲ) – (0)	$BiOCl + 2H^+ + 3e^- = Bi + Cl^- + H_2O$	0.1583
S(Ⅵ) – (Ⅳ)	$SO_4^{2-} + 4H^+ + 2e^- = H_2SO_3 + H_2O$	0.172
Sb(Ⅲ) – (0)	$SbO^+ + 2H^+ + 3e^- = Sb + H_2O$	0.212
Ag(Ⅰ) – (0)	$AgCl + e^- = Ag + Cl^-$	0.22233
As(Ⅲ) – (0)	$HAsO_2 + 3H^+ + 3e^- = As + 2H_2O$	0.248
Hg(Ⅰ) – (0)	$Hg_2Cl_2 + 2e^- = 2Hg + 2Cl^-$（饱和 KCl）	0.26808
Bi(Ⅲ) – (0)	$BiO^+ + 2H^+ + 3e^- = Bi + H_2O$	0.320
U(Ⅵ) – (Ⅳ)	$UO_2^{2+} + 4H^+ + 2e^- = U^{4+} + 2H_2O$	0.327
C(Ⅳ) – (Ⅲ)	$2HCNO + 2H^+ + 2e^- = (CN)_2 + 2H_2O$	0.330
V(Ⅳ) – (Ⅲ)	$VO^{2+} + 2H^+ + e^- = V^{3+} + H_2O$	0.337
Cu(Ⅱ) – (0)	$Cu^{2+} + 2e^- = Cu$	0.3419
Re(Ⅶ) – (0)	$ReO_4^- + 8H^+ + 7e^- = Re + 4H_2O$	0.368
Ag(Ⅰ) – (0)	$Ag_2CrO_4 + 2e^- = 2Ag + CrO_4^{2-}$	0.4470
S(Ⅳ) – (0)	$H_2SO_3 + 4H^+ + 4e^- = S + 3H_2O$	0.449
Cu(Ⅰ) – (0)	$Cu^+ + e^- = Cu$	0.521
I(0) – (–Ⅰ)	$I_2 + 2e^- = 2I^-$	0.5355
I(0) – (–Ⅰ)	$I_3^- + 2e^- = 3I^-$	0.536
As(Ⅴ) – (Ⅲ)	$H_3AsO_4 + 2H^+ + 2e^- = HAsO_2 + 2H_2O$	0.560
Sb(Ⅴ) – (Ⅲ)	$Sb_2O_5 + 6H^+ + 4e^- = 2SbO^+ + 3H_2O$	0.581
Te(Ⅳ) – (0)	$TeO_2 + 4H^+ + 4e^- = Te + 2H_2O$	0.593
U(Ⅴ) – (Ⅳ)	$UO_2^+ + 4H^+ + e^- = U^{4+} + 2H_2O$	0.612
＊＊Hg(Ⅱ) – (Ⅰ)	$2HgCl_2 + 2e^- = Hg_2Cl_2 + 2Cl^-$	0.63
Pt(Ⅳ) – (Ⅱ)	$[PtCl_6]^{2-} + 2e^- = [PtCl_4]^{2-} + 2Cl^-$	0.68
O(0) – (–Ⅰ)	$O_2 + 2H^+ + 2e^- = H_2O_2$	0.695
Pt(Ⅱ) – (0)	$[PtCl_4]^{2-} + 2e^- = Pt + 4Cl^-$	0.755
＊Se(Ⅳ) – (0)	$H_2SeO_3 + 4H^+ + 4e^- = Se + 3H_2O$	0.74
Fe(Ⅲ) – (Ⅱ)	$Fe^{3+} + e^- = Fe^{2+}$	0.771
Hg(Ⅰ) – (0)	$Hg_2^{2+} + 2e^- = 2Hg$	0.7973
Ag(Ⅰ) – (0)	$Ag^+ + e^- = Ag$	0.7996
Os(Ⅷ) – (0)	$OsO_4 + 8H^+ + 8e^- = Os + 4H_2O$	0.8
N(Ⅴ) – (Ⅳ)	$2NO_3^- + 4H^+ + 2e^- = N_2O_4 + 2H_2O$	0.803

续上表

电对	方程式	$\varphi^{\ominus}/V$
Hg(Ⅱ)-(0)	$Hg^{2+}+2e^-\!=\!Hg$	0.851
Si(Ⅳ)-(0)	$(quartz)SiO_2+4H^++4e^-\!=\!Si+2H_2O$	0.857
Cu(Ⅱ)-(Ⅰ)	$Cu^{2+}+I^-+e^-\!=\!CuI$	0.86
N(Ⅲ)-(Ⅰ)	$2HNO_2+4H^++4e^-\!=\!H_2N_2O_2+2H_2O$	0.86
Hg(Ⅱ)-(Ⅰ)	$2Hg^{2+}+2e^-\!=\!Hg_2^{2+}$	0.920
N(Ⅴ)-(Ⅲ)	$NO_3^-+3H^++2e^-\!=\!HNO_2+H_2O$	0.934
Pd(Ⅱ)-(0)	$Pd^{2+}+2e^-\!=\!Pd$	0.951
N(Ⅴ)-(Ⅱ)	$NO_3^-+4H^++3e^-\!=\!NO+2H_2O$	0.957
N(Ⅲ)-(Ⅱ)	$HNO2_++H^++e^-\!=\!NO+H_2O$	0.983
I(Ⅰ)-(-Ⅰ)	$HIO+H^++2e^-\!=\!I^-+H_2O$	0.987
V(Ⅴ)-(Ⅳ)	$VO_2^++2H^++e^-\!=\!VO^{2+}+H_2O$	0.991
V(Ⅴ)-(Ⅳ)	$V(OH)_4^++2H^++e^-\!=\!VO^{2+}+3H_2O$	1.00
Au(Ⅲ)-(0)	$[AuCl_4]^-+3e^-\!=\!Au+4Cl^-$	1.002
Te(Ⅵ)-(Ⅳ)	$H_6TeO_6+2H^++2e^-\!=\!TeO_2+4H_2O$	1.02
N(Ⅳ)-(Ⅱ)	$N_2O_4+4H^++4e^-\!=\!2NO+2H_2O$	1.035
N(Ⅳ)-(Ⅲ)	$N_2O_4+2H^++2e^-\!=\!2HNO_2$	1.065
I(Ⅴ)-(-Ⅰ)	$IO_3^-+6H^++6e^-\!=\!I^-+3H_2O$	1.085
Br(0)-(-Ⅰ)	$Br_2(aq)+2e^-\!=\!2Br^-$	1.0873
Se(Ⅵ)-(Ⅳ)	$SeO_4^{2-}+4H^++2e^-\!=\!H_2SeO_3+H_2O$	1.151
Cl(Ⅴ)-(Ⅳ)	$ClO_3^-+2H^++e^-\!=\!ClO_2+H_2O$	1.152
Pt(Ⅱ)-(0)	$Pt^{2+}+2e^-\!=\!Pt$	1.18
Cl(Ⅶ)-(Ⅴ)	$ClO_4^-+2H^++2e^-\!=\!ClO_3^-+H_2O$	1.189
I(Ⅴ)-(0)	$2IO_3^-+12H^++10e^-\!=\!I_2+6H_2O$	1.195
Cl(Ⅴ)-(Ⅲ)	$ClO_3^-+3H^++2e^-\!=\!HClO_2+H_2O$	1.214
Mn(Ⅳ)-(Ⅱ)	$MnO_2+4H^++2e^-\!=\!Mn^{2+}+2H_2O$	1.224
O(0)-(-Ⅱ)	$O_2+4H^++4e^-\!=\!2H_2O$	1.229
Tl(Ⅲ)-(Ⅰ)	$Tl^{3+}+2e^-\!=\!Tl^+$	1.252
Cl(Ⅳ)-(Ⅲ)	$ClO_2+H^++e^-\!=\!HClO_2$	1.277
N(Ⅲ)-(Ⅰ)	$2HNO_2+4H^++4e^-\!=\!N_2O+3H_2O$	1.297
**Cr(Ⅵ)-(Ⅲ)	$Cr_2O_7^{2-}+14H^++6e^-\!=\!2Cr^{3+}+7H_2O$	1.33
Br(Ⅰ)-(-Ⅰ)	$HBrO+H^++2e^-\!=\!Br^-+H_2O$	1.331
Cr(Ⅵ)-(Ⅲ)	$HCrO_4^-+7H^++3e^-\!=\!Cr^{3+}+4H_2O$	1.350
Cl(0)-(-Ⅰ)	$Cl_2(g)+2e^-\!=\!2Cl^-$	1.35827
Cl(Ⅶ)-(-Ⅰ)	$ClO_4^-+8H^++8e^-\!=\!Cl^-+4H_2O$	1.389

续上表

电对	方程式	$\varphi^{\ominus}/V$
Cl(Ⅶ) - (0)	$ClO_4^- + 8H^+ + 7e^- = 1/2Cl_2 + 4H_2O$	1.39
Au(Ⅲ) - (Ⅰ)	$Au^{3+} + 2e^- = Au^+$	1.401
Br(Ⅴ) - (-Ⅰ)	$BrO_3^- + 6H^+ + 6e^- = Br^- + 3H_2O$	1.423
I(Ⅰ) - (0)	$2HIO + 2H^+ + 2e^- = I_2 + 2H_2O$	1.439
Cl(Ⅴ) - (-Ⅰ)	$ClO_3^- + 6H^+ + 6e^- = Cl^- + 3H_2O$	1.451
Pb(Ⅳ) - (Ⅱ)	$PbO_2 + 4H^+ + 2e^- = Pb^{2+} + 2H_2O$	1.455
Cl(Ⅴ) - (0)	$ClO_3^- + 6H^+ + 5e^- = 1/2Cl_2 + 3H_2O$	1.47
Cl(Ⅰ) - (-Ⅰ)	$HClO + H^+ + 2e^- = Cl^- + H_2O$	1.482
Br(Ⅴ) - (0)	$BrO_3^- + 6H^+ + 5e^- = 1/2Br_2 + 3H_2O$	1.482
Au(Ⅲ) - (0)	$Au^{3+} + 3e^- = Au$	1.498
Mn(Ⅶ) - (Ⅱ)	$MnO_4^- + 8H^+ + 5e^- = Mn^{2+} + 4H_2O$	1.507
Mn(Ⅲ) - (Ⅱ)	$Mn^{3+} + e^- = Mn^{2+}$	1.5415
Cl(Ⅲ) - (-Ⅰ)	$HClO_2 + 3H^+ + 4e^- = Cl^- + 2H_2O$	1.570
Br(Ⅰ) - (0)	$HBrO + H^+ + e^- = 1/2Br_2(aq) + H_2O$	1.574
N(Ⅱ) - (Ⅰ)	$2NO + 2H^+ + 2e^- = N_2O + H_2O$	1.591
I(Ⅶ) - (Ⅴ)	$H_5IO_6 + H^+ + 2e^- = IO_3^- + 3H_2O$	1.601
Cl(Ⅰ) - (0)	$HClO + H^+ + e^- = 1/2Cl_2 + H_2O$	1.611
Cl(Ⅲ) - (Ⅰ)	$HClO_2 + 2H^+ + 2e^- = HClO + H_2O$	1.645
Ni(Ⅳ) - (Ⅱ)	$NiO_2 + 4H^+ + 2e^- = Ni^{2+} + 2H_2O$	1.678
Mn(Ⅶ) - (Ⅳ)	$MnO_4^- + 4H^+ + 3e^- = MnO_2 + 2H_2O$	1.679
Pb(Ⅳ) - (Ⅱ)	$PbO_2 + SO_4^{2-} + 4H^+ + 2e^- = PbSO_4 + 2H_2O$	1.6913
Au(Ⅰ) - (0)	$Au^+ + e^- = Au$	1.692
Ce(Ⅳ) - (Ⅲ)	$Ce^{4+} + e^- = Ce^{3+}$	1.72
N(Ⅰ) - (0)	$N_2O + 2H^+ + 2e^- = N_2 + H_2O$	1.766
O(-Ⅰ) - (-Ⅱ)	$H_2O_2 + 2H^+ + 2e^- = 2H_2O$	1.776
Co(Ⅲ) - (Ⅱ)	$Co^{3+} + e^- = Co^{2+}(2\ mol \cdot L^{-1}\ H_2SO_4)$	1.83
Ag(Ⅱ) - (Ⅰ)	$Ag^{2+} + e^- = Ag^+$	1.980
S(Ⅶ) - (Ⅵ)	$S_2O_8^{2-} + 2e^- = 2SO_4^{2-}$	2.010
O(0) - (-Ⅱ)	$O_3 + 2H^+ + 2e^- = O_2 + H_2O$	2.076
O(Ⅱ) - (-Ⅱ)	$F_2O + 2H^+ + 4e^- = H_2O + 2F^-$	2.153
Fe(Ⅵ) - (Ⅲ)	$FeO_4^{2-} + 8H^+ + 3e^- = Fe^{3+} + 4H_2O$	2.20
O(0) - (-Ⅱ)	$O(g) + 2H^+ + 2e^- = H_2O$	2.421
F(0) - (-Ⅰ)	$F_2 + 2e^- = 2F^-$	2.866
	$F_2 + 2H^+ + 2e^- = 2HF$	3.053

续上表

2. 在碱性溶液中(298 K)

电对	方程式	$\varphi^{\ominus}/V$
Ca(Ⅱ)-(0)	$Ca(OH)_2 + 2e^- = Ca + 2OH^-$	-3.02
Ba(Ⅱ)-(0)	$Ba(OH)_2 + 2e^- = Ba + 2OH^-$	-2.99
La(Ⅲ)-(0)	$La(OH)_3 + 3e^- = La + 3OH^-$	-2.90
Sr(Ⅱ)-(0)	$Sr(OH)_2 \cdot 8H_2O + 2e^- = Sr + 2OH^- + 8H_2O$	-2.88
Mg(Ⅱ)-(0)	$Mg(OH)_2 + 2e^- = Mg + 2OH^-$	-2.690
Be(Ⅱ)-(0)	$Be_2O_3^{2-} + 3H_2O + 4e^- = 2Be + 6OH^-$	-2.63
Hf(Ⅳ)-(0)	$HfO(OH)_2 + H_2O + 4e^- = Hf + 4OH^-$	-2.50
Zr(Ⅳ)-(0)	$H_2ZrO_3 + H_2O + 4e^- = Zr + 4OH^-$	-2.36
Al(Ⅲ)-(0)	$H_2AlO_3^- + H_2O + 3e^- = Al + OH^-$	-2.33
P(Ⅰ)-(0)	$H_2PO_2^- + e^- = P + 2OH^-$	-1.82
B(Ⅲ)-(0)	$H_2BO_3^- + H_2O + 3e^- = B + 4OH^-$	-1.79
P(Ⅲ)-(0)	$HPO_3^{2-} + 2H_2O + 3e^- = P + 5OH^-$	-1.71
Si(Ⅳ)-(0)	$SiO_3^{2-} + 3H_2O + 4e^- = Si + 6OH^-$	-1.697
P(Ⅲ)-(Ⅰ)	$HPO_3^{2-} + 2H_2O + 2e^- = H_2PO_2^- + 3OH^-$	-1.65
Mn(Ⅱ)-(0)	$Mn(OH)_2 + 2e^- = Mn + 2OH^-$	-1.56
Cr(Ⅲ)-(0)	$Cr(OH)_3 + 3e^- = Cr + 3OH^-$	-1.48
*Zn(Ⅱ)-(0)	$[Zn(CN)_4]^{2-} + 2e^- = Zn + 4CN^-$	-1.26
Zn(Ⅱ)-(0)	$Zn(OH)_2 + 2e^- = Zn + 2OH^-$	-1.249
Ga(Ⅲ)-(0)	$H_2GaO_3^- + H_2O + 2e^- = Ga + 4OH^-$	-1.219
Zn(Ⅱ)-(0)	$ZnO_2^{2-} + 2H_2O + 2e^- = Zn + 4OH^-$	-1.215
Cr(Ⅲ)-(0)	$CrO_2^- + 2H_2O + 3e^- = Cr + 4OH^-$	-1.2
Te(0)-(-Ⅰ)	$Te + 2e^- = Te^{2-}$	-1.143
P(Ⅴ)-(Ⅲ)	$PO_4^{3-} + 2H_2O + 2e^- = HPO_3^{2-} + 3OH^-$	-1.05
*Zn(Ⅱ)-(0)	$[Zn(NH_3)_4]^{2+} + 2e^- = Zn + 4NH_3$	-1.04
*W(Ⅵ)-(0)	$WO_4^{2-} + 4H_2O + 6e^- = W + 8OH^-$	-1.01
*Ge(Ⅳ)-(0)	$HGeO_3^- + 2H_2O + 4e^- = Ge + 5OH^-$	-1.0
Sn(Ⅳ)-(Ⅱ)	$[Sn(OH)_6]^{2-} + 2e^- = HSnO_2^- + H_2O + 3OH^-$	-0.93
S(Ⅵ)-(Ⅳ)	$SO_4^{2-} + H_2O + 2e^- = SO_3^{2-} + 2OH^-$	-0.93
Se(0)-(-Ⅱ)	$Se + 2e^- = Se^{2-}$	-0.924
Sn(Ⅱ)-(0)	$HSnO_2^- + H_2O + 2e^- = Sn + 3OH^-$	-0.909
P(0)-(-Ⅲ)	$P + 3H_2O + 3e^- = PH_3(g) + 3OH^-$	-0.87
N(Ⅴ)-(Ⅳ)	$2NO_3^- + 2H_2O + 2e^- = N_2O_4 + 4OH^-$	-0.85
H(Ⅰ)-(0)	$2H_2O + 2e^- = H_2 + 2OH^-$	-0.8277

续上表

电对	方程式	$\varphi^{\ominus}/V$
Cd(II) - (0)	$Cd(OH)_2 + 2e^- = Cd(Hg) + 2OH^-$	-0.809
Co(II) - (0)	$Co(OH)_2 + 2e^- = Co + 2OH^-$	-0.73
Ni(II) - (0)	$Ni(OH)_2 + 2e^- = Ni + 2OH^-$	-0.72
As(V) - (III)	$AsO_4^{3-} + 2H_2O + 2e^- = AsO_2^- + 4OH^-$	-0.71
Ag(I) - (0)	$Ag_2S + 2e^- = 2Ag + S^{2-}$	-0.691
As(III) - (0)	$AsO_2^- + 2H_2O + 3e^- = As + 4OH^-$	-0.68
Sb(III) - (0)	$SbO_2^- + 2H_2O + 3e^- = Sb + 4OH^-$	-0.66
*Re(VII) - (IV)	$ReO_4^- + 2H_2O + 3e^- = ReO_2 + 4OH^-$	-0.59
*Sb(V) - (III)	$SbO_3^- + H_2O + 2e^- = SbO_2^- + 2OH^-$	-0.59
Re(VII) - (0)	$ReO_4^- + 4H_2O + 7e^- = Re + 8OH^-$	-0.584
*S(IV) - (II)	$2SO_3^{2-} + 3H_2O + 4e^- = S_2O_3^{2-} + 6OH^-$	-0.58
Te(IV) - (0)	$TeO_3^{2-} + 3H_2O + 4e^- = Te + 6OH^-$	-0.57
Fe(III) - (II)	$Fe(OH)_3 + e^- = Fe(OH)_2 + OH^-$	-0.56
S(0) - (-II)	$S + 2e^- = S^{2-}$	-0.47627
Bi(III) - (0)	$Bi_2O_3 + 3H_2O + 6e^- = 2Bi + 6OH^-$	-0.46
N(III) - (II)	$NO_2^- + H_2O + e^- = NO + 2OH^-$	-0.46
*Co(II) - C(0)	$[Co(NH_3)_6]^{2+} + 2e^- = Co + 6NH_3$	-0.422
Se(IV) - (0)	$SeO_3^{2-} + 3H_2O + 4e^- = Se + 6OH^-$	-0.366
Cu(I) - (0)	$Cu_2O + H_2O + 2e^- = 2Cu + 2OH^-$	-0.360
Tl(I) - (0)	$Tl(OH) + e^- = Tl + OH^-$	-0.34
*Ag(I) - (0)	$[Ag(CN)_2]^- + e^- = Ag + 2CN^-$	-0.31
Cu(II) - (0)	$Cu(OH)_2 + 2e^- = Cu + 2OH^-$	-0.222
Cr(VI) - (III)	$CrO_4^{2-} + 4H_2O + 3e^- = Cr(OH)_3 + 5OH^-$	-0.13
*Cu(I) - (0)	$[Cu(NH_3)_2]^+ + e^- = Cu + 2NH_3$	-0.12
O(0) - (-I)	$O_2 + H_2O + 2e^- = HO_2^- + OH^-$	-0.076
Ag(I) - (0)	$AgCN + e^- = Ag + CN^-$	-0.017
N(V) - (III)	$NO_3^- + H_2O + 2e^- = NO_2^- + 2OH^-$	0.01
Se(VI) - (IV)	$SeO_4^{2-} + H_2O + 2e^- = SeO_3^{2-} + 2OH^-$	0.05
Pd(II) - (0)	$Pd(OH)_2 + 2e^- = Pd + 2OH^-$	0.07
S(II, V) - (II)	$S_4O_6^{2-} + 2e^- = 2S_2O_3^{2-}$	0.08
Hg(II) - (0)	$HgO + H_2O + 2e^- = Hg + 2OH^-$	0.0977
Co(III) - (II)	$[Co(NH_3)_6]^{3+} + e^- = [Co(NH_3)_6]^{2+}$	0.108
Pt(II) - (0)	$Pt(OH)_2 + 2e^- = Pt + 2OH^-$	0.14

续上表

电对	方程式	$\varphi^{\ominus}/V$
Co(Ⅲ)-(Ⅱ)	$Co(OH)_3 + e^- = Co(OH)_2 + OH^-$	0.17
Pb(Ⅳ)-(Ⅱ)	$PbO_2 + H_2O + 2e^- = PbO + 2OH^-$	0.247
I(Ⅴ)-(-Ⅰ)	$IO_3^- + 3H_2O + 6e^- = I^- + 6OH^-$	0.26
Cl(Ⅴ)-(Ⅲ)	$ClO_3^- + H_2O + 2e^- = ClO_2^- + 2OH^-$	0.33
Ag(Ⅰ)-(0)	$Ag_2O + H_2O + 2e^- = 2Ag + 2OH^-$	0.342
Fe(Ⅲ)-(Ⅱ)	$[Fe(CN)_6]^{3-} + e^- = [Fe(CN)_6]^{4-}$	0.358
Cl(Ⅶ)-(Ⅴ)	$ClO_4^- + H_2O + 2e^- = ClO_3^- + 2OH^-$	0.36
* Ag(Ⅰ)-(0)	$[Ag(NH_3)_2]^+ + e^- = Ag + 2NH_3$	0.373
O(0)-(-Ⅱ)	$O_2 + 2H_2O + 4e^- = 4OH^-$	0.401
I(Ⅰ)-(-Ⅰ)	$IO^- + H_2O + 2e^- = I^- + 2OH^-$	0.485
* Ni(Ⅳ)-(Ⅱ)	$NiO_2 + 2H_2O + 2e^- = Ni(OH)_2 + 2OH^-$	0.490
Mn(Ⅶ)-(Ⅵ)	$MnO_4^- + e^- = MnO_4^{2-}$	0.558
Mn(Ⅶ)-(Ⅳ)	$MnO_4^- + 2H_2O + 3e^- = MnO_2 + 4OH^-$	0.595
Mn(Ⅵ)-(Ⅳ)	$MnO_4^{2-} + 2H_2O + 2e^- = MnO_2 + 4OH^-$	0.60
Ag(Ⅱ)-(Ⅰ)	$2AgO + H_2O + 2e^- = Ag_2O + 2OH^-$	0.607
Br(Ⅴ)-(-Ⅰ)	$BrO_3^- + 3H_2O + 6e^- = Br^- + 6OH^-$	0.61
Cl(Ⅴ)-(-Ⅰ)	$ClO_3^- + 3H_2O + 6e^- = Cl^- + 6OH^-$	0.62
Cl(Ⅲ)-(Ⅰ)	$ClO_2^- + H_2O + 2e^- = ClO^- + 2OH^-$	0.66
I(Ⅶ)-(Ⅴ)	$H_3IO_6^{2-} + 2e^- = IO_3^- + 3OH^-$	0.7
Cl(Ⅲ)-(-Ⅰ)	$ClO_2^- + 2H_2O + 4e = Cl^- + 4OH^-$	0.76
Br(Ⅰ)-(-Ⅰ)	$BrO^- + H_2O + 2e^- = Br^- + 2OH^-$	0.761
Cl(Ⅰ)-(-Ⅰ)	$ClO^- + H_2O + 2e^- = Cl^- + 2OH^-$	0.841
* Cl(Ⅳ)-(Ⅲ)	$ClO_2(g) + e^- = ClO_2^-$	0.95
O(0)-(-Ⅱ)	$O_3 + H_2O + 2e^- = O_2 + 2OH^-$	1.24

摘自 David R. Lide, Handbook of Chemistry and Physics, 8-25-8-30, 78th. edition, 1997—1998

* 摘自 J. A. Dean Ed, Lange's Handbook of Chemistry, 13th. edition, 1985

* * 摘自其他参考书。

参考书目

[1] 关鲁雄. 化学基本操作与物质制备实验. 长沙：中南大学出版社，2002

[2] 刘毅敏. 医学化学实验. 北京：科学出版社，2010

[3] 冯清. 医学基础化学实验(双语版). 武汉：华中科技大学出版社，2007

[4] 贾振斌. 医用基础化学实验. 北京：科学出版社，2010

[5] 唐中坤. 医用化学实验(第二版). 北京：科学出版社，2010

[6] 胡庆红. 医用基础化学实验教程. 北京：人民卫生出版社，2011

[7] 武汉大学. 分析化学实验(第二版). 武汉：武汉大学出版社，2013

[8] 邓湘舟. 现代分析化学实验. 北京：化学工业出版社，2013

[9] P. V. Koppen. *General Chemistry Laboratory Manual*. University of California-Santa Barbara，Mcgraw Hill Education，2003

相关网址

1. 网上化学课程

　　中国化学课程网 http://chem. cersp. com/

　　网易名校公开课 http://v. 163. com/special/ocw/

　　美国德克萨斯大学网上课程 http://www. utexas. edu/world/lecture

2. 化学数据库

　　科学数据库 http://www. sdb. ac. cn

　　万方数据库 http://db. sti. ac. cn

　　基本科学指标数据库(ESI)http://www. webofknowledge. com

　　美国国家标准与技术研究院(NIST)物性数据库 http://webbook. nist. gov/chemistry

　　Cambridgesoft 公司的化学数据库 http://chemfinder. cambridgesoft. com

3. 专利数据库

　　中国知识产权网 http://www. cnipr. com

　　美国专利商标局(USPTO)http://www. uspto. gov/patft/

　　欧洲专利局(EPO)http://ep. espacenet. com

4. 信息资源

　　中国化学品网 http://www. ylrqcn. com/

　　中国医药信息网 http://www. cpi. ac. cn/

　　英国利物浦大学 Links for Chemists http://www. liv. ac. uk/chemistry/links

　　Sigma 公司 http://www. sigmaaldrich. com

　　世界电子图书馆 http://cn. ebooklibrary. org

　　美国化学会化学文摘(CAS)http://info. cas. org/ONLINE/online. htmL

　　德国专业信息中心(FIZ)http://www. fiz – karsruhe. de

　　Elsevier Science 公司期刊网 http://www. sciencedirect. com

　　Wiley 网上图书馆 http://www. interscience. wiley. com